London Mathematical Society Lecture Note Series, 156

Twistors in Mathematics and Physics

Edited by

T.N. Bailey
Lecturer in Mathematics, University of Edinburgh
R.J. Baston
University Lecturer, Mathematical Institute, University of Oxford

CAMBRIDGE UNIVERSITY PRESS

Cambridge

New York   Port Chester   Melbourne   Sydney

Published by the Press Syndicate of the University of Cambridge
The Pitt Building, Trumpington Street, Cambridge CB2 1RP
40 West 20th Street, New York, NY 10011, USA
10 Stamford Road, Oakleigh, Melbourne 3166, Australia

First published 1990

Printed in Great Britain at the University Press, Cambridge

British Library cataloguing in publication data available

Library of Congress cataloguing in publication data available

ISBN 0 521 39783 9

Preface

Our aim in editing *Twistors in Mathematics and Physics* has been to collect together review articles which reflect the wide diversity of ideas and techniques which constitute modern twistor theory. Whilst the origins and much continuing work in twistor theory are in the area of fundamental physics, there is an ever-growing body of 'twistor mathematics' which has now taken on a life of its own. This is reflected this by articles on representation theory and differential geometry, among other subjects.

The main objective in the 'twistor programme' for fundamental physics is a theory which unites Einstein's general relativity and the world of quantum physics—a theory in which the rôle of complex holomorphic geometry is fundamental; Penrose's article in this volume reviews its current status. Other contributors have covered the advances which have ocurred since the major successes of Penrose's *non-linear graviton* and Ward's construction of the *Yang–Mills instantons*—the most notable of these is probably Penrose's definition of *quasi-local mass* in general relativity, and topics such as the twistor description of vacuum space-times with symmetries and twistor particle theory are also covered.

Twistor mathematics is now a wide-ranging subject in itself, and articles in this volume cover differential geometry, integrable systems and several topics related to representation theory. The process by which Penrose originally encoded solutions of field equations on Minkowski space in terms of holomorphic functions on regions in twistor space, namely the *Penrose transform*, has been generalized to a complex homogeneous spaces, and so has applications in the theory of invariant operators (Verma modules) and the construction of unitary representations. The reader will find all these topics, and others, covered here.

Our hope is that this volume will be of use to workers in all areas of twistor theory and the many areas connected with it. We particularly hope that it will encourage the continued cross-fertilisation of ideas, particularly between pure mathematics and mathematical physics, which has always been one of the subject's particular strengths.

Contents

Twistor Theory After 25 Years — its Physical Status and Prospects

R. Penrose

Introduction

The primary objective of twistor theory originally was—and still is—to find a deeper route to the workings of Nature; so the theory should provide a mathematical framework with sufficient power and scope to help us towards resolving some of the most obstinate problems of current physical theory. Such problems must ultimately include: (1) removing the infinities of quantum field theory, (2) ascertaining the nature and origin of symmetry and asymmetry in the classification of particles and in physical interactions, (3) deriving, from some fundamental principle, the strengths of coupling constants and the masses of particles, (4) finding a quantum gravity theory capable of satisfactorily addressing the issues raised by space-time singularities and the structure of space-time in the small, (5) constructing a picture that makes sense of the puzzling non-locality and conceptual peculiarities inherent in the process of quantum measurement. Does twistor theory have anything of significance to contribute concerning these matters? Might it at least point us in some appropriate directions?

I shall comment on these issues individually in a moment. But as things stand, it must be said that the successes of twistor theory to date have been almost entirely in applications within *mathematics*, rather than in furthering our understanding of the nature of the physical world. I would think of twistor theory's physical role, so far, as being something perhaps resembling that of the Hamiltonian formalism. That formalism provided a change in the framework for classical Newtonian theory rather than a change in Newtonian theory itself. The Hamiltonian scheme (at least Hamilton's own part in its development) was motivated very much by a physical analogy between the behaviour of particles and of waves; but it was not until the advent of quantum physics that a change in physical theory was put forward—indeed one in which particles and waves became actually the same thing, rather than being merely analogous. When the mathematics for a quantum theory was required, Hamiltonian formalism was in place and provided the ideal vehicle, ready to accommodate the essential changes that were needed, in order that physical theory could be transported from classical to quantum. The ambitious role set out for twistor theory, then, is that, likewise, when enough of its mathematics has been developed, that theory also will be in place and, with relatively minor changes, will turn out to be just what is required for a much needed new physics.

In this article, I shall be concerned primarily with physical issues, and how I feel that twistor theory stands, or ought to stand, with regard to them. The *mathematical* applications of the theory are well covered by articles by other authors in this volume, and some of these applications have proved to be unexpectedly fruitful. With regard to *physical* applications and aside from developments connected with the fundamental issues referred to above, which I discuss in a moment), there has one noteworthy and *unanticipated* success: the concept of *quasi-local mass and (angular) momentum in general relativity*. For a great many years, relativists had resigned themselves to the idea that the mass-energy of the gravitational field cannot be localized, and only the *total* energy of an asymptotically flat space-time can be assigned an unambiguous meaning. Twistor theory now allows us to do a good deal better (Penrose 1982, Penrose and Rindler 1986), though various difficulties remain. A full and up-to-date account is to be found in Paul Tod's article (1990—this volume; see also Mason and Frauendiener (1990), this volume), and it will not be necessary for me to go into the details here.

Nonetheless, one is compelled to confess that, so far, rather little that is both tangible and new has come through with regard to twistor theory's *original* physical aspirations. Let us now try to see how the theory stands with regard to each in turn of the above-mentioned questions.

1   The infinities of quantum field theory

Of the fundamental physical problems referred to in the opening paragraph, it is only the issue of infinities of quantum field theory that has been significantly addressed, so far. The main progress in this direction has been in the theory of *twistor diagrams* (see Hodges 1990 and Huggett 1990, this volume) which has been evolved as the twistor analogue of Feynman graphs. The intention has been that a procedure essentially equivalent to the conventional Feynman theory could be developed—except that it is intended that *finite* answers are to be obtained in important cases where the Feynman graphs diverge. The initial work in this area proceeded to a considerable extent by guesswork, analogy, geometrical considerations, aesthetics, and wishfull thinking (Penrose and MacCallum 1972, Penrose 1975b; cf. also Sparling 1975, Qadir 1978), but then later work (Hodges 1983a,b, 1985a,b 1990, this volume, cf. also Huggett 1990 this volume) not only put the twistor diagram theory on a sound basis, but also led to actual changes in which certain of the infinities of the standard theory have indeed become replaced by finite expressions.

As an initial step of the original scheme, the usual momentum states of the conventional theory are replaced by the (finite-normed) *elementary states* that arise naturally in twistor theory, so that the calculated amplitudes have some chance of being actually finite, rather than involving delta-functions, as is the case with momentum states. This allows, as a general objective of the twistor diagram formulation, that amplitudes might be computed over *compact* (high-dimensional) contours, the integrands being supposed to be analytic expressions at all points of the contours, so that the answers would accordingly be always guaranteed to be finite whenever these two requirements can be satisfied. However, since some of the answers, as 'correctly' computed by Feynman graph methods, are actually divergent, this entails that some change must be introduced into the procedures from those that would be obtained by direct translation of the corresponding Feynman graphs. This applies, as Andrew Hodges noted a good many years ago, even to some 'tree diagrams' of the standard Feynman theory, which are infra-red divergent. He was able to circumvent this problem in an ingenious way (Hodges 1985a), by replacing the previous factors $Z^\alpha W_\alpha$ that had ocurred in twistor diagram expressions according to

$$Z^\alpha W_\alpha \longmapsto Z^\alpha W_\alpha + k$$

where k is some (dimensionless) numerical constant whose value would be ultimately fixed by theory or experiment. At first, k merely provides a number whose logarithm enters into a finite expression which replaces each infra-red divergent quantity (the divergence being recovered when $k \to 0$), but k seems also to play a key role in eliminating ultra-violet divergences (Hodges 1985a) and it has a separate importance in relation to twistor diagrams for *massive* particles.

The factors $Z^\alpha W_\alpha + k$ bear some resemblance to factors

$$Z^\alpha X^\alpha I_{\alpha\beta} + m,$$

which Hodges uses in the twistor diagrams describing massive particles ('projection operators' for the mass eigenvalue m). These diagrams also make use of the so-called 'universal bracket factor' $[...]_U$ (cf. Penrose 1979c, Hodges 1985b) which is 'defined' (formally) by the divergent expression

$$[x]_U = \ldots + (x)_{-2} + (x)_{-1} + (x)_0 + (x)_1 + (x)_2 + \ldots$$

where, for $n = 0, 1, 2, 3, \ldots$

$$(x)_{-n} = (x)^{-n}/n! \quad \text{(contour with boundary on x=0)}$$

and

$$(x)_{n+1} = -n!(-x)^{-n-1}/2\pi i \quad \text{(contour surrounding x=0),}$$

or by the formal expression, suggested by George Sparling (cf. Penrose 1979)

$$\oint [x]_U = \left\{ \oint \int_0^\infty + \oint \oint \right\} \frac{e^z}{z+x} dz$$

Although the divergence difficulties are cleverly circumvented in Hodges's particular 'mass-projection' expression, a proper general understanding of the universal bracket is still lacking.

Many other problems in twistor diagram theory remain to be solved, but some good progress has been made towards the goal of finding a complete formulation of the *standard model* of weakly or strongly interacting particles in twistor diagram terms (Hodges 1990, this volume). One particular problem is to obtain a fuller understanding of the (high-dimensional) contours that occur in twistor diagrams. As noted above, these are supposed always to be compact (perhaps with boundary) so that the integrals will always be finite. The extent to which it has been possible to satisfy this compactness requirement so far has been definitely encouraging. However, particularly when mass is present, the status of this requirement is still unclear and it seems to demand the use of 'blown-up twistor space' according to which the line I, in projective twistor space $\mathbf{PT}$, is replaced by a quadric surface. This is related to the 'googly twistor space' needed for the description of general relativity, and which will be described in outline below.

Another particularly important issue of twistor diagram theory is to understand, in purely twistorial terms, which twistor diagrams are to appear in any given process, and with what weighting (and sign) each diagram is to occur. A popular approach to the corresponding problem for Feynman diagrams, in modern quantum field theory, is to use the (formal) method of *path integrals*. However, an analogous procedure for twistor theory has not yet come to light. There is an apparently fundamental conflict between the twistor description of fields and that which is addressed by a path-integral approach. In the latter, paths are deliberately allowed in which the field equations are violated, whereas in twistor theory it is considered to be a virtue for the classical field equations to come out as solved *automatically* by the twistor descriptions! In twistor (diagram) theory 'off-shell' contributions in which field equations are violated come about in a different way (in effect, by the introduction of further twistors). The relation between this and path integrals has not yet come to light.

Another possible line of approach to the problem of 'twistor diagram generation' is through the ideas of a generalized conformal field theory involving 'pretzel twistor spaces' (Hodges, Singer and Penrose 1989, Penrose 1989), though this approach has not yet progressed very far. A 'pretzel twistor space' is a higher-dimensional analogue of a Riemann surface (with 'holes'), as occurs in standard conformal field theory (or string theory). Though perhaps superficially similar to a 'membrane' (or 'p-brane') theory (i.e. higher-dimensional string theory), this approach is fundamentally different in that the generalizations of Riemann surfaces are *complex manifolds*, either of three dimensions (projective case) or of four (non-projective), and in that the complex manifolds play the role of *twistor* spaces, in relation to space-time, rather than being regarded as being 'in' space-time. So long as general relativity is not involved, each such 'pretzel twistor space' would be a *flat twistor space* X, defined by the following properties, in the three-dimensional projective case:

1. X is a compact complex 3-manifold with boundary ∂X;

2. each component of ∂X is a copy of (i.e. is CR-equivalent to) the **PN** of standard twistor theory;

3. each point of $X - \partial X$ has a neighbourhood which is holomorphic to a neighbourhood of a line in $\mathbb{CP}^3$;

4. the canonical bundle of X admits a fourth root.

The reason for condition (4) is to enable the corresponding non-projective (four-dimensional) flat twistor space to be defined as the appropriate line bundle over the projective flat twistor space.

According to this proposal, a scattering process would be described by one of these flat twistor spaces (or by a linear superposition of processes described by different such spaces), where a positive or negative orientation would be assigned to each component of ∂X. Each positively oriented component would refer to an incoming particle state, and each negatively oriented component, to an outgoing particle state, where the in- and out-states (taken to be massless, in the first instance) would be described in the standard way by (1st) sheaf cohomology elements (restricted to **PN**). The procedure follows closely the one adopted in ordinary conformal field theory (cf. Segal 1990, Witten 1989). It is strongly motivated by the close analogy between the way (a) that the 'equator' S^1 (unit circle) divides the Riemann sphere into the 'northern hemisphere' S^+ and 'southern hemisphere' S^- and the way (b) that **PN** divides **PT** into **PT**$^+$ and **PT**$^-$. The splitting (a) of functions (i.e. H^0-elements) on S^1 into their positive and negative frequency parts according to whether they extend holomorphically into S^+ or S^-, is closely mirrored by the splitting (b) of solutions of the massless field equations into their positive and negative frequency parts according to whether the corresponding twistor functions (as H^1-elements) extend into **PT**$^+$ or **PT**$^-$. (This important fact realized one of the key original motivations behind twistor theory; cf. Penrose 1986a.)

The hope is that there should be some close relation between the construction of flat twistor spaces, their corresponding conformal field theories, and twistor diagrams. This would mirror the way that the early string theory showed how duality diagrams (Riemann surfaces with 'holes') could be used to make sense of the 'counting' of Feynman diagrams in strong interaction theory, and can also serve to replace certain collections of infinite Feynman diagrams by finite expressions. There does seem to be a corresponding role for twistor diagrams in relation to flat twistor spaces, but unfortunately this has not been explored very far as yet.

In conformal field theory, the in-states [out-states] are elements of Fermionic 'Fock spaces'

$$\mathcal{H} = \mathbb{C} \oplus H \oplus H \wedge H \oplus H \wedge H \wedge H \oplus H \wedge H \wedge H \wedge H \oplus \ldots$$

where each H is the space of positive-frequency [negative-frequency] functions (or sections of bundles) on a positively [negatively] oriented S^1 that constitutes a 'hole' boundary in the Riemann surface in question. In the case of a pretzel twistor space, the 'hole' boundaries are copies of **PN**, and instead of functions, we have first cohomology elements, representing wave-functions of massless fields. We cannot interpret the elements of the higher-order spaces

$$H \wedge H, \quad H \wedge H \wedge H, \quad H \wedge H \wedge H \wedge H, \quad \ldots$$

as representing many-particle states, since each different particle taking part in a scattering process is to be represented by a different **PN** hole. Instead, the elements of these higher-order spaces (functions of several twistors) must presumably represent massive particles, in accordance with the twistor particle programme, that will be briefly described in the next section.

2 Symmetry and asymmetry in particle interactions

One of the most striking things about the twistor formulation—for good or for bad—is that by choosing the twistor space **PT** to be primary, rather than the dual space **PT*** (or *vice versa*), we are led to an essentially *left-right asymmetric* description of physics. This would seem to be a desirable thing when we are trying to describe aspects of physics—notably weak interactions—for which such left-right asymmetry is known to be a fact of nature, but its desirability is more questionable for those interactions which are believed to be left-right *symmetric*. In particular, in the case of general relativity, we have a fundamental theory of space-time structure which is left-right symmetric, and this presents a severe challenge to any asymmetric twistorial description. It is a remarkable fact, however, that several new approaches to the description of standard general relativity have also been guided, for apparently quite independent reasons, into a left-right asymmetric formulation. These are approaches which relate, in one way or another, to what are known as 'Ashtekar variables' (Ashtekar 1988—see also Mason & Frauendiener 1990, this volume). The left-right asymmetry is expressed as an asymmetry between primed and unprimed 2-spinor indices—or, what amounts to the same thing, to an asymmetry between anti-self-dual and self-dual curvatures.

The basic 'twistorial' reason for believing in a left-right asymmetric approach to physics (i.e. a preferance either for **PT** or for **PT*** in the formulation) arises from the 'twistor-function' description of linear massless fields. A holomorphic function $f(Z^\alpha)$ (actually a representative 1-cocycle for an element of $H^1(\mathbf{PT}^+, \mathcal{O})$), which is homogeneous of degree $-n - 2$, describes a wave function for a massless particle of helicity $n/2$ (n being an arbitrary integer). It is a striking fact that in this way we can automatically incorporate the two essential requirements for a massless one-particle wave-function, namely satisfaction of both the massless field equation and the positive-frequency condition. The fact that we are using a *holomorphic* function of the twistor Z^α (i.e. a 'function of Z^α' rather than a 'function of Z^α and $\overline{Z}_\alpha$') is the twistorial version of the basic quantum-mechanical requirement that ordinary wave functions must be functions just of position (or just of momentum) not functions of both position and momentum. For ordinary wave functions we can, if we prefer, choose functions of momentum instead of functions of position for our descriptions of quantum states, so long as we are consistent about this. Likewise, we can, if we prefer, consistently use holomorphic functions of *dual* twistors W_α (i.e. holomorphic functions of $\overline{Z}_\alpha$, where we simply relabel $\overline{Z}_\alpha$ as W_α, i.e. *anti*-holomorphic functions of Z^α. However, such an alternative choice must be consistent: it would make no sense to use, say, a position description for particles of positive electric charge and a momentum description for particles of negative electric charge; and likewise it would make no sense to use, say, a dual twistor description for massless particles of positive helicity and a twistor description of massless particles of negative helicity. (Such a description might have seemed tempting in view of the fact that right-handed—i.e. self-dual—non-linear gravitons seem to have a natural description in terms of dual twistors and left-handed—i.e. anti-self-dual—non-linear gravitons, a natural description in terms of twistors.) A particularly awkward aspect of any such attempt to describe massless particles in this hybrid way arises from the fact that there is often the need to describe massless particles which are not simply entirely right-handed or left-handed, such as plane-polarized photons. Thus, at least if we are describing massless particles, it seems to be necessary to make a choice in our twistorial representation: *either* a description in terms of twistors Z^α must be used *or* a description in terms of dual twistors W_α.

So long as we are concerned only with *linear* massless fields (without sources), this does not imply any serious left-right asymmetry for what it is possible to achieve with the twistor formalism, but, as is apparent with the the 'non-linear graviton construction' for (anti-)self-dual gravitational fields (Penrose 1976) and the Ward construction for (anti-)self-dual Yang-Mills fields (Ward 1977), the situation seems very awkwardly different for non-linear fields. (This raises the issue of the 'googly problem' which I shall return to later.) If it is supposed that Nature's ways actually accord with some of the basic ideas of twistor theory, and that she thus prefers, say, a twistorial description—or else she prefers a dual-twistorial description—then it would be expected that some left-right asymmetry should be present with the actual physics of non-linear massless fields. Of course, we already know that weak interactions are left-right asymmetric, but the above considerations should apply also to the gravitational field. The standard Weinberg-Salam-Glashow-Ward theory of unified electromagnetic and weak interactions implies that there is an indirect left-right asymmetry in electromagnetism, but the above considerations seem to imply a 'twistor expectation' of a left-right asymmetry in gravitation also.

Even for linear fields, there is twistorial left-right (or, rather, a self-dual/anti-self-dual) asymmetry in the case of fields with *sources*. For example, in the case of a Coulomb field, the twistor-function for the self-dual part would have the form

$$f(Z^\alpha) \; = \; \frac{1}{(Q_{\alpha\beta} Z^\alpha Z^\beta)^2}$$

while that for the anti-self-dual part would be something like

$$g(Z^\alpha) \; = \; \log(Q_{\alpha\beta} Z^\alpha Z^\beta)$$

or

$$g'(Z^\alpha) \; = \; \log \frac{Q_{\alpha\beta} Z^\alpha Z^\beta}{A_\gamma Z^\gamma B_\delta Z^\delta}.$$

In none of these cases do we get a global representation of the space-time field as an H^1 element in twistor space, but in the self-dual case we get a global description as a relative H^1 element (Bailey 1985). In the anti-self-dual case this does not seem to be so, however, and the situation is more obscure. Moreover, when we go over to the 'non-linear' Ward representation in terms of a line-bundle over in the anti-self-dual case we get 'charge quantization' and a non-Hausdorff bundle (Penrose and Sparling 1979, Bailey 1985). There is no analogue (as yet) in the self-dual case.

I have phrased the above discussion in terms of left-right asymmetry (i.e. parity P), since this is the most obvious of the discrete symmetry operations which convert left-handed massless particle into right-handed ones. However the operation C of charge-conjugation (particle-antiparticle interchange) also achieves this (witness the case of a neutrino), as do the operations CT and PT (where T stands for time-reversal symmetry). All of these four symmetries are violated in weak interactions, and it would appear that such violations could be well accommodated by twistor theory. In the context of the rules governing twistor diagrams, one only needs an asymmetry under interchange of black spots (twistors Z^α) with white spots (dual twistors W_α). But in addition, T and CP are known to be violated in K_0-decay, and this could arise, twistorially, out of some sort of asymmetry between $\mathbf{PT}^+$ and $\mathbf{PT}^-$. There are (controversial!) reasons for believing that even CPT should be violated in quantum gravity theory (cf. Penrose 1981, 1986b, 1989 and below). Such further symmetry violations would seem much more natural in the context of twistor theory than they do in standard space-time descriptions, but a good deal more understanding is needed if we are to see what the exact role of twistor theory in symmetry violation actually is.

Let us next consider massive particles, and how the quantum description of such particles is best to be incorporated into twistor theory. We recall that the twistor theory of massless fields is closely bound up with the quantized expressions for momentum and angular momentum for a massless particle:

$$p_a = \pi_{A'}\overline{\pi}_A, \quad M^{ab} = i\omega^{(A}\overline{\pi}^{B)}\epsilon^{A'B'} - i\overline{\omega}^{(A'}\pi^{B')}\epsilon^{AB}$$

where, in the twistor (as opposed to dual twistor) representation of massless wave-functions, we make the replacements

$$\overline{\pi}_A \longmapsto -\frac{\partial}{\partial\omega^A}, \quad \overline{\omega}^{A'} \longmapsto -\frac{\partial}{\partial\pi_{A'}},$$

in accordance with the standard twisor quantization rule

$$\overline{Z}_\alpha \longmapsto -\frac{\partial}{\partial Z^\alpha}.$$

(Here I am taking $\hbar = 1$.) These give the standard momentum and angular momentum operators when applied to a twistor function of the one twistor variable Z^α, the squared mass $m^2 = p_a p^a$ being identically zero. The procedure which is adopted in twistor particle theory in order to handle particles of *non*-zero mass is to replace the above expressions by sums

$$p_a = \sum_{i=1}^{r} \pi_{iA'}\overline{\pi}_{iA} \quad M^{ab} = \sum_{i=1}^{r}\{i\omega_i^{(A}\overline{\pi}_i^{B)}\epsilon^{A'B'} - i\overline{\omega}_i^{(A'}\pi_i^{B')}\epsilon^{AB}\}$$

where instead of acting on functions of just one twistor, these (quantized) operators now act on functions of several twistors

$$Z_1^\alpha,\ldots,Z_r^\alpha,$$

with

$$Z_i^\alpha = (\omega_i^A, \pi_{iA'}).$$

A twistor (wave-)function for a massive particle is now to be a holomorphic function of $Z_1^\alpha,\ldots,Z_r^\alpha$ (although the possibility that certain of these twistor variables might better be taken as dual twistors should not be overlooked). According to the original twistor-particle scheme (which I sometimes refer to as 'naïve twistor particle theory') *leptons* were to be the particles described by functions of just *two* twistors, say Y^α and Z^α, and *hadrons* by functions of *three* twistors, say X^α, Y^α and Z^α. One of the reasons for this was that the n-twistor internal symmetry group (the group of linear transformations of Z_i^α and $\overline{Z}_{i\alpha}$ that leaves p_a and M_{ab} invariant) is a slight inhomogeneous extension of $\mathrm{SU}(2) \times \mathrm{U}(1)/\mathbb{Z}_2$ (or $\mathrm{U}(2)$) in the case of $n = 2$, and a rather larger inhomogeneous extension of $\mathrm{SU}(3)$ in the case $n = 3$. These were basically the symmetry groups that arose in the standard classification of leptons and hadrons, respectively—in the 'good old days' before 'charm' was discovered! The 3-twistor scheme for the hadrons of those days provided quite a strikingly good fit, for the most part, although there were some notable anomalous multiplets, such as the nonet (rather than the expected octet) which involves the η^0, and certain families of resonances which seemed to have the wrong symmetries (see Hughston 1979). The 2-twistor scheme for leptons seemed to present more serious problems, since it provided for only two quantum numbers (in addition to spin and mass)—identifiable with charge and lepton number—leaving no way of distinguishing the muon from the electron. An ingenious suggestion due to George Sparling was that the required additional quantum number might be, in effect, the sign of the quantum number for total spin! He noted that the squared total quantum-mechanical spin J_2 of a massive particle, since it takes the form

$$J_2 = j(j + 1)$$

(in units of $\hbar$), where j is the usual total spin quantum number, is invariant under

$$j \longmapsto -j - 1.$$

Thus, for an observed total spin value, there are really two possible values of the quantum number j, namely j and $-j - 1$. Sparling's suggestion was that we can allow for *negative* values of j , and the idea was that perhaps what distinguishes the muon from the electron was that one of these particles has $j = \frac{1}{2}$ and the other has $j = -\frac{3}{2}$. (There was some hint of support for this kind of idea in the expression for the spin operator in the 2-twistor scheme.) It may be that this suggestion has less plausibility now than it had at the time, owing to the further complication of the discovery of the τ-lepton. Nevertheless there does seem to be something in this idea, which shows up in the case of *hadrons*. It is (or was!) well known that if we plot the hadron resonances, for each particular set of values of its $\mathrm{SU}(3)$ quantum numbers, in a diagram with the j vertical and m^2 horizontal, then most of the resonances will lie on a family of remarkably straight lines rising off to the right—the *Regge trajectories*. In the case of the baryons, it is possible to divide these trajectories into two classes, those corresponding to natural parity ($\epsilon = 1$) and those corresponding to unnatural parity ($\epsilon = -1$). If we assume that (say) the natural parity baryons have positive j and the unnatural parity ones, negative j, and we now plot m^2 against j, we find that the pairs of Regge trajectories join together into single straight lines, the *extended* Regge trajectories (Penrose, Sparling and Tsou 1978)! There does seem to be some genuine physical content in this idea, since it has some predictive power and the later observations have been quite well in accordance with these predictions (Tsou 1981).

The ideas of (naïve?) twistor particle theory ought, in principle, to fit in well with the theory of twistor diagrams. For in order to know which interactions a specific particle is supposed to indulge in, i.e. which particular twistor diagrams are to be considered to contribute, there must be a way of specifying its appropriate *quantum numbers*. Twistor particle theory says that these quantum numbers ought to be expressible in terms of the eigenvalues of certain (scalar) twistor operators constructed from

$$X^\alpha, \ldots, Z^\alpha, \frac{\partial}{\partial X^\alpha}, \ldots, \frac{\partial}{\partial Z^\alpha},$$

as applied to a *twistor wave-function* $f(X^\alpha, ..., Z^\alpha)$, i.e. a function of a number of twistors, from which a particle's space-time wave function can be obtained by contour integration (see Penrose 1975a, Hughston 1979). The demand that f actually have a certain specified set of eigenvalues for these operators (i.e. that the particle whose twistor wave-function is f is supposed to be a particle of a specified type) can be given also in terms of certain types of twistor diagram. Here the diagram acts as a projection operator which projects out the part of f that has precisely the required eigenvalues.

It is now well understood that for twistor wave functions $f(Z^\alpha)$, of just one twistor variable Z^α, the function f is really to be thought of as a representative of *1st sheaf cohomology*, and the contour integration of f yields the evaluation of this cohomology element on lines in projective twistor space and provides a space-time wave-function for a particle of zero rest mass. When the number of twistors is greater than one, contour integration yields a wave-function for a particle which need not be massless, but it is not yet known, in the cases of twistor functions of more than one twistor, exactly how the function is to be interpreted in some cohomological sense. The most likely possibility seems to be in terms of relative cohomology, but more work needs to be done in order to decide what is really going on.

In the days before τ-leptons and charmed hadrons, etc. the twistor particle programme had seemed to fit in rather well with the observed particle classifications in a quite striking way. However things have got more complicated than they were, in the world of observational particle physics, and it would now seem that some essentially new ingredients are needed in order to make the twistor particle scheme work. If one had been strict about the original scheme, however, it would have been always clear that some new ingredient was needed. For the twistor mass operator commutes with everything else. This should imply that the mass of a particle, according to the scheme, could take absolutely any (positive) value, quite independently of the other quantum numbers. This is completely at variance with what is observed, since *observationally* the mass parameter seems to be fixed—or, at least, having only a very few descrete values allowed to it (e.g. those for the electron, the muon and τ-particle?). It would seem that the mass ought not to commute, exactly, with the other quantum numbers, but as yet, no plausible scheme for this has come to light. (Twistor particle theory is really not significantly worse than the standard schemes in this regard, there being no respectable theory at all for determing mass values for quarks, etc.)

It would seem that in order to understand how to treat mass properly, we must understand exactly how Einstein's gravity fits into the twistor scheme. Mass acts as the source for gravity, and it seems hardly likely that any deep understanding of mass can ever be obtained without general relativity being brought in in some essential way. (It would obviously be unreasonable to attempt to understand electric charge without bringing in Maxwell's theory!)

3   Coupling constants and particle masses

The problem of finding a theoretical determination of particles' mass values is part of the problem of understanding what fixcs (pure-number) coupling constants in physical interactions generally. Although twistor theory has not yet provided any tangible advances towards an understanding of these deep questions, it is not impossible that it might ultimately do so. Perhaps one of the more plausible places where such an advance might be expected, assuming that there is some relevant significance to the (naïve) twistor particle programme, is in an understanding of the Cabbibo and the Weinberg angles. Both of these would have to do with the process of 'twistor growing' (Penrose 1976b). The Cabbibo angle would arise in the way that the 2-twistor scheme (for leptons) would be mapped into the 3-twistor scheme (for hadrons), and the Weinberg angle, in the way that the 1-twistor scheme (for the case of the photon) would be mapped into the 2-twistor scheme (for leptons). In each case, there would be a 'mixing' of twistors, in the scheme involving the larger number of twistors, when it gets reduced to the scheme involving the smaller number of twistors. The degree of mixing, in each case, would give rise to the 'angle' in question. 'Twistor growing' would be the reverse of this process, where the number of twistors gets increased in the passage from one scheme to the next. (This procedure would seem to have to do with something like a cycle: Čech cohomology $\rightarrow$ Dolbeault cohomology $\rightarrow$ 'complexification' by 'freeing up' the complex conjugates into separate twistors $\rightarrow$ hyperfunctional Čech description.)

The fine structure constant α, or equivalently the charge on the electron, is the most noteworthy example of a pure-number coupling constant. Particle mass-values, scaled in terms of the Planck mass, would be the gravitational analogues. In each case, the question of renormalisation comes in to muddy the issue. Should one be attempting to determine theoretically the 'bare' or 'dressed' values of these numbers? The bare value might seem more fundamental, but for this to be meaningful, we should need the renormalization factors to be finite. Thus we require a *finite*, and not a merely renormalizable theory—which is in accordance with the aspirations of twistor diagram theory (and of other fundamental approaches).

This notwithstanding, it has never been clear to me that it should be a 'hopeless' matter to attempt to determine the 'dressed'—or rather the 'observed' values—directly. It has often been argued, since all the processes of physics contribute to these renormalization factors, that there would be no chance of having an *exact* theoretical determination of them until one had a complete theory of physics—and, even then, the renormalization factors would have to be enormously complicated. However it not at all clear to me that this need be the case. There might be an alternative way of looking at quantum field theory according to which renormalization becomes completely irrelevant. In relation to all this, there is a certain 'numerology' in the observed pure-number values that arise, suggestive of possible laws operative on the scale of the dressed values themselves. Perhaps most noteworthy example of such numerology is Wyler's value of the fine structure constant, which I re-express as:

$$\alpha = (4\pi/3)^2(120\pi^3)^{1/4} = 137.0360824....$$

(the current observed value of α being 137.035895 ± 0.000061, which leaves Wyler's number only marginally outside the experimental range). If we take this numerology seriously, and think in terms of the 'elementary quark charge', namely one third of that of the electron, where we may take the factor of 4π as being 'natural' in the context of Gauss's law, etc., then we may regard the factor of $(4\pi/3)^2$ as being 'accounted for'—so what has to be explained is the $(120\pi^3)^{1/4}$. From the point of view of twistor theory, the unnatural-looking fourth root in this expression might somehow be related to the fourth root $Z^\alpha \to \{1 + f_{-4}(Z^\alpha)\}^{1/4}Z^\alpha$ that occurs in the 'googly photon' construction (Penrose 1979a), according to which the self-dual part of the Maxwell field can be coded in terms of deformations of the fibres in the map

$$\mathsf{T} - \{0\} \longrightarrow \mathsf{PT}$$

that preserve the volume form

$$d^4Z = \epsilon_{\alpha\beta\gamma\delta}dZ^\alpha \wedge dZ^\beta \wedge dZ^\gamma \wedge dZ^\delta.$$

This is in contradistinction to the standard Ward construction for the anti-self-dual part, where the Euler operator

$$\Upsilon = Z^\alpha\frac{\partial}{\partial Z^\alpha}$$

is preserved instead. It would seem that in the correct twistorial description of interacting electromagnetic fields these right- and left-handed parts of the photon would have to be combined together, but it has been hard to see how to do this in such a way that the right-handed part does not come out as being 'charged' with respect to the left-handed part. In order to disentangle the two parts of the photon, it seems to be necessary to 'go to infinity' and thereby destroy conformal invariance in the description. As yet, an elegant way of formulating all this has not emerged.

The problem is intimately connected with the corresponding one for gravitation. As things stand, there is only a rather inelegant way of coding the self-dual part of the gravitational field into a deformation of twistor space, coding transformations like $Z^\alpha \to \{1 + f_{-6}(Z^\alpha)\}^{1/6} Z^\alpha$ into a 'blown-up' version of **T**. Again, an elegant combination of such a 'googly graviton' (positive-helicity) with the standard 'leg-break' (negative-helicity) non-linear graviton construction (Penrose 1976a) would be needed (see later).

If such a combined construction were found, then it is conceivable that it might have a bearing on the problem of how the masses of particles are determined from their other quantum numbers. One should note that sourced fields (e.g. Coulomb, Schwarzschild) should, strictly, always involve both self-dual and anti-self-dual parts combined; and the quanta that are involved (photons, gravitons), being 'off shell', do not have well-defined helicities. Thus, there must be some more comprehensive treatment involving both 'leg-break' and 'googly' constructions in combined form, when sources are to be correctly incorporated. It is possible that global considerations in relation to this putative combined construction would lead to severe constraints on the strengths of the sources for these fields. Thus, it is conceivable that α, or basic particle masses, could be determined by the theory (and a corresponding comment would apply to the strong and weak coupling constants). Until such a combined construction becomes available, however, this must remain a matter of pure speculation.

4   Quantum gravity

Of course, quantum gravity was always very much in mind as one of the initial motivations for developing twistor theory. If something drastic happens to the structure of space-time at the very tiny scale of the Planck length, then some radical theory is needed in order to provide the necessary mathematical descriptions. Perhaps the very concept of a space-time point loses its meaning at such scales, in which case it is then necesary to find an alternative geometric description. This description would have to be consistent with the normal ones that we adopt on larger scales—and it could well provide better pictures, at those levels (concerning particles or atoms, or perhaps even molecules) where quantum effects become so important that they confuse our ordinary space-time view. But our description should also allow for the possibility that the very concept of a space-time point can evaporate in situations where quantum gravity effects become important, such as at the Planck length or at space-time singularities. The aspirations of twistor theory are very much in accordance with such goals, and the idea was mooted quite early on (Penrose 1975b) that 'at the quantum gravity level', instead of having a space-time manifold with a 'fuzzy *metric*' at the Planck scale (which had been a more conventional view), twistor theory would have 'fuzzy *points*', yet a well-defined concept of null direction. Some recent work (Mason 1990) has added some substance to this idea, but its proper implementation would have to wait for a more complete union between classical twistor theory and general relativity.

The conviction that classical Einstein theory must find an appropriate (and *elegant*) formulation within twistor theory received its main impetus from the *non-linear* ('leg-break') *graviton* construction (Penrose 1976a). Before that time, it had seemed that twistor theory was an essentially flat- (or conformally flat-) space theory, and that somehow gravitational effects were to enter in discrete 'quantized' elements which effected the 'glueing together' of pieces of Minkowski space (Penrose 1968). It had seemed that at the level of classical general relativity, twistor theory might lose its usefulness, and that gravitational effects might only be incorporated in terms of 'jumps' from one flat space to another. The central idea of the non-linear graviton—that twistor space could itself become 'curved' in a global way, the local structure being identical with that of flat twistor space, but that local curvature information for the space-time could be encoded in this global twistor structure—provided a new viewpoint and a new stimulus. There was now the hope that general relativity at the *classical* level could perhaps be completely transcribed into twistor terms. Indeed, it has become clearer, as time has gone on, that such a transcription is vital, if twistor theory is ever to move significantly further towards its goals, not only in providing a better foundation for quantum gravity, but in making significant headway towards resolving the other major obstacles to progress in physical theory, as described in the other sections of this account.

One argument indicating that the full (vacuum, at least) Einstein equations, rather than merely their linear approximation, needs to find an appropriate twistor description, if twistors are to have a chance of adequately describing quantized gravity, is that gravitational quanta ought not to be described merely by linear spin-2 massless states, but by (complex) solutions of the *full* vacuum equations. (See Penrose 1976; also, in the full Einstein theory, the gravitational radiation field at null infinity $\mathcal{I}$ involves non-linear complications in relation to the Bondi-Metzner-Sachs group, that do not occur with the linearized radiation field, cf. Penrose and Rindler 1986.)

At the time of writing, nearly fifteen years after the original non-linear graviton construction came to light, there is still no clear message as to how Einstein's great theory could be fully encompassed by a twistor formulation. There have been numerous ideas, but all remain rather incomplete as of now. Recent work on the 'ambitwistor' formulation of Einstein's equations has been encouraging (LeBrun 1990, this volume, Baston and Mason 1987). The points of (projective) ambitwistor space $\mathbf{P}\mathcal{A}$ represent complex null geodesics in complexified space-time, where $\mathbf{P}\mathcal{A}$ arises as a deformation of the (doubly) projective space of pairs (Z^α, W_α) for which $Z^\alpha W_\alpha = 0$. Here the Einstein equations are described in terms of the possibility of extending $\mathbf{P}\mathcal{A}$ into $\mathbf{PT} \times \mathbf{PT}^*$ to various infinitesimal orders. From the point of view of twistor wave functions, an ambitwistor description is not really what is wanted, since this would lead to picture more akin to a *classical* function on phase space (like $\Psi(x^a, p_a)$) rather than a quantum wave function (like $\psi(x^a)$ or $\tilde{\psi}(p_a)$). The non-linear graviton description is closer to what is required in this respect, since this indeed provides a 'non-linearization' of a twistor function $f_2(Z^\alpha)$, rather than of something like $F_{2,2}(Z^\alpha, W_\alpha)$ which would be the case for ambitwistors (suffixes indicating degrees of homogeneity). However, the problem as to how best to encode the 'googly' part of the non-linear graviton in terms of a non-linear version of a twistor function $f_{-6}(Z^\alpha)$ is not yet fully resolved.

The present status of the googly problem seems to be that there are indeed 'deformations' of twistor space that can encode the required information, but that it is not yet clear whether this is the appropriate mathematical procedure to adopt, since a natural construction for the corresponding self-dual complex space-time in terms of the googly-deformed twistor space has not yet presented itself. The nature of the proposed deformations is as follows.

I consider just the 'pure' googly problem, in which there is no anti-self-dual field or gravitational sources, and we have to try to encode the information of a self-dual gravitational field into a 'twistor space' $\mathcal{T}$ for which the projective twistor space $\mathbf{P}\mathcal{T}$ is flat, i.e. simply (part of) $\mathbf{PT}$, except that we need to blow up $\mathbf{PT}$, by replacing the line $\mathbf{I}$ (representing Minkowski infinity) by a quadric surface $\mathbf{PI}$, to obtain the blown-up (flat) projective twistor space $\mathbf{PT_I}$. Here, the blown-up non-projective flat twistor space $\mathbf{T_I}$ is defined as the space of

$$\left(\omega^A \pi_{B'}, \pi_{A'}\pi_{B'}\right),$$

where $\pi_{A'} \neq 0$, together with the limit points of that space. These limit points (given by $\omega^A \to \lambda^{-1}\omega^A, \pi_{A'} \to \lambda\pi_{A'}$ as $\lambda \to 0$) constitute the three-dimensional space $\mathbf{I}$, whose projective version is the quadric $\mathbf{PI}$. We now apply transformations $Z^\alpha \to \{1 + f_{-6}(Z^\alpha)\}^{1/6} Z^\alpha$ in a neighbourhood $\mathcal{K}$ of $\mathbf{I}$, giving transition function relations between overlaps of pairs of open sets of a covering the complement of $\mathbf{I}$ in $\mathcal{K}$. These transformations are the ones which locally preserve the map $\mathcal{T} \to \mathbf{P}\mathcal{T}$ and the form (regular on $\mathbf{I}$ because of the $\pi_{S'}\pi_{T'}$ factors):

$$\pi_{S'}\pi_{T'}\epsilon_{\alpha\beta\gamma\delta}dZ^\alpha \wedge dZ^\beta \wedge dZ^\gamma \wedge dZ^\delta.$$

The idea is that the resulting space $\mathbf{T_I}$ locally extends to $\mathbf{I}$, although it does not do so globally. (The construction provides something akin to an 'orbifold', but where there is *functional* freedom at the 'vertex' I.) The cohomology class defined by f_{-6} apparently comes in the form of an element of relative second cohomology, and it is, in effect, 'exponentiated' to give the required local deformation of T_I. When used as an ordinary twistor function to determine a massless spin-2 self-dual radiation field, f_{-6} would give the required self-dual gravitational radiation field at $\mathcal{I}^+$ as 'seen' from (future) timelike infinity i^+.

Of course, even if this programme were entirely successful, since it is based on the behaviour of the space-time at null infinity $\mathcal{I}$, it would still leave open the question of how best to describe *finite* space-time regions. We shall have to be able to describe such regions if, in particular, we are ever to obtain a twistorial understanding of space-time *singularities*—and thus come to terms with what is generally considered to be one of the major tasks of quantum gravity theory.

Indeed, one of the hopes for twistor theory is that it might be possible to provide a coherent description of the physics actually *at* a space-time singularity, where the normal physical concepts, being dependent on ordinary space-time ideas, become inappropriate. It is conceivable that a twistor space can still be used even when the concept of a space-time point has evaporated. One can imagine that some kind of construction analogous to that of the non-linear (leg-break) graviton whereby the space-time points arise as something like 'holomorphic curves' in a curved twistor space $\mathcal{T}$. Such 'curves' would then cease to exist where they would otherwise represent singular points, but the twistor space $\mathcal{T}$ could still be perfectly well defined. For such an idea to work it would certainly be necessary to find a description of (at least vacuum) solutions of the Einstein equations that need neither be self-dual nor anti-self-dual.

These remarks could apply to the 'generic' type of singularity that is expected to arise inside a black hole (or at the big crunch of a collapsing universe) where we anticipate that the Weyl curvature becomes infinite. However, the big bang, if we assume the *Weyl curvature hypothesis*, would have vanishing Weyl curvature, so we have conformal flatness and a simple flat twistor-space description there (cf. Penrose and Rindler 1986). The second law of thermodynamics is associated with a general tendency for the Weyl tensor to get larger with time. Thus, there is a tendency for the twistorial description of space-time to get more and more complicated as time unfolds!

If twistors are ever to provide a good description of a quantum gravity theory capable of coping with the physical problems provided by the existence of classical space-time singularities, then such a description must come to terms with this striking fact that the space-time singularities of our observed universe are indeed *time-asymmetric*. Accordingly, the theory of quantum gravity that twistor theory (or indeed any other theory, if successful agreement with our universe is to be obtained) must be time-asymmetric in its implications. According to my own views (cf. Penrose 1989), there would have to be an actual change in the rules of quantum theory as they are presently understood. The quantized description of gravity would not be a 'standard' one; it would gain its time-asymmetry from the fact that an objective *wave-function collapse* would need to be part of the the resulting theory. As I have stated elsewhere (Penrose 1986b, 1989) it is not implausible to suppose that non-linear effects will ultimately alter the very basis of quantum theory, leading to reduction of the state vector, as soon as the space-time geometry itself becomes subject to quantum superposition (cf. also Károlyházy 1974, Károlyházy, Frenkel and Lukács 1986, Komar 1969). However, mere non-linearity of the ordinary sort (such as the introduction of a non-linear term into Schrödinger's equation) would seem to be quite insufficient. For reasons indicated in the next section, I believe that a radical new viewpoint is needed with regard to the structure of space and time, and that twistor theory could supply an essential input towards finding such a viewpoint.

5   Quantum measurement

One of the most puzzling features of state-vector reduction is its non-local nature. As we know from experiments of the EPR (Einstein-Podolsky-Rosen) type, a measurement in one region of space-time can seemingly cause the state in a distant region to 'jump', the second space-time region being spacelike-separated from the first. The separation between the regions can be many meters, such as in the famous experiment of Aspect and his associates (cf. Aspect and Grangier 1986), or (presumably) even millions of light-years. If we are ever to obtain any kind of objective physical picture of what is going on in such situations, it would seem that any such picture would have to be constructed from essentially non-local ingredients. One of the early motivations for the introduction of a twistor-type formalism was, indeed, the hope that some non-local description of space-time might lead to a more comprehensible picture of reality in relation to puzzling non-local quantum phenomena like those of EPR.

Initially, twistor non-locality was evident simply because the space-time realization of a (projective) twistor is indeed a non-local object: an entire light ray, in the case of a null twistor, and something with also a spatial spread (a Robinson congruence) in the case of a non-null twistor. Conversely, a space-time point has a non-local description in projective twistor space: as a projective line. However, as our understanding of the correspondence between twistor space and space-time descriptions has grown, this non-locality has emerged in more subtle guises. It has been one of the most striking and characteristic features of the twistor representation of physical fields that local field quantities in space-time (Maxwell field strength, anti-self-dual Yang Mills curvature, free massless fields of general spin, the Dirac field, anti-self-dual space-time curvature, etc.) are represented in twistor terms as global structures (holomorphic bundles, 1st cohomology classes, global deformations, etc.) for which the *local* structure in twistor space remains identical with that in the field-free case. Here I refer to *holomorphic* local structure only and not to 'real' information such as the CR-structure of the imbedding of **PN** within **PT**. It is the hope that this pleasing and powerful feature will remain, for the twistor description of *all* physics at the basic level.

However, despite this fundamentally non-local character of such twistor descriptions, it is still not clear what light twistor theory sheds on the puzzling non-locality of EPR-type phenomena. In twistor theory, all basic physical phenomena seem to have a non-local twistor description. Even local space-time activities are described non-locally in twistor terms, so it becomes perhaps less surprising when something else that is twistorially non-local (such as a two-photon state in an Aspect-type EPR experiment) should turn out to have a *non*-local space-time description. In this connection I should perhaps mention the curious 'elemental states' described by Andrew Hodges which are, on the other hand, local in twistor space (H^0-elements) but their space-time translations are non-local (H^1- or H^2-elements). These are expressions like $(A_\alpha Z^\alpha)_n$ or $(B^\alpha W_\alpha)_n$ and they can be viewed as being essential ingredients of twistor diagrams. In space-time terms they provide kinds of 'cohomological wave-functions'.

All this perhaps begs the question of the real puzzle presented by quantum non-locality, which comes in only with the process of quantum *measurement*. From the point of view of twistor theory, there is another feature of quantum measurement—or of state-vector reduction R—apart from its non-locality, indeterminism, non-linearity, discontinuity and time-asymmetry, which distinguishes it from unitary evolution U, and this is the fact that it is *non-holomorphic*. As far as I can make out, all the processes of quantum mechanics that come under the heading of U can be expressed in holomorphic terms. Indeed, holomorphicity has proved to be an extremely useful tool, even in conventional space-time descriptions, particularly in relation to 'causality' and the positive frequency condition (dispersion relations, the analytic S-matrix; also the Riemann surfaces of string and conformal field theories). The rôle of holomorphicity is much more central in twistor descriptions, where global holomorphicity seems to take over even the very foundational rôle of differential equations in space-time physics. The fundamental ingredient of R, on the other hand, is the convertion of a probability amplitude λ into an actual probability according to the decidedly *non*-holomorphic operation

$$\lambda \to \lambda\overline{\lambda}.$$

. The understanding of this holomorphicity/non-holomorphicity in twistorial terms must involve the introduction of new conceptions. I am inclined to believe that not only must gravity be involved in quantum measurement, but also that there is something odd about the way that *time* is involved in the R process. Perhaps there is even some kind of a hint of a beginning of a space-time bifurcation going on, incorporating a partial reversal of time-directionality—which might go some way towards explaining the presence of complex conjugation here. Moreover, the very fact that we have probabilistic behaviour in the Rpart of standard quantum theory, and discontinuous 'quantum jumps', suggests that something radically new is needed in the translation from an essentially holomorphic twistorial description into a real space-time one. It seems to me to be not at all impossible that something like a 'quantum jump' might occur, not localizable in space-time terms, when distortions of twistor space might necessitate jumping from one family of objects in twistor space, representing space-time points, to another (say from one family of holomorphic curves in some curved projective twistor space $\mathcal{T}$ to a separate family of holomorphic curves in the same $\mathcal{T}$, according to the 'leg-break' non-linear graviton; cf. Penrose 1986b). All this would require the *full* Einstein theory to be translatable into twistor terms, if it is indeed true that the problem of quantum measurement, i.e. the process Ris indeed an essentially *quantum gravity* process. As I have stated above, I believe that a (non-linear) modification of the rules of standard quantum theory will be needed if Ris ever to be understood as physical process, and so that the classical world can emerge out of the quantum one. To date, twistor theory has been very conservative in this respect. All we have set ourselves up to do, in twistor diagram theory for example, has been to compute complex-number amplitudes for different physical processes. There has been no attempt to 'explain' why the procedure $\lambda \to \lambda\bar{\lambda}$ should give rise to probabilities, nor even to tell us under what physical circumstances this procedure should actually be applied. These are the problems of quantum measurement. Since, according to the above aspirations, this understanding should be part of the correct quantum gravity theory, it is indeed among twistor theory's ultimate tasks to address such issues also.

6   Where next?

How do I expect that twistor theory might develop in the future, if it is to make any real headway towards resolving any of these grandiose issues? It is, of course, risky to attempt predictions of this kind. For the developments that might be most clearly discernable at present are those, on the whole, that have already been most thoroughly explored. It is likely that the most important future developments will come through some new ideas that cannot now be adequately guessed. There are, however, some promising lines of development in addition to the ones that I alluded to in earlier sections, that have not in fact been explored very far as yet, mainly through lack of time.

A recurrent theme of the earlier sections has been the importance of finding an appropriate twistorial description of the full (at least vacuum) Einstein equations. The lack of such a description seems to be the major stumbling block to fundamental progress in almost all the areas that I have considered above. Moreover, the difficulty in finding one is, in my opinion, not due to an inherent inadequacy in twistor theory, but is indicative of the depth of the physical problems involved in finding the appropriate union between quantum theory and general relativity. Since the very conception of twistor theory was rooted in quantum–space–time ideas, it should not be surprising if it turns out that in order for twistors to catch up with classical general relativity, it will be necessary for them also to go beyond it.

There are some very recent developments concerning the description of space-times, particularly in relation to the Einstein equations, that have arisen through an attempt to bring hypersurface twistor theory and Ashtekar variables into closer relationship (Mason and Penrose 1989). Hypersurface twistors are defined with respect to an (analytic) hypersurface $\mathcal{H}$ (usually either spacelike or one of the null hypersurfaces $\mathcal{I}^-$ or $\mathcal{I}^+$, the spacelike case being considered here) in an (analytic) space-time $\mathcal{M}$. The points of the projective hypersurface twistor space $\mathsf{P}\mathcal{T}(\mathcal{H})$ are *α-curves* in the complexification $\mathbb{C}\mathcal{H}$ of $\mathcal{H}$, where an α-curve is a null curve with tangent vector of the form $p^a = \tilde{\pi}^A \pi^{A'}$ for which

$$p^a \nabla_a \pi_{B'} = 0.$$

In order that p^a be actually tangent to the α-curve, we must have

$$\tilde{\pi}^A = t^{AA'} \pi_{A'},$$

where t^a is normal to $\mathbb{C}\mathcal{H}$ in $\mathbb{C}\mathcal{M}$. The non-projective hypersurface twistor space $\mathcal{T}(\mathcal{H})$ is a line bundle over $\mathsf{P}\mathcal{T}(\mathcal{H})$ (usually without the zero section) where the fibres are given by the scalings for $\pi_{A'}$. Now, associated with any such $\mathsf{P}\mathcal{T}(\mathcal{H})$ (or $\mathcal{T}(\mathcal{H})$) is a complex anti-self-dual conformal 4-manifold $\mathcal{M}(\mathcal{H})$, referred to as *heaven-on-earth*, which is obtained from $\mathsf{P}\mathcal{T}(\mathcal{H})$ by the standard non-linear graviton ('leg-break') construction, the points of $\mathcal{M}(\mathcal{H})$ being holomorphic curves in $\mathsf{P}\mathcal{T}(\mathcal{H})$. A complex 3-submanifold of $\mathcal{M}(\mathcal{H})$, canonically identifiable with $\mathbb{C}\mathcal{H}$, consists of those points of $\mathcal{M}(\mathcal{H})$ which are described by the holomorphic curves in $\mathsf{P}\mathcal{T}(\mathcal{H})$ given by the families of α-curves passing through actual points of $\mathbb{C}\mathcal{H}$

The space $\mathsf{P}\mathcal{T}(\mathcal{H})$, or equivently $\mathcal{M}(\mathcal{H})$, provides a good part of the intrinsic and extrinsic geometry of $\mathcal{H}$, within $\mathcal{M}$, that would be necessary for the initial value problem for general relativity. However, it is lacking in two essential ingredients. One of these is the choice of *conformal scale* for $\mathcal{M}$, the construction of $\mathsf{P}\mathcal{T}(\mathcal{H})$ (and $\mathcal{M}(\mathcal{H})$) being entirely conformally invariant. The other is the actual location of $\mathbb{C}\mathcal{H}$ within $\mathcal{M}(\mathcal{H})$ (i.e. the specification of which holomorphic curves in $\mathsf{P}\mathcal{T}(\mathcal{H})$ actually represent points of $\mathbb{C}\mathcal{H}$). Lionel Mason and I have found that a convenient way of specifying both bits of information is in terms of a rank-two holomorphic affine bundle $\mathcal{B}$ over $\mathsf{P}\mathcal{T}(\mathcal{H})$. The fibre of $\mathcal{B}$ which lies above a point Z of $\mathsf{P}\mathcal{T}(\mathcal{H})$ is the space of solutions $\nu_{B'}$, given χ, of the equation

$$\pi^{A'} \nabla_{AA'} \nu_{B'} = \nabla_{AB'} \chi$$

on the α-plane in $\mathcal{M}(\mathcal{H})$ which meets $\mathbb{C}\mathcal{H}$ in the α-curve representing Z. (Note that $\mathcal{M}(\mathcal{H})$ contains α-planes since it is anti-self-dual, a separate α-plane passing through each α-curve in $\mathbb{C}\mathcal{H}$.) The ∇-operator and metric on $\mathcal{M}(\mathcal{H})$ are chosen to agree, at $\mathbb{C}\mathcal{H}$, with that on $\mathbb{C}\mathcal{M}$, and are such that the scalar curvature of $\mathcal{M}(\mathcal{H})$ vanishes. The scalar χ on $\mathcal{M}(\mathcal{H})$ is chosen so that

$$\chi = 0 \quad \text{on} \quad \mathbb{C}\mathcal{H}$$

where

$$\Box\chi = 0$$

in $\mathcal{M}(\mathcal{H})$, and where $\nabla^a\chi$ defines a *lapse function* on $\mathbb{C}\mathcal{H}$, the normal

$$t^a = \nabla^a\chi,$$

to $\mathbb{C}\mathcal{H}$ in $\mathcal{M}(\mathcal{H})$ being identifiable with that in $\mathbb{C}\mathcal{M}$.

The spinor $\nu^{A'}$ is taken to have homogeneity -1 in $\pi_{A'}$ (which allows the bundle $\mathcal{B}$ to be defined over $\mathbf{P}\mathcal{T}(\mathcal{H})$ rather than just over $\mathcal{T}(\mathcal{H})$ and χ is taken to have homogeneity 0. There is an additional structure to $\mathcal{B}$ defined by the bundle map $\mathcal{B} \to \mathbf{P}\mathcal{T}(\mathcal{H}) \times \mathbb{C}$, given by

$$\nu^{A'} \to \pi_{A'}\nu^{A'} + \chi$$

(the expression on the right being constant on each α-plane of $\mathcal{M}(\mathcal{H})$), so there is a sub-bundle $\mathcal{B}_0$ defined as the kernel of this map (cf. Eastwood 1990). The holomorphic curves in $\mathbf{P}\mathcal{T}(\mathcal{H})$, over which lies a holomorphic cross-section of $\mathcal{B}_0$ define the points of $\mathbb{C}\mathcal{H}$. The necessary initial data at $\mathbb{C}\mathcal{H}$ for the evolution of Einstein's equations are then determined by the remaining structure of $\mathcal{B}$.

Thus far, the Einstein equations have played no special role in this construction, but it turns out that the constraint equations for the Einstein equations can be expressed quite concisely in terms of another bundle over $\mathcal{T}\mathcal{H}$ (or over $\mathbf{P}\mathcal{T}(\mathcal{H})$), closely related to $\mathcal{B}$, whose fibres consist of the local (dual) twistors U_α (of homogeneity -1) on the α-curves with local-twistor derivative along the α-curve, equal to

$$(0, \tilde{\pi}^A\nabla_A^{A'}\chi).$$

This bundle is an 'affinized' version of the cotangent bundle of $\mathcal{T}(\mathcal{H})$, and it is related to ambitwistor space $\mathcal{A}$. It is hoped that the evolution of the Einstein's equations should find a reasonably concise expression in terms of it. Further work is in progress.

What seems really to be required is a twistor space that applies globally to the description of an entire space-time, and which is freed from any dependence on a hypersurface like $\mathcal{H}$. The space $\mathcal{A}$ is not really the proper analogue of flat twistor space T (it has too many dimensions, and is really the anologue of the hypersurface CN in $T \times T^*$), and $\mathcal{T}(\mathcal{I}^-)$ or $\mathcal{T}(\mathcal{I}^+)$ though canonical, if $\mathcal{I}^-$ or $\mathcal{I}^+$, does not carry all the necessary information. Even if the 'googly' information is incorporated into $\mathcal{T}(\mathcal{I}^\pm)$, say according to the description indicated earlier (where sources, black holes, etc. are assumed to be absent), then we still have the difficulty of extracting the information about local space-time regions. This seems a formidable problem, and some way of encoding the Einstein evolution (perhaps incorporating the ideas indicated above) seems essential.

If any such twistorial description of general relativity turns out to be successful, then it is to be hoped that it would tie in with the expression for *quasi-local mass* (and angular momentum, etc.). See Penrose (1982), Penrose and Rindler (1986), Tod (1990 this volume). That expression is still in need of a better theoretical foundation, so that ambiguities in the definition and other matters may be cleared up. (See Mason and Frauendiener 1990, this volume, for some new ideas concerning its relation to the Hamiltonian for general relativity.)

One point concerning the further development of twistor theory, which is exemplified by the above ideas for a 'googly' construction, is that for many purposes it is necessary to consider the *non*-projective twistor space T or $\mathcal{T}$ as fundamental, rather than the projective twistor space PT or $P\mathcal{T}$. This has already been a feature of the twistor particle programme for a long time, since the helicity operator, which defines homogeneity for a twistor function of just one twistor variable, is merely one among a list of different twistor operators defining particle quantum numbers. Moreover, in Hodges's introduction of the 'k' term in twistor diagram theory, we have another example of an essential rôle for non-projective twistor space. It seems to me to be likely that any adequate description of Einstein vacuum space-times will also be an essentially non-projective one. This would certainly appear to have to be the case if the optimistic hope of describing general solutions of the Einstein vacuum equations in terms of a single twistor space can ever be realized.

It is clear that in many respects twistor theory remains far from its goals; yet there are others in which it has achieved things that had not originally been anticipated, and there have certainly been enough positive signals along the route to suggest that the theory has a good deal more to contribute to fundamental physics. However, some of the most obstinate problems seem almost as intractable as ever. What is the best route to further progress? Perhaps more account should be taken of conventional successful theories than has been done so far, and some new and useful input can be thereby obtained. Indeed, one of the criticisms of twistor theory (and of twistor theorists) in the past has been that they have paid too little attention to the ideas that are currently popular or successful in particle physics and quantum field theory.

As a counter to such criticisms, I should draw attention to the fact that in the important developments of twistor diagram theory that have been made over recent years by Andrew Hodges, the standard models of electroweak theory and QCD have been the guiding influences, so that I think it can now be said that these models can essentially be completely described in twistor terms. It is true that path integrals, as such, have not yet had a great deal of influence on twistor theory, but there are certain fundamental difficultes in bringing these ideas together, as was indicated earlier. The currently popular ideas of conformal field theory have certainly had their impact on twistor thinking, as I have mentioned above, but there are still some severe difficulties in implementing these ideas.

Another line of thought that has been very popular in recent years, as part of the impact of modern quantum field theory on pure mathematics, is the idea of a *topological* quantum field theory. Here one is concerned with the quantized version of a 'field theory', such as $(2 + 1)$-gravity (cf. Witten 1988/9), in which there is no local field information and all the 'physics' is stored in global structure. This kind of situation is of interest *mathematically*, where one may be concerned with such questions as the global topology of manifolds or with knots or links, but its interest from the physical point of view would seem to be severely limited if the ambient space of the theory is to be interpreted as 'space-time' (as is normally intended to be the case). However, twistor theory sugggests a different tack. We are now familiar with the fact that local space-time field information is not stored locally in twistor space but globally. Thus, if we could re-interpret the ambient space of a topological quantum field theory as *twistor space* rather than space-time, then the above problem with physical interpretation would be turned to our advantage. Among the new difficulties that we now must face, however, is the fact that twistor space is *complex*, not real, so there is a whole host of new problems in connection with how one must interpret the 'complexification' of a real topological situation. A start has been made to the understanding of these issues in the concept of 'holomorphic linking', which seems to have some tantalising connections with twistor diagram theory (Penrose 1988. 1990b). There is even the hope that some conection might ultimately be made with an early, but almost forgotten, motivation behind twistor theory, that of *spin-networks* (Penrose 1971, 1972). The idea here was that there should be an underlying *combinatorial* foundation to physics, which has seemed to be at odds with the underlying *complex-holomorphic* foundation to twistor theory. However, these foundations may not be so far apart from one another as they seem. Combinatorial and holomorphic concepts can often enrich one another in important ways. In the idea of holomorphic linking there is a clear aspect of this, and it is conceivable that the long-awaited relationship between twistor diagrams and spin networks may come to light through this kind of idea.

Finally, the fact should should not be ignored that despite all the mar-vellous new insights that have come to us through quantum field theory, it is *general relativity* that is now *the most accurate theory known to science*! (Observations of the binary pulsar PSR 1913 + 16 have shown GR to be accurate to about one part in 10^{14} (T. Damour, private communication; see Weinberg and Taylor 1984) whereas the accuracy of QED, as exemplefied by the calculated accuracy of the magnetic moment of the electron, is known to only one part in 10^{11}. It is the challenge of coming to terms with Einstein's marvellous theory, in the form in which he actually gave it to us, that has engaged the attention of twistor theorists more than has any other aspect of physics.

References

Ashtekar, A (1988) *New perspectives in Canonical Gravity.* Bibliopolis, Naples.

Bailey, T.N. (1985) Twistors and fields with sources on worldlines. Proc. Roy. Soc. **A397**, 143-155

Baston, R.J. and Mason, L.J. (1987) Conformal gravity, the Einstein equations and spaces of complex null geodesics. Class. Quan. Grav. **4** 815-826.

Eastwood, M.G. (1990) The Penrose transform without cohomology, in Mason and Hughston (1990).

Ginsberg, M.L. (1983) Scattering theory and the geometry of multitwistor spaces. Trans. Amer. Math. Soc. **276**, 789-815.

Ginsberg, M.L. and Huggett, S.A. (1979) Sheaf cohomology and contour integrals. In Hughston and R.S. Ward (1979), pp.287-292.

Hodges, A.P. (1982) Twistor diagrams. Physica **114A**, 157-75.

Hodges, A.P. (1983a) Twistor diagrams and massless Moller scattering. Proc. Roy. Soc. Lond. **A385**, 207-228.

Hodges, A.P. (1983b) Twistor diagrams and massless Compton scattering. Proc. Roy. Soc. Lond. **A386**, 185-210.

Hodges, A.P. (1985a) A twistor approach to the regularization of divergences. Proc. Roy. Soc. London **A397**, 341-74.

Hodges, A.P. (1985b) Mass eigenstates in twistor theory. Proc. Roy. Soc. London **A397**, 375-96.

Hodges, A.P., Penrose, R. and Singer M.A. (1989) A twistor conformal field theory for four space-time dimensions, Phys Lett. **B216**, 48-52.

Hodges, A.P. (1990) String amplitudes and twistor diagrams, an analogy. In Quillen, Segal and Tsou (1990).

Hodges, A.P. (1990b) this volume.

Hughston, L.P. (1979) Twistors and Particles. Lecture Notes in Physics No. 97 (Springer-Verlag, Berlin).

Hughston, L.P. and Ward R.S. (1979) Advances in Twistor Theory. (Pitman Notes in Mathematics, Pitman, San Francisco).

Isham, C.J., Penrose, R. and Sciama D.W., eds. (1975) Quantum Gravity, an Oxford Symposium, (Oxford University Press, Oxford).

Károlyházy, F. (1974) Gravitation and quantum mechanics of macroscopic bodies, Magyar Fizikai Polyoirat **12**, 24.

Károlyházy, F., Frenkel, A. and Lukács, B. (1986) On the possible role of gravity on the reduction of the wave function. In *Quantum Concepts in Space and Time* eds. R.Penrose and C.J.Isham (Oxford University Press, Oxford).

Komar, A.B. (1969) Int. J. Theor. Phys. 2, 257.

LeBrun, C.R. (1985) Ambitwistors and Einstein's equations Class. Quant. Grav. **2**, 555-63

Mason, L.J. (1990) Insights from twistor theory, in 'Conceptual problems in quantum gravity', ed. J. Stachel, Birkhauser.

Mason, L.J. and Frauendiener (1990) The Sparling three-form, Ashtekar variables and quasi-local mass (this volume).

Mason, L.J. and Hughston, L.P. (1990) Further Advances in Twistor Theory, Vol. 1: *The Penrose transform and ith applications.* (Pitman Notes in Mathematics 231, Longman, Harlow Essex, UK and John Wiley New York, USA.)

Penrose, R. (1968) Twistor Quantization and curved space-time Int. J. Theor. Phys. **1**, 61-99.

Penrose, R. (1971) Angular momentum: an approach to combinatorial space-time, in Quantum theory and Beyond, ed. Ted Bastin (Cambridge University Press, Cambridge).

Penrose, R. (1972) On the nature of quantum gravity. In *Magic without Magic*, ed. J.R.Klauder (Freeman, San Francisco).

Penrose, R. (1975a) Twistors and particles: an outline. In *Quantum Theory and the Structures of Time and Space*, eds. L. Castell, M. Drieschner and C.F. von Weizsäcker (Carl Hanser Verlag, Munich).

Penrose, R. (1975b) Twistor theory: its aims and achievements. In Isham, Penrose and Sciama 1975.

Penrose, R. (1976) Non-Linear gravitons and curved twistor theory Gen. Rel. Grav. **7**, 31-52.

Penrose, R. (1976b) How to grow (or at least half grow) new twistors from old. Twistor Newsletter **2**, reprinted in Mason and Hughston (1990).

Penrose, R. (1979) The backhanded (googly) photon, In Hughston and Ward (1979).

Penrose, R. (1979b) A googly graviton. In Hughston and Ward (1979).

Penrose, R. (1979c) The Universal Bracket Factor. In Hughston and Ward (1979).

Penrose, R. (1981) Time-asymmetry and quantum gravity. In *Quantum Gravity 2*, eds D.W.Sciama, R.Penrose and C.J.Isham (Oxford University Press, Oxford).

Penrose, R. (1982) Quasi-local mass and angular momentum in general relativity. Proc. Roy. Soc London **A381**, 53-63.

Penrose, R. (1986a) On the origins of twistor theory Gravitation and Geometry *I. Robinson Festschrift*, ed. W. Rindler and A. Trautman, Bibliopolis, Naples.

Penrose, R. (1986b) Gravity and state-vector reduction. In *Quantum Concepts in Space and time eds. R.Penrose and C.J.Isham* (Oxford University Press, Oxford) 126-146.

Penrose, R. (1988) Topological QFT and twistors: holomorphic linking. Twistor Newsletter **27**; Holomorphic linking, postcript, *ibid*.

Penrose, R. (1989) *The Emperor's New Mind: Concerning Computers, Minds, and the Laws of Physics* (Oxford University Press, Oxford).

Penrose, R. (1990) Non-Hausdorff Riemann surfaces and complex dynamical systems, (in Mason and Hughston 1990).

Penrose, R. (1990b) Twistor particles, strings and links. In Quillen, Segal and Tsou (1990).

Penrose, R. and MacCallum, M.A.H. (1972) Twistor theory: an approach to the quantization of fields and space-time, Phys. Repts. **6C**, 241-315.

Penrose, R. and Rindler, W. (1986) *Spinors and Space-Time, Vol. 2: Spinor and Twistor Methods in Space-Time Geometry.* (Cambridge University Press, Cambridge).

Penrose, R. and Sparling, G.A.J. (1979) The anti-self-dual Coulomb field's non-Haussdorff twistor space, Twistor Newsletter **9**. In Mason and Hughston (1990).

Penrose, R., Sparling, G.A.J. and Tsou, S.T. (1978) Extended Regge Trajectories. J. Phys. **A11**, L231.

Perjes, Z. (1977) Perspectives of Penrose theory in particle physics. Rept. Math Phys. **12**, 193-211.

Perjes, Z. (1982) Introduction to twistor particle theory. In *Twistor Geometry and Non-Linear Systems*, ed. H.D. Doebner & T.D. Palev (Springer-Verlag, Berlin) Lecture Notes in Mathematics **970**,53-72.

Perjes, Z. (1983) Twistor internal symmetry groups Group theoretical methods in physics, ed. M.A. Markov, Nauka, English translation by Harwood Acad. Publ. Co. p.631-645.

Perjez, Z. and Sparling, G.A.J. (1979) The twistor structure of hadrons. In Hughston and Ward (1979) pp. 192-7.

Qadir, A. (1978) Penrose Graphs. Phys. Repts. **39C**, 131-67.

Quillen, D.G., Segal, G.B. and Tsou S.T (1990) The Interface of Mathematics and Particle Physics (I.M..A Conference Series, Clarendon Press, Oxford).

Segal, G.B. (1990) The definition of conformal field theory (*to appear*).

Singer, M.A. (1990) Twistors and four-dimensional quantum field theory. In Quillen, Segal and Tsou (1990).

Tsou S.T. (1981) Extended Regge trajectories updated. Twistor Newsletter **12**. In Mason and Hughston 1990.

Ward, R.S. (1977) On self-dual gauge fields, Phys. Lett. **61A**, 81-2.

Weisberg, J.M. and Taylor, J.H. (1984). Phys.Rev.Lett. **52**, 1348-50.

Witten, E. (1988) Quantum field theory, Grassmanians and algebraic curves. Commun.Math.Phys. **113**, 529-600.

Witten, E. (1988/89) 2+1 dimensional gravity as an exactly solvable system. Nucl.Phys. **B311**, 46-78.

Between Integral Geometry and Twistors

S.G. Gindikin

A remarkable feature of the twistor theory created by Roger Penrose is that it has turned out to be related to different areas of mathematics. Individual mathematicians are attracted by different aspects of the theory and for very different reasons. For me the starting point was integral geometry, and I would like to recall here the circumstances under which my friends and I more than 10 years ago became interested in twistor theory.

In 1977 Henkin and I began to consider the remarkable results of Martineau on linearly concave domains. A domain D in $\mathbb{C}^n$ (or, better, in $\mathbb{CP}^n$) is said to be linearly concave if it is a union of the (holomorphic) hyperplanes contained within it. Then the domain $D_1 = \mathbb{C}^n \setminus \bar{D}$ is a domain of holomorphy (linearly convex domain) and for a class of sufficiently regular boundaries there is a remarkable duality: functionals on the space $H(D_1)$ of holomorphic functions on D_1 correspond to holomorphic functions on the dual domain D^* the points of which are hyperplanes belonging to D (the Fantappie indicators of functionals).

On the other hand, these functionals correspond to $(n-1)$-dimensional $\bar{\partial}$-cohomology with coefficients in Ω^n. What is the relation between the two different objects corresponding to functionals, i.e. cohomology and functions? A natural conjecture was that one has to integrate (Dolbeault) cohomology over hyperplanes in D, i.e. to consider a kind of Radon transform for $\bar{\partial}$-cohomology in D. Indeed, we have been able to prove that this version of the Radon transform coincides with the Fantappie indicator.

Next we undertook a systematic investigation of the Radon transform for $\bar{\partial}$-cohomology. Most instructive for us was the problem of finding an inversion formula, i.e. of reconstructing the cohomology class from its Radon transform. How can one write an analytic formula in this situation where many forms represent the same cohomology class? It turned out that one has to consider the same differential form that appears in the inversion formula for the real Radon transform (in an odd-dimensional space) and which is obtained from the Radon transform by a differential operator. One gets a holomorphic form on the flag manifold F the points of which are pairs consisting of a point $z \in D$ and a hyperplane π lying in D passing through z (i.e. $\pi \in D^*$). Note that F is a Stein manifold. The decisive point is that the integration is not carried out as in the real case but after restricting the resulting form to a section (naturally, non-holomorphic!) of the natural bundle $F \to D$. For different sections one gets different representatives of the cohomology class.

There arose in that way a kind of holomorphic continuation of $\bar{\partial}$-cohomology to the Stein manifold F which in analytical problems turned out to be more convenient and effective than the original cohomology. It became possible to use in the study of higher $\bar{\partial}$-cohomology the language of holomorphic functions and forms.

In the real integral geometry an important role is played by a generalized Radon transform when one integrates over planes of codimension greater than 1. Such an approach goes back to John who by associating to functions in $\mathbf{R}^3$ their integrals over lines obtained functions in four variables which are solutions of an ultra hyperbolic equation.

Starting from that construction, we considered a generalization of the notion of linearly concave domain, which we called $(n-1-q)$-concave domains D. Such domains are unions of the q-dimensional planes contained within them. There then arises a dual (Stein) domain D^* in the Grassmannian $G_{n,k}$, the points of which are the q-planes belonging to D. The Andreotti-Norguet results imply that the qth $\bar{\partial}$-cohomology for D (with coefficients, say, in Ω^n or in some other natural bundle) is infinite dimensional. We studied the q-dimensional Radon transform for this cohomology, i.e. integrals over q-planes in D. The result is a holomorphic function on D^* (or a section of a holomorphic bundle according to the type of coefficients). For $q < n - 1$ one does not obtain all functions (sections), but only those that are solutions of a system of differential equations with constant coefficients (depending on q but not on the specific form of the domain D) generalizing the John equations. In order to prove the main theorem on the isomorphism between the qth $\bar{\partial}$-cohomology in D and the space of holomorphic solutions of that system of equations in D^* we had to apply sophisticated analytic techniques. Using the Gelfand-Graev-Shapiro x-operator we have been able to write out the inversion formula for arbitrary q in the same form as the above mentioned formula for $q = n - 1$.

It should be noted that Andreotti and Norguet had already considered the operation of integrating $\bar{\partial}$-cohomology over cycles of the same dimension. Our results can be interpreted as an effective description of that operator (kernel, image) for $(n - 1 - q)$-linearly concave domains.

In 1978 we announced our results in [1] with detailed publications in [2]. When we were preparing our papers for publication, we learned (from notes of Wells' lectures) about the Penrose transform and took a great interest in possible application of our results to twistor theory. We realized that we actually had a ready proof of the fact that the Penrose transform yields an isomorphism between the space of first $\bar{\partial}$-cohomology for positive twistors and the space of holomorphic solutions of corresponding massless equations in the future tube (which was not apparently known at the time).

Immediately after having become acquainted with twistor theory we considered the isomorphism problem for $\bar{\partial}_b$-cohomology spaces (tangent Cauchy-Riemann cohomology) and the space of solutions of massless equations on the conformal compactification of Minkowski space. Here one encounters additional analytic difficulties arising because there is no Grothendieck-Dolbeault lemma on a CR-manifold. We obtained several versions of the isomorphism theorem for different functional and distribution spaces. A specialization of those results to Maxwell's equations as well as other "twistor" results have been included in [3]. A summary of our joint activity in twistor theory was given in the survey paper [4]. We would also like to mention the survey paper on integral geometry [5] which makes very clear the reverse influence of twistor theory on real integral geometry.

Among my papers related to twistor theory and published after [4] article [6] should be mentioned. It defines the Fantappie indicator for q^{th} $\bar{\partial}$-cohomology in $(n - 1 - q)$-linearly concave domains for $q < n - 1$ and also includes explicit integral formulae reconstructing the Radon transform from the $\bar{\partial}$-cohomology boundary values. The formula has a holomorphic kernel in the sense that a formula with a holomorphic kernel reconstructs the above holomorphic 'continuation' of a cohomology class on the Stein flag domain F, a bundle over D. Restricting that form to sections one obtains representatives of a cohomology class.

For the Penrose transform one obtains an integral formula reconstructing the Penrose transform from the $\bar{\partial}$-cohomology boundary values on the null twistor surface.

Studying other aspects of integral geometry I was unexpectedly brought back to twistor theory but in its curved version. In 1978 I M Gelfand, Z Ya Shapiro and I were investigating the general inversion problem for the integral transformation $f \to \hat{f}$ associating to each function f on a complex manifold X its integrals $\hat{f}$ over curves from some family of curves M on X. We were interested in local inversion formulae, i.e. formulae in which one integrates a form obtained from f by applying a differential operator over a submanifold of curves going through the point at which the value of the function is reconstructed. It turned out that all such forms are restrictions of a universal form $x\hat{f}$ defined on the infinite dimensional manifold of all curves on X (and which is obtained from $\hat{f}$ by applying a differential operator). A local inversion formula exists if and only if the restriction of that form to M can be computed through the restriction of $\hat{f}$ to M. There arises a non-linear charactericity type condition on M. A manifold M of curves satisfying that condition is called admissible.

I N Bernstein and I studied admissible manifolds of curves in [8, 9]. As it turned out, curves from an admissible family can be canonically continued to rational curves and those families have to be full in some natural sense.

Our main result is an effective description of admissible subfamilies in an admissible family of curves (e.g. in the family of lines or second order curves or sections of a vector bundle on the projective line).

The admissibility condition is equivalent to the existence of a generalized conformal structure on M with an integrability condition. In particular, for the 4-parameter family of curves on a 3-manifold one obtains a self-dual conformal metric. The admissible subfamilies theorem makes it possible to construct examples of such metrics automatically. For example, one can describe all 4-parameter families of second order curves in CP^3 (an 8-parameter family) on which the incidence relation induces a conformal 4-metric. It is interesting to compare these constructions with the Penrose non-linear graviton construction. Penrose considers perturbations of a flat 4-metric generated by a perturbation of the complex structure on the twistor manifold. Then our restriction of some generalized multidimensional conformal structures to 4-dimensional submanifolds means, in the twistor language, a special kind of birational transformation.

After the problem of self-dual 4-metrics we proceeded to the investigation of the problem of right-flat metrics (i.e. self-dual solutions of the vacuum Einstein equations). We again tried to construct examples by restricting multidimensional structures rather than by perturbations. We defined such generalized metric structures on manifolds of rational curves on 3-dimensional (twistor) manifolds. In particular, the problem of right flat metrics on M, dim $M = 4$, is equivalent to the problem of constructing on M a bundle $F(t)$ of closed simple forms (i.e. $dF(t) = 0$, $F(t) \wedge F(t) = 0$) which depends quadratically on the parameter $t \in C$. The language of bundles of differential forms is discussed in detail in [10]. Explicit constructions of right flat metrics are given in [11]. Generalizations to hyperKähler metrics can be found in [12].

The important common aspect of the above problems of integral geometry both for curves and right flat (hyperKähler) metrics is the leading role played by families of rational curves. These and other non-linear equations can be formulated as a compatibility condition for a system of linear differential equations of the first order with rational spectral parameter. The families of rational curves approach is developed in [13].

In twistor theory we believe it is important to consider, instead of curves, submanifolds of greater dimension. In integral geometry there is progress in that direction (see, e.g. [14]). It would be interesting to analyse similar problems for differential equations. In our view, for problems considered here, various types of generalized conformal structures are very important. This notion is discussed in a separate article [15] in this volume.

In conclusion we mention that [16, 17, 18, 19] are survey papers on the above problems.

References

[1] Gindikin, S.G. & Henkin, G. *Transformation de Radon pour la $\bar{\partial}$-cohomologie des domains q-lineairment concaves*, C. R. Acad. Sci., 1978, AB 298, N4 209–212.

[2] Gindikin, S.G. & Henkin, G. *Integral geometry for $\bar{\partial}$-cohomology in q-linearly concave domains in* $\mathbb{CP}^n$, Funct. Anal. Appl. **12** (1978) 4, 6–23 (Russian).

[3] Gindikin, S.G. & Henkin, G. *Complex geometry and Penrose representation for solutions of the Maxwell equation*, Teoret i matemat. phyzika **43** (1980), 18–31 (Russian).

[4] Gindikin, S.G. & Henkin, G. *Radon transformation and complex integral geometry*, Modern Problems of Mathematics, **17**, Moscow, VINITI (1981), 57–111 (Russian).

[5] Gelfand, I., Gindikin, S & Graev, M. *Integral geometry in affine and projective spaces*, Modern Problems of Mathematics **16** VINITI, USSR Acad. of Sci. Moscow (1980) 52–226 (Russian).

[6] Gindikin, S.G., *Integral formulas and integral geometry for $\bar{\partial}$-cohomology in* $\mathbb{CP}^n$, Funct. Anal. Appl., **18** (1984) 2, 26–39 (Russian).

[7] Gelfand, I., Gindikin, S. & Shapiro, Z. *Local problem of integral geometry in the space of curves*, Funct. Anal. Appl., **13** (1979), 2, 11–31 (Russian).

[8] Gindikin, S., *Integral geometry and twistors*, Lecture Notes Math., **970** (1980) 2–42.

[9] Bernstein, J. & Gindikin, S., *Four papers on integral geometry and Einstein equation* Reports Dep. Math., Univ. of Stockholm **23** (1986), 1–61.

[10] Gindikin, S.G., *Bundles of differential forms and the Einstein equation*, Nuclear Physics **36** (1982), 537–548 (Russian).

[11] Gindikin, S.G., *Some solutions of the self-dual Einstein equation* Funct. Anal. Appl., **19** (1985), 58–60 (Russian).

[12] Gindikin, S.G., *On one construction of hyperkähler metrics*, Funct. Anal. Appl., **20** (1986), 82–83 (Russian).

[13] Gindikin, S.G., *Reduction of manifolds of rational curves and related problems of differential equations theory*, Funct. Anal. Appl., **18** (1984), 14–39 (Russian).

[14] Gindikin, S.G., *Horospheres and twistors*, Lect Notes in Physics **313** (1988), 3–10.

[15] Gindikin, S.G., *Generalised conformal structure*, this volume.

[16] Gindikin, S.G., *Integral geometry as geometry and as analysis*, Contemporary Mathematics, **63** (1987) 75–107.

[17] Gindikin, S.G., *Differential geometry problems in integral geometry* Diff. Geom. and its Appl., Proc of the Conf. 1986, Brno 53–61.

[18] Gindikin, S.G., *Twistors, rational curves, quaternionic structures, infinite dimensional Lie algebras and quantum field theory*, World Sci. 1988 11–27, Proc of the Varna Summer School, 1987.

[19] Gindikin, S.G., *Twistors, integral geometry, generalized conformal structures*, Geometry and Physics, 1989 (in print).

Generalized Conformal Structures

S.G. Gindikin

A complex conformal structure is defined by a choice of quadratic cones in tangent spaces. Similar structures appear in classical mathematics where quadratic cones are replaced by other cones. Such structures have recently arisen in some problems of mathematical physics and integral geometry. We consider that fact to be of fundamental importance and for that reason deem it necessary to discuss those examples in more detail.

1 Basic Definitions and Examples

1.1 V-conformal structures

A class of complex generalised structures is defined by a choice of algebraic cone $V \subset \mathbf{C}^n$. We say that a *V-conformal structure* is given on a complex manifold M, dim $M = n$, if in each tangent space $T_z M$, $z \in M$, one has chosen a cone $V_z \subset T_z M$ which is linearly equivalent to V. The usual conformal structure under that definition corresponds to a non-degenerate quadratic cone V.

In the sequel all real objects are assumed to be C^∞ and all complex objects to be holomorphic.

Consider a set $\Sigma = \Sigma(v)$ of maximal linear subspaces σ, dim $\sigma > 1$ lying on V, and let $\Sigma_1, \ldots, \Sigma_l$ be its connected components. We say that a V-conformal structure on M is Σ_i-*integral* if

$$\bigcup_{\sigma \in \Sigma_i} \sigma = V$$

and for each point $z \in M$ and each $\sigma \in \Sigma_i(V_z)$ there exists (locally) a unique submanifold $S \ni z$ such that $T_z S = \sigma$ and $T_w S \in \Sigma_i(V_w)$ for all $w \in S$.

Note: Sometimes it is convenient not to exclude the case dim $\sigma = 1$. In that case an integral surface always exists (locally), although, unless V is reduced to a finite number of lines, there is no uniqueness. On the other hand, for dim $\sigma > 1$ the uniqueness condition is true in a very general situation (see, e.g. [3]).

Real generalised conformal structures correspond to real forms $V_{\mathbf{R}}$ of complex cones V, where $V_{\mathbf{R}}$ is defined by an involution ρ which preserves V and leaves invariant the points of a real subspace $\mathbf{R}^n_\rho$ in $\mathbf{C}^n \cong \mathbf{R}^{2n}$; $V_{\mathbf{R}} = V \cap \mathbf{R}^n_\rho$.

Both the cone $V_{\mathbf{R}}$ and its generators σ can be imaginary. In that case $V_{\mathbf{R}}$ is given by a system of (homogeneous) algebraic equations which have no real solutions. A cone V_x (either real or imaginary) linearly equivalent to $V_{\mathbf{R}}$ is fixed in every tangent space $T_z M_{\mathbf{R}}$. In this paper we deal mainly with complex generalised conformal structures although their real forms also deserve special attention.

1.2   Twistor manifolds

Under the Σ_i-integrability condition, denote by $\mathcal{T}_i$ the manifold of (local) integrable submanifolds S (i.e. $T_z S \in \Sigma_i(V_z)$ for all $z \in S$). We call $\mathcal{T}_i$ a Σ_i-twistor manifold for that V-conformal structure. Then

$$\dim \mathcal{T}_i = \operatorname{codim} \sigma + \dim \Sigma_i$$

and any point $z \in M$ corresponds to a submanifold E_z, $\dim E_z = \dim \Sigma_i$ in $\mathcal{T}_i$ (i.e. the submanifold of integrable submanifolds going through z). Evidently, any V-conformal structure is fully reconstructed from the family E_z. Indeed, take the family M of submanifolds E_z and consider submanifolds S_u for $u \in \mathcal{T}_i$:

$$z \in S_u \Leftrightarrow u \in E_z$$

Taking $\sigma = T_z S_u$ for such S_u that $z \in S_u \Leftrightarrow u \in E_z$ and considering the closure of $\bigcup \sigma$ one gets the cone V_z.

Consider the geometric structure corresponding to the incidence relation between $\mathcal{T}_i M$:

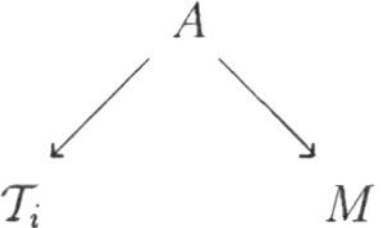

where A consists of pairs (u, z), $u \in \mathcal{T}_i$, $z \in M$ for which $u \in E_z$ (or, equivalently, $z \in S_u$) with natural projections on $\mathcal{T}_i$ and M. Such structures, called double fibrations, have been introduced in integral geometry [4]. As we have just seen in some cases they are equivalent to (integrable) generalised conformal structures. In our view the infinitesimal language of generalised conformal structures is often more effective then the global language of double fibrations.

1.3   Flat structures

Let $M = \mathbf{C}^n$, and let V_z denote the shift of the cone $V \subset \mathbf{C}^n$ into the point $z \in \mathbf{C}^n$. Such a V-conformal structure as well as those equivalent to it is called *flat*. Affine automorphisms of a flat structure consist of compositions of shifts and linear automorphisms of the cone V. In some cases (e.g., for the usual conformally flat structure) the group of automorphisms include other elements (there are generalised inversions). A natural question is that of compactifying a manifold with a flat structure. Other natural problems are:

(W) *The Weyl Problem.* Are there any locally non-flat manifolds with a V-conformal structure? If so, how can one define their curvature as an obstacle to the local trivialisation?

(L) *The Liouville Problem.* Are there any local automorphisms of a flat V-conformal structure that cannot be extended to a global one?

(P) *The Penrose Problem.* Is a V-conformal structure which is Σ_i-integrable for all i necessarily a flat one?

Evidently, any flat structure is Σ_i-integrable. The problem (L) is also interesting in the non-flat case.

1.4   Two dimensional examples

A few words on the basic examples. For $n = 2$ a complex conformal structure is given by any pair of intersecting lines which, in the non-degenerate cases, do not coincide. It has two real forms. The most interesting of them is the structure corresponding to a pair of imaginary lines. In that case the complex conformal structure coincides with the structure of a one-dimensional complex curve on a two-dimensional real manifold. The corresponding flat structure is that of $\mathbf{C}^1$ or $\mathbf{R}^2$. As for the Liouille problem, there evidently exist an infinite-dimensional family of local automorphisms of the flat structure which can not be extended to global ones. It is for that reason that one has one-dimensional complex analysis or, equivalently, a theory of conformal mappings on a plane.

For a real form corresponding to a pair of real lines on $M_{\mathbf{R}}$ there arises (locally) a coordinate network (of tangent directions) and the automorphisms transform each coordinate separately. Thus two real forms of the same structure yield substantially different theories.

1.5   Higher dimensional examples

In the multidimensional case, the preceding example gives rise to two series of structures. First, there are conformal structures corresponding to non-degenerate quadratic cones. The central result here is the Liouville theorem stating that for $n > 2$ (both in the complex case and for all real forms) local automorphisms of a conformally flat structure can be extended to global ones, and that, consequently, the group of local conformal automorphisms is finite-dimensional.

However, there is another way of generalising the basic example. Take for a cone $V \subset \mathbf{C}^{2k}$ a pair of subspaces of dimension k which intersect only at zero. Then, if $V_{\mathbf{R}}$ consists of imaginary subspaces, the $V_{\mathbf{R}}$-conformal structure is an almost complex structure; Σ consists of two elements and the condition of Σ-integrability coincides with the integrability of the almost complex structure. There is (fortunately?) no analogue of the Liouville theorem for that structure which explains the fact that there is a non-trivial local multidimensional complex analysis.

1.6   *The classical twistor picture*

Consider in more detail a conformal structure on $\mathbf{C}^4$. Choose coordinates in such a way that the cone V is defined by the equation

$$z_1 z_2 - z_3 z_4 = 0.$$

There are two families of generators: Σ_1 consists of two-dimensional subspaces

$$\begin{aligned} z_1 \tau_0 + z_3 \tau_1 &= 0 \\ z_4 \tau_0 + z_2 \tau_1 &= 0 \end{aligned} \quad \left(\tau \in \mathbf{C}^2, \tau \neq (0,0) \right),$$

and Σ_2 consists of subspaces

$$\begin{aligned} z_1 \tau_0 + z_4 \tau_1 &= 0 \\ z_3 \tau_0 + z_2 \tau_1 &= 0 \end{aligned} \quad \left(\tau \in \mathbf{C}^2, \tau \neq (0,0) \right).$$

Any two planes of the same family intersect in a single point, while any two planes from different families have a common line. Both families are parameterised by points of the projective line $\mathbf{C}P_\tau^1$. Integral surfaces for Σ_i are of the form

$$\begin{aligned} z_1 \tau_0 + z_3 \tau_1 &= u_1, \\ z_4 \tau_0 + z_2 \tau_1 &= u_2, \end{aligned}$$

i.e. the twistor manifold $\mathcal{T}_1$ is obtained from the projective space $\mathbf{C}P^3$ with homogeneous coordinates $(\tau_0, \tau_1, u_1, u_2)$ by deleting the line $\{\tau_0 = \tau_1 = 0\}$.

As a result we have interpreted the points of the conformally flat space $\mathbf{C}^4$ as those lines in $\mathbf{C}P^3$ that do not intersect the line $\{\tau = 0\}$, the conformal distance between lines vanishing if and only if the lines intersect. Considering all lines in $\mathbf{C}P^3$ one has a conformal compactification of $\mathbf{C}^4$: one adds to $\mathbf{C}^4$ 'at infinity' the cone V of lines intersecting the line $\{\tau = 0\}$. With the use of Plücker coordinates that compactification is canonically identified with the quadric in $\mathbf{C}P^5$ (the Plücker-Klein quadric), that is, the usual complexified, compactified Minkowski space.

The twistor manifold $\mathcal{T}_2$ is interpreted as the dual projective space. Then, considering a real form $V_{\mathbf{R}}$ of signature $(2,2)$ one obtains lines in $\mathbf{R}P^3$. An interesting question is that of a geometrical interpretation of other real forms of a conformally flat structure (see, e.g. [5]).

1.7 Integrability

Consider now a 4-dimensional complex manifold with conformal structure. It is conformally flat if and only if the Weyl tensor (or the conformal curvature) W vanishes. In the 4-dimensional case W splits into two parts: self-dual one W_+ and an anti-self-dual one W_-. Penrose has shown [6] that the condition of $\Sigma_1(\Sigma_2)$-integrability is equivalent to the vanishing of W_+ (respectively W_-). Thus, if both conditions of Σ_1 and Σ_2-integrability are satisfied simultaneously, then a conformal metric is flat.

1.8 Other examples

Conformal and complex structures are not the only classical examples of generalised conformal structures. The contact structure corresponds to the case when V is a hyperplane in an odd-dimensional space, however, in that case there is an important extra condition of non-degeneracy (maximal non-integrability). Under that condition all structures are locally equivalent (the Darboux theorem).

It is very interesting that almost simultaneously with the Liouville theorem a number of similar theorems appeared, although the analogy had not apparently been realised for a long time. Thus, the Darboux theorem stating that any transformation of the projective plane taking lines into lines is projective may be interpreted in such a way. Here one has to consider a 3-dimensional flag manifold F consisting of a point and a line going through it. There is a canonical generalised conformal structure on F: V consists of a pair of lines, one of which is tangent to the curve of flags containing the same point, while the second one is tangent to the curve of flags containing the same *line*. The plane containing both those lines defines a contact structure. The Darboux theorem is an analogue of the Liouville theorem for that structure on F.

Another example is provided by the Poincaré theorem on the extension of the locally bi-holomorphic mappings of a complex ball in a neighbourhood of its boundary point to the global mapping of the ball. Here one has to consider a CR-structure on the boundary of the ball, which can be interpreted as some (real) generalised conformal structure with a non-degeneracy condition (the Levi form has to be positive).

It is, undoubtedly, very interesting to carry out a systematic investigation of generalised conformal structures, and, in particular, to single out situations when some analogue of the Liouville theorem is true. It is a rather widespread situation and the Liouville theorem, in a sense, can be regarded as an obstacle to the existence of a non-trivial analytic theory. However, in this paper we do not aim at constructing a general theory. Our goal is to consider some specific examples which, in our view, imply that generalised conformal structures arise rather frequently in various areas of mathematics.

2 $\mathcal{P}$-structures and rational curves

A natural idea is to investigate not only structures related to a specific cone V, but series of structures related to cones of different dimensions so that one could investigate relations of manifolds with such structures of different dimensions (as it is the case for usual conformal structures). A customary situation in geometry is that the same structures in lesser dimensions generate different series of multidimensional structures (recall that in dimension 2 conformal and complex structures coincide). We now consider a series of structures which for $n = 4$ coincide with the conformal structure (§1.6). First, we recall that on a quadratic cone in $\mathbf{C}^4$ there are two families of two-dimensional planes which are parametised by points of a projective line. This property we take as a starting point for a generalisation procedure.

2.1 $\mathcal{P}$-cones

A cone V is called a $\mathcal{P}$-*cone* if there are two families of flat generators Σ and Π on it of dimension $n - l$, $(l > 1)$, and 2, respectively, such that

1. $V = \bigcup\limits_{\sigma \in \Sigma} \sigma = \bigcup\limits_{J_1 \in \Pi} J_1;$

2. Any $\sigma \in \Pi$ and $\pi_1 \in \Pi$ intersect, and a generic 2-plane $\pi \in \Pi$ intersects all planes $\sigma \in \Sigma$ in a single line (so that using that plane π_1 one can parametrise all planes $\sigma \in \Sigma$ by points of a projective line (the projectivisation of a 2-plane π)).

2.2 *Canonical form*

Theorem 2.1 *Each $\mathcal{P}$-cone $V \subset \mathbf{C}^n$ can by a linear transformation be transformed into a cone which is a union of subspaces $\sigma(\tau)$ parameterised by points of the projective line $\mathbf{C}P^1_\tau$:*

$$
\begin{aligned}
z^1(\tau) &= z^1_0 \tau_0^{k_1} + z^1_1 \tau_0^{k_1 - 1} \tau_1 + \ldots + \; z^1_{k_1} \tau_1^{k_1} = 0 \\
&\;\;\vdots \qquad\qquad\qquad\qquad\qquad\qquad\qquad \vdots \\
z^l(\tau) &= z^l_0 \tau_0^{k_l} + z^l_1 \tau_0^{k_l - 1} \tau_1 + \ldots + \; z^l_{k_l} \tau_1^{k_l} = 0
\end{aligned}
$$

where $\sum k_j + l = n$, and z^i_j are coordinates in $\mathbf{C}^n$.

A few words on the proof. First, we check that the cone is indeed a $\mathcal{P}$-cone. Evidently, $\sigma(\tau)$ forms the first system of linear spaces making up Σ. We now describe the system of 2-dimensional planes which form Π. Let $y^1(\tau), \cdots, y^l(\tau)$ be a system of homogeneous polynomials in $\tau = (\tau_0, \tau_1)$ of degrees $k_1 - 1, \cdots, k_l - 1$ respectively. Let $z^i(\tau) = y^i(\tau)(\lambda_0 \tau_1 - \lambda_1 \tau_0)$ and $z^i_j(\lambda)$ be coefficients in $z^i(\tau)$ appearing before the monomial $\tau_0^{k_i - j} \tau_1^{j}$; they are linear forms in λ. The points $\{z^i_j(\lambda)\}$ generate a 2-plane $\pi(y)$ on the cone V (they are parameterised by coefficients y_j^i of the polynomials $y^i(\tau)$). Clearly, $\{z_j^i(\lambda)\}$ is the point at which $\sigma(\lambda)$ intersects $\pi(y)$. Those planes $\pi(y)$ that are in general position correspond to a system of polynomials $y^i(\tau)$ with no common linear factor. If all $y^i(\tau)$ do have a common factor $\mu_0\tau_1 - \mu_1\tau_0$ then $\pi(y) \subset \sigma(\mu)$. Hence, V is a $\mathcal{P}$-cone which will be denoted by $V^{(k)}, k = (k_1, \ldots, k_l)$.

We now give an outline of the proof for the reverse statement. Consider a $\mathcal{P}$-cone $V \subset \mathbf{C}^n$. Consider a vector bundle over the projective line $\mathbf{C}P^1_\tau$ with fibers $\mathbf{C}^n/\sigma(\tau)$ over the points τ (recall that $\sigma \in \Sigma$ are parameterised by points of the projective line $\mathbf{C}P^1_\tau$). Then the points of $\mathbf{C}^n$ naturally correspond to the sections of the bundle with the points of $\sigma(\tau)$ corresponding to sections vanishing at τ, i.e. the points of the cone V correspond to sections vanishing at some point. By the Grothendieck theorem any vector bundle over the projective line is of the form $\mathcal{O}(k_1) \oplus \ldots \oplus \mathcal{O}(k_l)$ and the bundle has sections only if $k_j \geq 0$. Then the cone of vanishing sections is precisely of the form described above. However, it is possible, *a priori*, that the sections corresponding to the points $z \in \mathbf{C}^n$ form only a proper subspace in the space of all sections. A straightforward analysis of sections of the cone $V^{(k)}$ shows that a proper section is not a union of 2-dimensional planes π.

2.3   *$\mathcal{P}$-structures*

We now consider $V^{(k)}$-conformal structures. Structures of that type will be called *$\mathcal{P}$-structures*. The above considerations imply that a flat structure is realised in the space of sections of the vector bundle $\mathcal{O}(k_1) \oplus \ldots \oplus \mathcal{O}(k_l)$ on the projective line. The twistor manifold is the total space of that bundle.

One can assume that $k_1 \geq k_2 \geq \ldots \geq k_l$ and there is no need to exclude the case $l = 1$ when $\dim V = n$ and there is no geometric way to distinguish flat generators. For $k_2 > 0$ one has $\dim \sigma(\tau) > 1$ and the problem of Σ-integrability satisfies the uniqueness condition. This can be interpreted as a version of the Desargues condition in the projective geometry [3]. If a structure is both Σ and Π-integrable then it is flat and there is an analogue of the Liouville theorem.

Consider for $k_2 > 0$ a Σ-integrable $V^{(k)}$-conformal structure on the manifold M, $\dim M = n$. Then, according to §1.2, M is realised as a manifold of rational curves E_z on the twistor manifold $\mathcal{T}$, $\dim \mathcal{T} = l + 1$. The projective parameterisation of $\sigma \in \Sigma$ induces the projective line structure on E_z. That statement does not require the existence of π. It is equivalent to the fact that the family of rational curves E_z is full in the following sense. Vectors in $T_z M$ correspond to sections of the normal bundle NE_z to the curve E_z (infinitesimal deformations of the curve). It is required that the subspace of such sections should coincide with the full subspace of sections of a subbundle $\widetilde{NE_z} \subset NE_z$. Other reformulations of that condition are also possible. The condition is a local one: it is fully determined by the character of intersections of curves in a small neighbourhood on $\mathcal{T}$ and local curves can be uniquely continued to global ones preserving that condition. In particular, that condition is invariant under bi-rational transformations and frequently, after an appropriate transformation of such a kind, a full family of rational curves goes into a family of all rational curves on some modified manifold.

Full n-parameter families M of rational curves E_z on an n-dimensional manifold $\mathcal{T}$ can be treated in a similar fashion. Then submanifolds S_u on M are curves and one has the case $k_2 = 0$. In the language of the curves, S_u, the condition that the family M is full is written as a differential condition of second order (the generalised Desargues axiom). In the holomorphic case it is equivalent to the fact that there is a unique curve S_u going out of the point $z \in M$ in each direction. A similar statement for $k_2 > 0$ can be proved (as is the case in the classical projective geometry when the Desargues axiom becomes a theorem).

Thus the problem of Σ-integrable $V^{(k)}$-conformal structures is equivalent to the problem of full families of rational curves. In particular, a self-dual 4-metric is realised on a full 4-parameter family of rational curves on a 3-dimensional manifold [6]. Therefore the problem of constructing such metrics (i.e. solutions of the equation $W_+ = 0$) is equivalent to the problem of constructing such families of curves.

2.4 Constructing families of rational curves

What are the ways to construct full families of rational curves? The simplest examples are those of sections of vector bundles on a projective line (which correspond to flat $\mathcal{P}$-structures), and lines and second order curves in the projective space. Penrose [6] has shown that if one deforms the complex structure of the total manifold of a vector bundle on the projective line in a neighbourhood of a fixed section, then the new manifold will also include a full family of rational curves. It is very instructive that in this construction one does not deform a (flat) $\mathcal{P}$-structure itself, but a more simple object, namely the complex structure on the twistor manifold.

Besides deforming flat structures, another traditional way of constructing curved structures is to restrict multidimensional flat structures to submanifolds. For general $\mathcal{P}$-structures the situation is more complicated than for usual conformal structures which on any submanifolds generate structures of the same type. The reason is that any section of a quadratic cone by a subspace is itself a quadratic cone. On the other hand, a conical section of the cone $V^{(k)}$ is not necessarily of the same type.

Therefore one has to investigate sections of cones $V^{(k)}$. As mentioned .above, no proper sections are $\mathcal{P}$-cones in the exact sense. However, one can modify the notion of a section. Consider the intersection of a subspace with $\sigma(\tau)$ in general position and let us call the closure of a union of such intersections a *regularised section*. Such regularised sections may be $\mathcal{P}$-cones. In such a case we call their planes *regular*.

Thus, one has a problem of linear algebra: how can one describe regular planes? It turns out that each regular plane is an intersection of regular hyperplanes, and that a hyperplane is regular if and only if it contains a plane $\sigma(\tau)$ for some τ. Thus the regularity condition is written as an explicit system of algebraic equations.

Let M be a manifold with a $V^{(k)}$-conformal Σ-integrable structure. Let us call a submanifold $N \subset M$ regular if $T_z N$ is a regular plane for all $z \in N$. Then, evidently, a Σ-integrable $\mathcal{P}$-structure is induced on N. The regularity condition for the submanifold N is written as a system of algebraic differential equations of the first order, obtained from the regularity condition on tangent planes. It turns out that the system can be explicitly integrated using a generalisation of the Hamilton-Jacobi approach [3] which yields a full description of regular submanifolds. It is a matter of principle that the description is given in twistor language.

The manifold M is realised as a full manifold of rational curves on the twistor manifold $\mathcal{T}$. One has to describe sub-families of curves corresponding to regular submanifolds N, i.e. full subfamilies N in M. First we formulate the answer to that question for regular submanifolds in general position [3, 8].

Theorem 2.2 *Let $\Gamma_1, \cdots, \Gamma_p$, $S_1, \cdots, S_q$ be submanifolds in $\mathcal{T}$, codim $\Gamma_j > 1$, codim $S_i = 1$ and let $m_1, \ldots, m_q$ be integers. Let $N(\Gamma, S, m)$ be a submanifold of curves intersecting all Γ_j and tangent to S_i of order m_i for all $1 \leq i \leq q$. Then $N(\Gamma, S, m)$ is a regular submanifold in M (a full subfamily of rational curves), and each regular submanifold in general position can be represented in such a form.*

If one does not require that a submanifold should be in general position, then the construction is more complicated. One has to consider a tower of σ-processes $\widetilde{\mathcal{T}} \to \mathcal{P}$, submanifolds $S_1, \ldots, S_q$ on $\widetilde{\mathcal{T}}$ and integers $m_1, \ldots, m_q$. Then one takes a subfamily N of curves which are lifted to $\widetilde{\mathcal{T}}$ and their tangent to S_j of order m_j. Then the subfamily N is full and it is the most general construction of full subfamilies. Note that the conditions of intersecting Γ_j are equivalent to the condition that one can lift a curve as the result of a σ-process along Γ_j.

Thus we have obtained a very general and constructive way of constructing full families of rational curves starting from the simplest ones: sections of vector bundles on a line or second order curves. We stress that the procedure differs from the Penrose approach in that the twistor manifold does not change while the set of curves is diminished (in some cases this can be interpreted as a modification of the twistor manifold by σ-processes).

In that way the problem of restricting $\mathcal{P}$-structures is completely solved. There is, however, another problem, viz. that of immersing a manifold with a $\mathcal{P}$-structure into a flat manifold with $\mathcal{P}$-structure of a higher dimension. Compared to conformal structures that problem is not trivial even in the local situation. It is very interesting to describe invariants that can be regarded as obstacles to such local immersions. Here we are interested only in local considerations, although global questions are also very interesting.

2.5 Self-duality

The above facts show that $\mathcal{P}$-structures give rise to an elegant and peculiar differential geometry. It is very important that full manifolds naturally arise in a number of important analytical problems. One example of exactly integrable non-linear systems connected with manifolds of rational curves is provided by the problem about conformal self-dual 4-metrics. Such manifolds arise in all cases when a non-linear system can be written as a compatibility condition for a system of linear equations with a rational spectral parameter [3].

Another example is provided by integral geometry. Let M be an n-parameter family of holomorphic curves on an n-dimensional complex manifold $\mathcal{T}$. To each C_0^∞-function φ on $\mathcal{T}$ one associates integrals over the curves with respect to some fixed measures. Thus one obtains a function $\hat{\varphi}$ on M. The problem is to reconstruct φ from $\hat{\varphi}$. The inversion formula should be a local one, i.e. a value of φ at a point should be expressed only through integrals over curves close to it. It turns out to be possible to find such a formula if and only if M is a full system of rational curves [8, 9]. There is a generalisation of that problem for $\dim \mathcal{T} < \dim M$ and in that case a local inversion formula also exists only for full systems of rational curves.

Among the cones $V^{(k)}$ one of the most simple ones corresponds to the case $k_1 = \ldots = k_l = 1$. This lies in the space $\mathbf{C}^{2l}$ with coordinates z_0^i, z_1^i, $1 \leq i \leq l$, and is given by the system of equations of order 2:

$$z_0^i z_1^j - z_1^i z_0^j = 0, \quad 1 \leq i, j \leq l.$$

The case $l = 2$ (Plücker) has been considered in §1.6. In the flat case M is realised as a manifold of lines in $\mathbf{C}P^{l+1}$.

One has $\dim \sigma(\tau) = l = n/2$. If l is even $(l = 2p)$ then there is an involution on $\mathbf{C}^n$ that preserves $V^{(1)}$ and has no non-zero stable points. Then the real form $V_R^{(1)}$ of the cone $V^{(1)}$ is imaginary, and on the manifold $M_\mathbf{R}$ with $V_R^{(1)}$-conformal Σ-integrable structure there arises a family of complex structures depending on the parameter $\tau \in \mathbf{C}P^1$. Such a structure is naturally called quaternionic since in the quaternionic algebra there is a family of imaginary units parameterised by points of $\mathbf{C}P^1$. In particular, a flat structure is realised on $\mathbf{H}^p$ where the quaternionic coordinates h_j can be written in the form

$$h_j = \begin{pmatrix} z_j & w_j \\ -\overline{w}_j & \overline{z}_j \end{pmatrix}$$

and the complex coordinates for the structure depending on the parameter τ are

$$\left(\tau_0 z_j - \tau_1 \bar{w}_j, \ \ \tau_0 w_j + \tau_1 \bar{z}_j\right).$$

The Liouville theorem for that structure implies that there is a finite-dimensional group of its automorphisms, and that therefore there is no natural local version of quaternionic analysis. However, quaternionic manifolds exist and can be constructed using manifolds of rational curves [10].

2.6   *Generalized metrics and bundles of forms*

A natural idea is to assign to $\mathcal{P}$-structures some kind of a generalised metric structure similar to metrics assigned to conformal structures. We give the construction for exen $l = 2p$ taking for a simplicity's sake the case $k_1 = \ldots k_l = k = 2p$. Consider $V^{(k)}$-conformal structure, $n = (k+1)l$, on M, which means that there is a system of first order equations depending on a parameter:

$$
\begin{aligned}
\omega^1(\tau) \ \ &= \omega_0^1 \tau_0^k + \ldots + \ \ \omega_k^1 \tau_1^k = 0 \\
&\ \ \vdots \qquad\qquad\qquad\quad \vdots \qquad\qquad \omega_j^i \text{ are 1-forms} \qquad (1) \\
\omega^{2p}(\tau) \ \ &= \omega_0^{2p} \tau_o^k + \ldots + \ \ \omega_k^{2p} \tau_1^k = 0
\end{aligned}
$$

which in every tangent space defines a system of subspaces $\sigma_z(\tau)$ whose union is the cone $V_z^{(k)}$ linearly equivalent to $V^{(k)}$. A flat structure corresponds to the case

$$\omega_j^i = dz_j^i.$$

Consider a bundle of 2-forms $F^{(k)}$ of order $2k$ in τ:

$$F^{(k)}(\tau) = \omega^1(\tau) \wedge \omega^2(\tau) + \ldots + \omega^{2p-1}(\tau) \wedge \omega^{2p}(\tau). \qquad (2)$$

That bundle will be interpreted as a metric continuation of the $V^{(k)}$-conformal structure, and the class of such new structures will be called $\mathcal{R}$-structures. That object is a metric in the naïve sense: unlike cones, $V^{(k)}$, it changes when $\omega_j^i(\tau)$ are multiplied by functions. More precisely, the gauge group for a $V^{(k)}$-conformal structure includes $GL(2p)$, i.e. non-degenerate transformations of the system (1) at each point and $SL(2)$, i.e. projective transformations of the parameter τ. When one considers $F^{(k)}$, the group $GL(2p)$ is reduced to $Sp(p)$, i.e. transformations preserving $F^{(k)}$.

The bundles $F^{(k)}$ may be described axiomatically:

1. The $(p+1)$-th exterior power of $F^{(k)}$ vanishes;

2. The p-th exterior power of $F^{(k)}$ is non-degenerate.

We stress that the degree of $F^{(k)}$ in τ is rigidly connected with the dimension of the manifold M.

Consider the differential equation

$$dF^{(k)}(\tau) = 0 \quad \text{for all } \tau \in \mathbf{C}^2 \tag{3}$$

Together with the algebraic equations on $F^{(k)}$ one has a system of non-linear differential equations. The condition on $F^{(k)}$ to be closed is a stronger one than the integrability condition for the system (1) (Σ-integrability of a $V^{(k)}$-conformal structure). This is verified directly by the Frobenius theorem.

For $k = 1$ we construct from (1) the non-degenerate metric

$$g = (\omega_0^1 \omega_1^2 - \omega_1^1 \omega_0^2) + \ldots + (\omega_0^{2p-1} \omega_1^{2p} - \omega_1^{2p-1} \omega_0^{2p})$$

where one takes symmetric products of 1-forms. For $p = 1$ this is the metric associated to the conformal structure on a 4-dimensional manifold. For $p > 1$ we associate metrics to generalised conformal structures. The bundle $F^{(1)}$ and metric g can be expressed in terms of one another, but the bundle $F^{(k)}$ can be defined for $k > 1$, as well, which will be used quite often.

Proposition 2.3 *Under condition (3) the metric g is hyperKahler and each hyperKahler metric can be expressed in terms of the forms ω_j^i in such a way that (3) is satisfied. (See [11])*

We remind the reader that a metric is called hyperKahler if its holonomy group is reduced to the symplectic group. If a metric g has a real form, which is a Riemannian metric on a real manifold $M_{\mathbf{R}}$ them, as we have mentioned in the preceding section, there is a family of complex structures on $M_{\mathbf{R}}$ depending on parameter τ. The hyperKahler condition means that the metric is Kahler with respect to all those complex structures.

HyperKahler metrics play an important role in mathematical physics. In particular, for $p = 1$ they provide self-dual solutions for the vacuum Einstein equations (right-flat metrics). It is important to be able to construct examples of hyperKahler metrics and, in particular, right-flat ones. A natural idea to do that is to restrict $\mathcal{R}$-structures to submanifolds. Since each $\mathcal{R}$-structure includes a $\mathcal{P}$-structure it is reasonable to start from results about restricting $\mathcal{P}$-structures.

Let M be a manifold with an $\mathcal{R}$-structure of dimension $n = 2p(k+1)$. We are interested in submanifolds N, dim $N = 2p(s + 1)$, $s < k$, which satisfy the conditions of §2.4 and on which there is an induced $\mathcal{R}$-structure. First we have to define what the word 'induced' means. The reason is that if one simply reduces $F_M^{(k)}$ to N one gets a bundle of closed forms of degree $2k$ in τ while we need a bundle of forms of degree $2s$. It turns out that the condition on N to be regular (in the sense of §2.4) is equivalent to the fact that after restricting (1) to N one can make a gauge transformation (depending on τ)

which will result in a system of the type (1) such that the corresponding bundle $F_N^{(s)}$ is of a degree $2s$. Otherwise one can divide $F_M^{(k)}|N$ by a function $f(\tau, z)$ which is a homogeneous polynomial in τ of degree $2(k-s)$, the result being again a polynomial bundle. Unfortunately, after such a procedure the bundle $F_N^{(s)}$ is not necessarily a closed one, and the problem is to formulate conditions under which that property can be preserved. We have been unable to solve it in full, but some series of examples have been constructed.

Let the structure on M be flat. Then the twistor manifold $\mathcal{T}$ (cf. §1.6) is obtained from the projective space $\mathbb{C}P^{2p+1}$ with homogeneous coordinates $(\tau_0, \tau_1, u_1, \ldots, u_{2p})$ by deleting the line $\tau_0 = \tau_1 = 0$. Consider the construction of Theorem 2.2 which involves only the submanifolds $\Gamma_1, \ldots, \Gamma_{2(k-1)}$ of codimension greater than 1.

Proposition 2.4 *(See [11]) Suppose that each submanifold Γ_j lies in a plane of the form $a_j\tau_0 + b_j\tau_1 = 0$. Then an $\mathcal{R}$-structure is induced on $N(\Gamma)$ such that*

$$F_N^{(s)}(\tau) = \frac{F_M^{(k)}|N}{\displaystyle\prod_{j=1}^{2(k-s)} (a_j\tau_0 + b_j\tau_1)}.$$

The corresponding metrics (for which $j = 1$) have singularities. For $p = 1$ there is another way of constructing submanifolds with $\mathcal{R}$-structures [12], which yields regular metrics, but it is not generalised to the case $p > 1$. An interesting feature of that approach is that it involves global constructions.

2.7   Remarks

A few concluding remarks are in order on $\mathcal{P}$-structures. In our view there is an important reason for the fact that they appear in various analytic problems. It seems that a right guideline has been given by the following observation of A. B. Goncharov. Consider instead of $V^{(k)}$, dual cones $V^{(k)*}$ consisting of planes dual to $V^{(k)}$ as well as their projectivisations $Q^{(k)}$. It turns out that $Q^{(k)}$ are algebraic varieties of minimal degree amoungstt all varieties of the same codimension l. For codimension $l = 1$, varieties of minimal degree are quadrics. Enriques has shown there are no other series of manifolds of minimal degree; there is only the Veronese surface in $\mathbb{C}P^5$ and the cone over it. This might explain the special place occupied by conformal structures and $\mathcal{P}$-structures among other generalised conformal structures.

The construction of $\mathcal{P}$-cones admits a natural generalisation if one considers families of generators depending on a parameter $\tau \in \mathbb{C}P^r$, $r > 1$. However, the fact that such cones are closely related to bundles over $\mathbb{C}P^r$ and there is no way to classify bundles on $\mathbb{C}P^r$ for $r > 1$, one feels that there is apparently no way to manage the situation for $r > 1$. This may well reflect difficulties of integrating non-linear equations connected with several spectral parameters.

3   Infinitesimal Structures on Complex Symmetric Manifolds

3.1   *Structures associated to Hermitian symmetric spaces*

We have seen that flat $V^{(k)}$-conformal structures for $k = (1, \ldots, 1)$ are realised on manifolds $G_{2,l+2}$ of lines in the projective space $\mathbb{C}P^{l+1}$ or, equivalently, 2-dimensional subspaces in $\mathbb{C}^{l+2}$. The group of (holomorphic) automorphisms of $G_{2,l+2}$ is locally isomorphic to the group of automorphisms of that structure.

That construction admits a natural generalisation to other Grassmannians $G_{p,m}$ (a manifold of $(p-1)$-dimensional planes in $\mathbb{C}P^{m-1}$). One can assume that $p \leq m - p$, since there is a canonical isomorphism $G_{p,m} = G_{m-p,m}$. As a cone $V_{p,m}$ we take the cone of p-dimensional subspaces in $\mathbb{C}^m$ intersecting a fixed p-dimensional subspace in a $(p-1)$-dimensional subspace. That cone is isomorphic to the cone of rectangular $p \times (m-p)$ matrices of rank 1. It has two linear systems in it. One consists of matrices whose rows are scalar multiples of a given vector, the other is obtained by a similar construction from columns. Those cones were first considered by Segre, while the corresponding generalised conformal structures (called Grassmann structures) have been introduced by M. A. Akivis [13]. All the results formulated for $p = 2$ are valid for them as well.

Grassmannians form one of the series of compact Hermitian symmetric manifolds (there are three more series and two special manifolds among those that are irreducible). A natural question then is to find out if there are more generalised conformal structures associated to other compact Hermitian symmetric spaces. Such structures have been described by A. B. Goncharov [14]. See also [1, 2].

Any such manifold M is of the form G/L where G is a complex semisimple Lie group, L is its parabolic subgroup whose nilpotent radical is abelian. Let P be its reductive part. Consider its linear action on the tangent space $T_z M$. Then the cone V_M is an orbit of minimal dimension (which coincides with the set of highest vectors for different parabolic subgroups in P). For different manifolds M the cones V_M can be described explicitly. An interesting fact is that they are always cones over compact Hermitian symmetric manifolds of lesser dimension. Therefore, orbits of minimal dimension define a generalised conformal structure on M. Some major facts concerning that structure are the following.

1. The structure is a flat one (which is the consequence of the fact that the radical is an abelian one).

2. The group of (holomorphic) automorphisms of M is locally isomorphic to the group of automorphisms of a V_M-conformal structure on M, and

local automorphisms of the structure can be extended to global ones (the Liouville theorem).

3. Let there be linear elements on V_M of dimension greater than 1. This excludes manifolds of rank 1 and the series containing the symplectic group G. Then any V_M-conformal structure that is Σ_i-integrable for all i is a flat one (the Penrose theorem).

This provides a remarkable geometric characterisation for symmetric Hermitian space of rank greater than 1.

3.2   *Psuedo-Hermitian symmetric structures*

Generalised conformal structures are associated not only to Hermitian symmetric spaces but to pseudo-Hermitian as well. These include complex semi-simple groups G considered as homogeneous spaces with respect to the action of the group $G \times G$ by left and right shifts.

Define a generalised conformal structure on G. Let V_G be the cone of nilpotent elements in its Lie algebra. Applying shifts to V_G one obtains a V_G-conformal structure on G. Linear elements on V_G are maximal nilpotent subalgebras, while integral submanifolds are given by orispheres, i.e. (two-sided) shifts of maximal nilpotent subgroups. The group of automorphisms of a V_G-conformal structure on G is locally isomorphic to $G \times G$, and local automorphisms can be extended to global ones (the Liouville theorem). There is also an analogue of the Darboux theorem from projective geometry: transformations preserving orispheres are group shifts. On the other hand, a V_G-structure on G is Σ-integrable (there are orispheres) but not flat. This is an essential difference from the Hermitian case. Therefore, no analogue of the Penrose theorem is valid in this case.

The twistor manifold is the manifold of orispheres $\mathcal{T}$, $\dim \mathcal{T} = \dim G$, which is a homogeneous space with respect to $G \times G$. According to the general scheme, there are submanifolds E_g on $\mathcal{T}$ corresponding to elements $g \in G$ (i.e. orispheres going through g). For $u \in \mathcal{T}$ one takes tangent spaces $\sigma_g \subset T_u\mathcal{T}$ to $E_g \ni u$ and the cone V_u^G which is their union. That cone can be naturally interpreted as a dual to V_G. For different points the cones V_u^G are linearly equivalent and one gets a V^G-conformal structure on $\mathcal{T}$ (dual to the V_G-structure on G).

Thus, we have obtained one more class of generalised conformal structures. For all groups G the cones V^G have a remarkably simple stratification structure. There is a family Σ_1 of one-dimensional subspaces lying in the same 2-dimensional space. For each of them there is a one-parameter family of 2-dimensional subspaces going through it and lying in the same 3-dimensional subspace (let Σ_2 be the set of those subspaces) etc.. For each k-dimensional subspace from Σ_k there is a one-parameter family of subspaces going through it and lying in the same $(k+1)$-dimensional subspace. The process ends when $\dim E_g$ and $\dim \sigma_g$ coincide.

The transform of a function to its integrals over the orispheres plays an important role in group representation theory. Its inversion is essential in the derivation of the Plancherel formula. The inversion is based only on the described V_G-conformal structure and its equivalence at each point to a flat one up to the third order infinitesimals. Thus as in the group case one has an explicit local inversion formula on an arbitray manifold G with a Σ-integrable generalized conformal structure whose dual on the twistor manifold has the indicated properties. It would be interesting to study such manifolds and to elucidate the place of the group ones among them. Note that the stratification of the cone V_G is related to the existence of root systems on the groups.

I would like to end with an optimistic hope that the generalized conformal structures should be important in modern mathematics.

References

[1] R.J. Baston. Almost Hermitian symmetric manifolds I: local twistor theory. Preprint (1990).

[2] R.J. Baston. Almost Hermitian symmetric manifolds II: Differential invariants. Preprint (1990).

[3] Gindikin, S., *Reduction of Manifolds of notional curves and related probelms of differential equations theory* Funct. Anal. Appl **18** (1984), 4, 14–39 (Russian).

[4] Gelfand, I., Graeze, M. & Shapizo, Z., *Differential forms and integral geometry*, Funct. Anal. Appl. **3** (1969), 3, 24–40 (Russian).

[5] Gindikin, S., *The Complex Universe of Roger Penrose*, The Math. Intelligencer, **5** (1983), 1, 27–35.

[6] Penrose, R., *Nonlinear gravitons and curved twistor theory*, Gen Rel. Grav. **7** (1976), 31–52.

[7] Gindikin, S., *Bundles of differential forms and the Einstein equation*, Nuclear Physics **36** (1982) 2(8), 537–548 (Russian).

[8] Bernstein, J. & Gindikin, S. *Four papers on integral geometry and Einstein equations*, Reports Dept Math, Univ Stockholm **2B** (1986), 1–61.

[9] Gelfand, I., Gindikin, S. & Shapizo, Z., *Local problem of integral geometry in the space of curves* Funct. Anal. Appl. **13** (1979), 2, 11–31.

[10] Gindikin, S., *Twistors, Rational curves, quaternionic structures, Infinite dimensional Lie algebras and quantum field theory* World Sci. (1988) 11–27. Proc. of Varna Summer School 1987.

[11] Gindikin, S., *On one construction of hyperKähler metrics* Funct. Anal. Appl **20** (1986), 3, 82–32 (Russian).

[12] Gindikin, S., *Some solutions of the self-dual Einstein equation*, Funct. Anal. Appl **13** (1985) 3, 58–60 (Russian).

[13] Akivis, M., *Nets and almost Grassmanian structures* Proc Nat. Acad., U.S.S.R **252** (1980), 2, 267–270.

[14] Goncharov, A.B., *Generalised conformal structures on Manifolds*, Selecta Mathematica Sovetica **v6** (1987), 4, 308–340.

[15] Gindikin, S., *Differential geometry problems in Integral geometry* Differential geometry and its applications, Proc of the Conference, 1986, Brno, Czechoslovakia, 53–69.

Riemannian Twistor Spaces and Holonomy Groups

F.E. Burstall

1 Introduction

The essence of the twistor programme is to encode the differential geometry of a manifold by holomorphic data on some auxiliary complex space (a *twistor space*). Thus problems in (pseudo) Riemannian geometry are converted into (hopefully soluble) problems in complex analysis or algebraic geometry. As examples of such constructions, let us mention:

- The Penrose fibration $\mathbb{C}P^3 \to \mathbb{H}P^1 = S^4$ encodes a large part of the geometry of the 4-sphere. For instance, instanton solutions to the Yang-Mills equations on S^4 pull back to holomorphic bundles on $\mathbb{C}P^3$ [30]. This forms the basis of the Atiyah-Drinfeld-Hitchin-Manin classification of the instantons [1]. Again, horizontal curves in $\mathbb{C}P^3$ project onto branched minimal surfaces in S^4. Using this Bryant [7] was able to establish the existence of embedded minimal surfaces of arbitrary genus in S^4.

- In a similar vein, consider the homogeneous fibrations

$$\frac{\mathrm{U}(n+1)}{\mathrm{U}(r) \times \mathrm{U}(1) \times \mathrm{U}(n-r)} \longrightarrow \frac{\mathrm{U}(n+1)}{\mathrm{U}(1) \times \mathrm{U}(n)}$$

of flag manifolds over projective spaces. Again horizontal holomorphic curves in the flag manifolds project onto branched minimal surfaces (harmonic maps) in $\mathbb{C}P^n$. Further, all minimal 2-spheres in $\mathbb{C}P^n$ arise in this way. This is the starting point of the classification theorem for harmonic 2-spheres in complex projective spaces [9, 14, 16, 18].

- Finally we briefly consider an example with a different flavour. The space of geodesics in $\mathbb{R}^3$ may be identified with the holomorphic line bundle $T^{1,0}\mathbb{C}P^1$. Then magnetic monopoles on $\mathbb{R}^3$ can be shown to correspond to certain algebraic curves (*spectral curves*) in $T^{1,0}\mathbb{C}P^1$ [20]. Somewhat more transparently, minimal surfaces in $\mathbb{R}^3$ correspond to algebraic curves (essentially without restriction) in $T^{1,0}\mathbb{C}P^1$ and this provides a geometrical interpretation of the Weierstrass representation formulae [20, 27].

In this article we shall concentrate on the construction of twistor spaces rather than their applications. We shall describe fibrations of complex manifolds over Riemannian manifolds that generalise those in the first two of the preceding examples. For applications to the theory of harmonic maps, the reader is referred to the survey articles [10, 11, 29].

2   The Bundle of Almost Complex Structures

The first two examples listed above have much in common: in both cases the twistor space is a complex manifold fibred over the Riemannian manifold of interest. The fibration is not holomorphic (even when this makes sense) but the fibres are complex submanifolds of the twistor space. Let us now see how we might build such fibrations over more general Riemannian manifolds.

So let N be a $2n$-dimensional Riemannian manifold. We may at least construct such a fibration of an *almost complex* manifold over N as follows: let $\pi\colon J(N) \to N$ be the bundle of almost Hermitian structures of N. Thus the fibre at $x \in N$ is

$$J_x(N) = \{j \in \operatorname{End}(T_xN)\colon\ j^2 = -1,\ j \text{ skew-symmetric}\}.$$

This bundle is associated to the orthonormal frame bundle of N with typical fibre $J(\mathbf{R}^{2n}) = \mathrm{O}(2n)/\mathrm{U}(n)$ which is a Hermitian symmetric space (in fact it is two disjoint copies of the compact irreducible Hermitian symmetric space $\mathrm{SO}(2n)/\mathrm{U}(n)$). In particular, the typical fibre has an $\mathrm{O}(2n)$-invariant complex structure and thus the vertical distribution $\mathcal{V} = \ker d\pi$ inherits an almost complex structure $J^{\mathcal{V}}$. The Levi-Civita connection on the orthonormal frame bundle induces a horizontal distribution $\mathcal{H}$ on $J(N)$ so that we have a splitting

$$TJ(N) = \mathcal{V} \oplus \mathcal{H}$$

with $d\pi$ giving an isomorphism between $\mathcal{H}$ and $\pi^{-1}TN$. This enables us to define a tautological almost complex structure $J^{\mathcal{H}}$ on $\mathcal{H}$ by

$$J^{\mathcal{H}}_j = j$$

and adding this to $J^{\mathcal{V}}$ gives us an almost complex structure $\mathcal{J} = J^{\mathcal{V}} \oplus J^{\mathcal{H}}$ on $J(N)$. By construction, the fibres of π are almost complex submanifolds with respect to $\mathcal{J}$.

Before going any further, let us remark that if we make a conformal change of metric on N, the bundle $J(N)$ remains unchanged although the horizontal distribution $\mathcal{H}$ will vary. However, despite this, it can be shown that the almost complex structure $\mathcal{J}$ is independent of the choice of metric within a conformal class. Thus our construction may be viewed as one in conformal geometry but we shall not pursue this here.

Having got our almost complex structure, it is natural to ask whether or not it is integrable so that $J(N)$ is an honest complex manifold. For this, of course, it is necessary and sufficient that the Nijenhuis tensor $N^{\mathcal{J}}$ of $\mathcal{J}$ vanish. The obstruction to this vanishing lies in the curvature tensor of N [22] :

Theorem 2.1 *Let* $j \in J(N)$ *with* $\sqrt{-1}$*-eigenspace* $T^+ \subset T_{\pi(j)}N^{\mathbf{C}}$. *Let* R *denote the Riemann curvature tensor of* N. *Then* $N^{\mathcal{J}}$ *vanishes at* j *if and only if*

$$R(T^+, T^+)\, T^+ \subset T^+. \tag{1}$$

Thus $\mathcal{J}$ is integrable if (1) holds for all maximally isotropic subspaces T^+ of $TN^{\mathbf{C}}$. This is a condition on the curvature tensor that can be analysed in terms of the representation theory of $O(2n)$ on the space of curvature tensors and one concludes:

Corollary 2.2 $\mathcal{J}$ *is integrable if and only if the Weyl tensor of* R *vanishes identically (i.e.* N *is locally conformally flat).*

Thus $J(N)$ is a complex manifold only in extremely restricted circumstances. The moral to be drawn from this is that $J(N)$ is 'too big' in general for $\mathcal{J}$ to be integrable. It is therefore appropriate to seek subbundles of $J(N)$ picked out by the geometry of N in the hope that some of these are complex manifolds. One way to do this is is to restrict attention to those elements of $J(N)$ that are compatible with the holonomy of N. It is to this that we now turn.

3   Reduction to the Holonomy Group

So let our $2n$-dimensional manifold N have holonomy group K and let $P \to N$ denote the holonomy bundle i.e the reduction of the orthonormal frame bundle of N to K. The typical fibre $J(\mathbf{R}^{2n})$ of $J(N)$ decomposes into a disjoint union of K-orbits and, correspondingly, $J(N)$ decomposes into a disjoint union of subbundles, each one associated to P with such an orbit as typical fibre. We now investigate whether any of these subbundles are complex manifolds with respect to $\mathcal{J}$.

There are two aspects to this question: firstly, we are doomed to failure unless our subbundle is $\mathcal{J}$-invariant i.e is an almost complex submanifold of $(J(N), \mathcal{J})$. Since our subbundles are associated to the holonomy bundle P, their inclusion into $J(N)$ preserves horizontal distributions which are therefore $\mathcal{J}$-invariant. Thus, for our subbundle to be $\mathcal{J}$-invariant, it is necessary and sufficient that the fibres be $\mathcal{J}$-invariant or, equivalently, that the corresponding K-orbit in $J(\mathbf{R}^{2n})$ be a holomorphic submanifold. Secondly, once we have a $\mathcal{J}$-invariant subbundle, we must ascertain whether the induced almost complex structure thereon (also called $\mathcal{J}$) is integrable. According to O'Brian-Rawnsley [22], this happens precisely when the Nijenhuis tensor of $\mathcal{J}$ vanishes on the subbundle. In view of (2.1), this is a condition that can be analysed in terms of the representation theory of K on the space of curvature tensors of metrics with holonomy K. This last topic is dealt with in some detail in the book [25] which is our source for most of the information about holonomy groups that we need below.

We begin by assuming that N is orientable so that $K \subseteq \mathrm{SO}(2n)$. Of course, $\mathrm{SO}(2n)$ acts transitively on connected components of $J(\mathbf{R}^{2n})$ and so we have two orbits both of which are complex submanifolds. The corresponding subbundles are denoted $J_+(N)$ and $J_-(N)$ with $J_+(N)$ consisting of those almost Hermitian structures j compatible with the orientation in the sense that $x_1 \wedge j x_1 \wedge \ldots \wedge x_n \wedge j x_n$ is a non-negative multiple of the volume form for any vectors $x_1, \ldots, x_n \in T_{\pi(j)}N$. We may now examine the integrability of $\mathcal{J}$ on $J_+(N)$ and $J_-(N)$ separately. If $2n \geq 6$, we get nothing new and $\mathcal{J}$ is integrable precisely when the Weyl tensor vanishes [15, 22]. However, when N is 4-dimensional, it is a celebrated result of Singer-Thorpe [26] that the Weyl tensor splits into two parts under the action of $\mathrm{SO}(4)$ and then each part constitutes the obstruction to the integrability of $\mathcal{J}$ on one of $J_+(N)$ or $J_-(N)$ [2]. In summary, we have

Theorem 3.1 *If $2n \geq 6$, $\mathcal{J}$ is integrable on $J_+(N)$ or $J_-(N)$ if and only if N is locally conformally flat. If $2n = 4$, $\mathcal{J}$ is integrable on $J_+(N)$ if and only if N is anti-self-dual and on $J_-(N)$ if and only if N is self-dual.*

To go further, we must consider N with holonomy strictly contained in $\mathrm{SO}(2n)$. To simplify matters, we shall suppose that N is simply-connected, oriented and irreducible. If N is not a symmetric space, the classification of Berger [4] tells us that the only possiblities for K are $\mathrm{SO}(2n)$, $\mathrm{U}(n)$, $\mathrm{SU}(n)$, $\mathrm{Sp}(1)\mathrm{Sp}(\frac{n}{2})$, $\mathrm{Sp}(\frac{n}{2})$ or $\mathrm{Spin}(7)$ (this last acting on $\mathbf{R}^8$ via the spin representation). Before discussing each of these in turn, let us establish some generalities about K-orbits in $J(\mathbf{R}^{2n})$.

Firstly, an analysis of the complex structure of $J(\mathbf{R}^{2n})$ (c.f. [22]) provides us with the following criterion for the holomorphicity of a K-orbit.

Lemma 3.2 *The K-orbit of $j \in J(\mathbf{R}^{2n})$ is a complex submanifold of $J(\mathbf{R}^{2n})$ if and only if the Lie algebra $\mathrm{k} \subset \mathrm{so}(2n)$ of K satisfies*

$$[[\mathrm{k}, j], j] = [\mathrm{k}, j]$$

The second fact we use about a holomorphic K-orbit is

Proposition 3.3 *If the K-orbit of j is a holomorphic submanifold then the stabiliser of j is the centraliser of a torus in K and, in particular, contains a maximal torus in K.*

To prove this, observe that a connected component of the Kähler manifold $J(\mathbf{R}^{2n})$ is an adjoint orbit of $\mathrm{SO}(2n)$ and the inclusion may be regarded (after identifying $so(2n)$ with its dual) as the moment map for the (isometric, holomorphic and hence symplectic) action of $\mathrm{SO}(2n)$. Now a holomorphic K-orbit acquires a K-invariant Kähler structure from $J(\mathbf{R}^{2n})$ and hence, by the functoriality of the moment map construction, is an adjoint orbit of K. This suffices to establish (3.3).

For our first application of these results, let us suppose that N is a Kähler manifold with $K = \mathrm{U}(n)$. We start by considering the situation in a typical fibre: let $j_0 \in J(\mathbf{R}^{2n})$ be the complex structure corresponding to the Kähler structure. Then j_0 lies in the centre of $u(n)$ and so lies in any maximal toral subalgebra of $u(n)$. It is now easy to conclude from (3.2) and (3.3) that the $\mathrm{U}(n)$ orbit of j is a holomorphic submanifold of $J(\mathbf{R}^{2n})$ if and only if j commutes with j_0. This provides us with a simple geometric interpretation of the holomorphic orbits: let T be the $\sqrt{-1}$-eigenspace of j_0; if j commutes with j_0 then we have an orthogonal decomposition of T into eigenspaces of j so that

$$T = T' \oplus T''$$

with $j = j_0$ on T' and $j = -j_0$ on T''. Conversely, an such splitting of T determines a j commuting with j_0 by the above prescription. Thus the holomorphic $\mathrm{U}(n)$-orbits are just the Grassmannians $G_r(T)$ of r-dimensional complex subspaces of T for $r = 0, 1, \ldots, n$.

The corresponding subbundles are just the bundles of Grassmannians $G_r(T^{1,0}N)$ embedded in $J(N)$ by the map

$$W \longmapsto j_W = \left\{ \begin{array}{ll} \sqrt{-1} & \text{on } W \oplus (\overline{W}^{\perp} \cap T^{0,1}N) \\ -\sqrt{-1} & \text{on } \overline{W} \oplus (W^{\perp} \cap T^{1,0}N) \end{array} \right.$$

These Grassmannian bundles with the almost complex structure $\mathcal{J}$ inherited from $J(N)$ were considered by O'Brian-Rawnsley [22] who showed that for $1 \le r \le n-1$, $\mathcal{J}$ is integrable on $G_r(T^{1,0}N)$ if and only if the Bochner tensor of N vanishes. Of course, $G_0(T^{1,0}N)$ and $G_n(T^{1,0})$ are just the images of the sections of $J(N)$ defined by the Kähler structure and its negative and so always have integrable $\mathcal{J}$.

As an example of this development, let us take N to be a complex projective space $\mathbf{C}P^n$. In this case the Bochner tensor vanishes so that $\mathcal{J}$ is integrable on each $G_r(T^{1,0}\mathbf{C}P^n)$. In fact, $G_r(T^{1,0}\mathbf{C}P^n)$ is naturally isomorphic to the flag manifold $\mathrm{U}(n+1)/\mathrm{U}(r) \times \mathrm{U}(1) \times \mathrm{U}(n-r)$ and the bundle projection is just the homogeneous fibration discussed in the introduction.

Finally we remark that the Grassmannian bundles always acquire an integrable complex structure from the holomorphic frame bundle of N. For this complex structure the bundle projection is holomorphic and so does not coincide with our $\mathcal{J}$ unless the fibres are zero-dimensional (i.e. $r = 0$ or n).

If N is Ricci-flat Kähler so that $K = \mathrm{SU}(n)$ we get nothing new: the only holomorphic orbits are the Grassmannian orbits we have already discovered. However, in this case, the Bochner tensor constitutes the whole of the curvature tensor and so unless N is flat, $\mathcal{J}$ is not integrable for $1 \le r \le n-1$.

Examples

Let us illustrate the situation by considering the two 4-dimensional symmetric spaces. Firstly, let $N = S^4$ with holonomy $\mathrm{SO}(4)$. Of course S^4 is conformally flat so that both $J_+(S^4)$ and $J_-(S^4)$ are both complex manifolds with respect to $\mathcal{J}$. Indeed, both these spaces are isometrically biholomorphic to $\mathrm{SO}(5)/\mathrm{U}(2) \cong \mathbb{C}P^3$ and the projections are just the Penrose fibration and its post-composition with the antipodal map.

Now let $N = \mathbb{C}P^2$ which has holonomy $\mathrm{U}(2)$ and is self-dual but not conformally flat. From (3.1) we conclude that $\mathcal{J}$ is integrable on $J_-(\mathbb{C}P^2)$ but not on $J_+(\mathbb{C}P^2)$. In fact, we may deduce this result from the above discussion since it is easy to show that $j \in J_-(\mathbb{R}^4)$ if and only if $[j_0, j]$ vanishes and $j \ne \pm j_0$. This provides an isomorphism of $J_-(\mathbb{R}^4)$ with $G_1(T) \cong \mathbb{C}P^1$ and so we may identify $J_-(\mathbb{C}P^2)$ with the flag manifold $P(T^{1,0}\mathbb{C}P^2)$. The integrability of $J_-(\mathbb{C}P^2)$ now follows from the fact that the Bochner tensor of $\mathbb{C}P^2$ vanishes (for a Kähler surface, this is the same as being self-dual). In $J_+(\mathbb{C}P^2)$ on the other hand, the only subbundles on which $\mathcal{J}$ is integrable are the two copies of $\mathbb{C}P^2$ corresponding to the Kähler structure and its negative.

Consider now the case of quaternionic Kähler manifolds, that is manifolds with holonomy contained in $\mathrm{Sp}(1)\mathrm{Sp}(k)$ [24] where $2n = 4k$. Geometrically, this means that there is a parallel subbundle $\mathcal{C}$ of $\mathrm{End}(TN)$ locally spanned by sections I, J, K which satisfy the familiar identities

$$I^2 = J^2 = K^2 = -1, \ IJ = -JI = K \ \text{etc.} \tag{2}$$

Following [24], we identify the representation of $\mathrm{Sp}(1)\mathrm{Sp}(k)$ on a complexified tangent space T as follows

$$T \cong H \otimes_{\mathbb{C}} E$$

where H and E are the 2 and $2k$ complex dimensional representations of $\mathrm{Sp}(1)$ and $\mathrm{Sp}(k)$ respectively. In this setting the real structure on T is the tensor product of the quaternionic structures j_H and j_E on H and E.

Let us now suppose that $K = \mathrm{Sp}(1)\mathrm{Sp}(k)$ and consider K-orbits in $J(\mathbb{R}^{4k})$. We begin by recalling that H and E carry invariant complex symplectic forms ω_H and ω_E whose tensor product gives the metric on T. Using (3.2) and (3.3), one may now deduce that there are three kinds of holomorphic orbit: the first and most important of which arise by choosing a complex line $H^+ \subset H$ and considering the almost complex structure with $\sqrt{-1}$-eigenspace

$$T^{1,0} = H^+ \otimes E.$$

This is Hermitian since ω_H necessarily vanishes on H^+. These almost complex structures comprise a single orbit $P(H) \cong \mathrm{Sp}(1)/\mathrm{U}(1)$ which can also be described as the elements in a fibre of $\mathcal{C}$ with square -1. This is just the sphere of radius 2 (with respect to the trace norm) in $\mathcal{C}$.

Dual to this orbit is one composed of almost complex structures with $\sqrt{-1}$-eigenspace

$$T^{1,0} = H \otimes E^+$$

where $E^+ \subset E$ is a Lagrangian subspace. $\mathrm{Sp}(k)$ acts transitively on such E^+ and we get a single orbit isomorphic to $\mathrm{Sp}(k)/\mathrm{U}(k)$. Finally, there is a class of holomorphic orbits obtained by mixing these constructions. Pick $H^+ \subset H$ and $E^+ \subset E$ with E^+ isotropic for ω_E but not maximally isotropic (Lagrangian). Then we have a j_E invariant decomposition

$$E = E^+ \oplus E^0 \oplus E^-$$

with E^0 stable under j_E and $j_E E^+ = E^-$. Then the orbit of the j with $\sqrt{-1}$-eigenspace

$$(H^+ \otimes E^0) \oplus (H \otimes E^+)$$

is holomorphic and is isomorphic to $\mathrm{Sp}(1) \times \mathrm{Sp}(k)/\mathrm{U}(1) \times \mathrm{U}(r) \times \mathrm{Sp}(k-r)$ where $\dim_{\mathbf{C}} E^+ = r$. These three kinds of orbit exhaust the holomorphic orbits of $\mathrm{Sp}(1)\mathrm{Sp}(k)$ in $J(\mathbf{R}^{4k})$.

Turning to the integrability of $\mathcal{J}$ on the corresponding subbundles, the first orbit considered corresponds to the radius 2 sphere bundle of $\mathcal{C}$ and it is a theorem of Salamon [24] that $\mathcal{J}$ is always integrable on this bundle. As for the remaining subbundles, the analysis in [25] of the curvature of quaternionic Kähler metrics can be used to show that $\mathcal{J}$ is integrable precisely when N is locally symmetric and so locally isometric to quaternionic projective space.

In case that $K = \mathrm{Sp}(k)$, the bundle $\mathcal{C}$ is flat and so spanned by global parallel sections I, J, K satisfying (2). Such manifolds are called *hyper-Kähler* manifolds and have been the object of much recent study (c.f. [25]). From our point of view though, we get nothing new. $\mathrm{Sp}(k)$ acts trivially on H and so the sphere bundle of $\mathcal{C}$ decomposes into the images of the global parallel sections $aI + bJ + cK$ where $a^2 + b^2 + c^2 = 1$ and, of course, $\mathcal{J}$ is integrable on all of these. As for the remaining subbundles, $\mathcal{J}$ is not integrable on any of these unless N is flat.

Finally, we must consider the case $K = \mathrm{Spin}(7)$ acting on $\mathbf{R}^8$ by the spin representation. It turns out that there are two holomorphic orbits in $J(\mathbf{R}^8)$ isomorphic to $\mathrm{SO}(7)/\mathrm{U}(3)$ and the 5-quadric $\mathrm{SO}(7)/\mathrm{SO}(2) \times \mathrm{SO}(5)$. However, the form of the curvature tensor prevents $\mathcal{J}$ from being integrable on either of the corresponding subbundles.

This concludes the analysis of $J(N)$ when N is not locally symmetric. We have seen that with the notable exception of the quaternionic Kähler manifolds, honest complex submanifolds of $J(N)$ are pretty hard to come by.

However, the situation is quite different for symmetric spaces. Indeed, using (2.1), (3.2) and (3.3), one can prove

Lemma 3.4 *If N is locally symmetric and irreducible, then the K-orbit of $j \in J(\mathbf{R}^{2n})$ is a complex submanifold if and only if the Nijenhuis tensor of $\mathcal{J}$ vanishes on the corresponding subbundle of $J(N)$.*

Thus in this situation our subbundle has integrable $\mathcal{J}$ as soon as it is an almost complex manifold and one would expect to find a good supply of complex submanifolds of $J(N)$. This is indeed the case as we shall now see.

4 Flag Manifolds and Symmetric Spaces

We now suppose that N is an irreducible $2n$-dimensional Riemannian symmetric space. Thus we may realise N as a coset space $N = G/K$ with $G^\tau \subset K \subset (G^\tau)_0$ for some involution τ of G. Now K is (a covering of) the holonomy group of N and similarly the coset fibration $G \to G/K$ covers the holonomy bundle $P \to N$. In this setting, $J(N)$ is associated to G:

$$J(N) \cong G \times_K J(\mathbf{R}^{2n})$$

and if K/H is a K-orbit in $J(\mathbf{R}^{2n})$ then the corresponding subbundle is $G \times_K K/H = G/H$ and the projection is just the coset fibration. Thus we see that the subbundles of $J(N)$ that we are considering are just the orbits of G in $J(N)$. We may now rephrase (3.4) as follows:

Proposition 4.1 *Let $j \in J(N)$. Then $G \cdot j$ is an almost complex submanifold of $J(N)$ on which $\mathcal{J}$ is integrable if and only if j lies in the zero-set of the Nijenhuis tensor $N^{\mathcal{J}}$.*

This focusses our attention on the zero-set of $N^{\mathcal{J}}$ which we denote by Z. In favourable circumstances, the structure of this set can be completely described. We begin by assuming that N is of compact type so that G is compact and semi-simple. We also assume that N is *inner* i.e. that τ is an inner involution of G or, equivalently, that $\operatorname{rank} G = \operatorname{rank} K$. The class of inner symmetric spaces include the even-dimensional spheres, the Hermitian symmetric spaces, the quaternionic Kähler symmetric spaces and indeed all symmetric G-spaces for $G = \mathrm{SO}(2n+1)$, $\mathrm{Sp}(n)$, E_7, E_8, F_4 and G_2. Moreover, all inner symmetric spaces are necessarily even-dimensional and so fit into our framework.

Our assumption that N be inner can also be motivated by (3.3): the stabiliser in G of $j \in Z$ is the centraliser of a torus in K and so contains a maximal torus of K; if N is inner, this torus is also maximal in G from which it follows that the stabiliser is the centraliser of a torus in G. Thus, for N inner, our orbit is of the form $G/C(T)$ for some torus $T \subset G$ and is therefore a *flag manifold* (c.f. [31]). According to Borel [5], these exhaust the compact Kählerian G-spaces for G semi-simple. We now have the following remarkable theorem:

Theorem 4.2 ([12]) *Let $N = G/K$ be a simply-connected inner Riemannian symmetric space of compact type. Then Z consists of a finite number of connected components on each of which G acts transitively. Moreover, any G-flag manifold is realised as such an orbit for some N.*

Remark

A similar result for a certain subset of Z has been proved by Bryant [8].

The proof of (4.2) requires a detour into the geometry of flag manifolds and reveals an interesting interaction between the complex geometry of flag manifolds and the real geometry of inner symmetric spaces. For this, we begin by noting that a coset space of the form $G/C(T)$ admits several invariant Kählerian complex structures in general [6]. To fix attention on just one of these, we use a complex realisation of $G/C(T)$ as follows: having fixed a complex structure, the complexified group $G^{\mathbf{C}}$ acts transitively on $G/C(T)$ by biholomorphisms with parabolic subgroups as stabilisers. Conversely, if $P \subset G^{\mathbf{C}}$ is a parabolic subgroup then the action of G on $G^{\mathbf{C}}/P$ is transitive and $G \cap P$ is the centraliser of a torus in G. Let us examine the infinitesimal situation: let $F = G/C(T)$ be a flag manifold and let $o \in F$. We have a splitting of the Lie algebra of G

$$\mathbf{g}^{\mathbf{C}} = \mathbf{h} \oplus \mathbf{m}$$

with $\mathbf{m} \cong T_o F$ and $\mathbf{h}$ the Lie algebra of the stabiliser of o in G. An invariant complex structure on F induces an ad $\mathbf{h}$-invariant splitting of $\mathbf{m}^{\mathbf{C}}$ into $(1,0)$ and $(0,1)$ spaces

$$\mathbf{m}^{\mathbf{C}} = \mathbf{m}^{+} \oplus \mathbf{m}^{-}$$

with $[\mathbf{m}^{+}, \mathbf{m}^{+}] \subset \mathbf{m}^{+}$ by integrability. One can show that $\mathbf{m}^{+}$ and $\mathbf{m}^{-}$ are nilpotent subalgebras of $\mathbf{g}^{\mathbf{C}}$ and in fact $\mathbf{h}^{\mathbf{C}} \oplus \mathbf{m}^{-}$ is a parabolic subalgebra of $\mathbf{g}^{\mathbf{C}}$ with nilradical $\mathbf{m}^{-}$. If P is the corresponding parabolic subgroup of $G^{\mathbf{C}}$ then P is the stabiliser of o and we obtain a biholomorphism between the complex coset space $G^{\mathbf{C}}/P$ and the flag manifold F.

Conversely, let $P \subset G^{\mathbf{C}}$ be a parabolic subgroup with Lie algebra $\mathbf{p}$ and let $\mathbf{n}$ be the conjugate of the nilradical of $\mathbf{p}$ (with respect to the real form $\mathbf{g}$). Then $H = G \cap P$ is the centraliser of a torus and we have orthogonal decompositions (with respect to the Killing inner product)

$$\mathbf{p} = \mathbf{h}^{\mathbf{C}} \oplus \overline{\mathbf{n}}, \ \mathbf{g}^{\mathbf{C}} = \mathbf{h}^{\mathbf{C}} \oplus \mathbf{n} \oplus \overline{\mathbf{n}}$$

which define an invariant complex structure on G/H realising the biholomorphism with $G^{\mathbf{C}}/P$.

The relationship between a flag manifold $F = G^{\mathbf{C}}/P$ and an inner symmetric space comes from an examination of the central descending series of $\mathbf{n}$. Recall that this is a filtration $0 = \mathbf{n}_{k+1} \subset \mathbf{n}_k \subset \ldots \subset \mathbf{n}_1 = \mathbf{n}$ of $\mathbf{n}$ defined by

$$\mathbf{n}_i = [\mathbf{n}, \mathbf{n}_{i-1}].$$

We orthogonalise this filtration using the Killing inner product by setting

$$\mathbf{g}_i = \mathbf{n}_{i+1}^{\perp} \cap \mathbf{n}_i$$

for $i \geq 1$ and extend this to a decomposition of $\mathbf{g}^{\mathbf{C}}$ by setting $\mathbf{g}_0 = \mathbf{h}^{\mathbf{C}} = (\mathbf{g} \cap \mathbf{p})^{\mathbf{C}}$ and $\mathbf{g}_{-i} = \overline{\mathbf{g}_i}$ for $i \geq 1$. Then

$$\mathbf{g}^{\mathbf{C}} = \sum \mathbf{g}_i$$

is an orthogonal decomposition with

$$\mathbf{p} = \sum_{i \leq 0} \mathbf{g}_i, \ \mathbf{n} = \sum_{i > 0} \mathbf{g}_i \,.$$

The crucial property of this decomposition is that

$$[\mathbf{g}_i, \mathbf{g}_j] \subset \mathbf{g}_{i+j}$$

which can be proved by demonstrating the existence of an element $\xi \in \mathbf{h}$ with the property that, for each i, $\operatorname{ad} \xi$ has eigenvalue $\sqrt{-1} i$ on $\mathbf{g}_i$. This element ξ (necessarily unique since $\mathbf{g}$ is semi-simple) was shown to exist by Burstall-Rawnsley [12] who called it the *canonical element* of $\mathbf{p}$. Since $\operatorname{ad} \xi$ has eigenvalues in $\sqrt{-1}\mathbf{Z}$, $\operatorname{Ad} \exp \pi \xi$ is an involution of $\mathbf{g}$ which we exponentiate to obtain an inner involution τ_ξ of G and thus an inner symmetric space G/K where $K = (G^{\tau_\xi})_0$. Clearly, K has Lie algebra given by

$$\mathbf{k} = \mathbf{g} \cap \sum_i \mathbf{g}_{2i}$$

and so contains H whence we obtain a homogeneous fibration $G/H \to G/K$ of our flag manifold over our inner symmetric space. Moreover, this fibration is essentially unique: the only ambiguity in the prescription is that several points in the symmetric space might have the same stabiliser K (e.g. antipodal points on a sphere). However, the number of such points is finite and so we only get a finite number of such fibrations. We call these fibrations the *canonical fibrations* of F. To summarise:

Theorem 4.3 *Let $F = G^{\mathbf{C}}/P$ be a flag manifold. Then there is a unique inner symmetric space G-space N associated to F together with a finite number of homogeneous fibrations $F \to N$.*

Let us emphasise that this construction depends on nothing but the conjugacy class of $\mathbf{p} \subset \mathbf{g}^{\mathbf{C}}$ and the choice of compact real form $\mathbf{g}$. Equivalently, it depends solely on the choice of invariant complex structure on F.

We have now seen that every flag manifold fibres over an inner symmetric space. Conversely, it is straightforward to show [12] that every inner symmetric space is the target of the canonical fibrations of at least one flag manifold. Let us now see how this story relates to the geometry of $J(N)$.

So let $p\colon F \to N$ be a canonical fibration. By construction, the fibres of p are complex submanifolds of F and this allows us to define a fibre map $i_p\colon F \to J(N)$ as follows: at $f \in F$ we have an orthogonal splitting of $T_f F$ into horizontal and vertical subspaces both of which are invariant under the complex structure of F. Then dp restricts to give an isomorphism of the horizontal part with $T_{p(f)}N$ and therefore induces an almost Hermitian structure on $T_{p(f)}N$: this is $i_p(f) \in J_{p(f)}N$. Such a construction is possible whenever we have a Riemannian submersion of a Hermitian manifold with complex submanifolds as fibres. In the case at hand we have:

Proposition 4.4 $i_p\colon F \to J(N)$ *is a G-equivariant holomorphic embedding.*

This implies that $i_p(F)$ is an almost complex submanifold of $J(N)$ on which $\mathcal{J}$ is integrable. Thus as a corollary of (4.1) and (4.4) we have

Corollary 4.5 $i_p(F)$ *is a G-orbit in $Z \subset J(N)$.*

In particular, this guarantees that Z is non-empty. Moreover, it turns out that the converse to (4.5) is true.

Theorem 4.6 ([12]) *If $j \in Z \subset J(N)$ then $G \cdot j$ is a flag manifold canonically fibred over N. In fact, $G \cdot j = i_p(F)$ for some canonical fibration $p\colon F \to N$ of a flag manifold F.*

For this, the main observation is the following: at $\pi(j)$, we have the symmetric decomposition

$$\mathbf{g} = \mathbf{k} \oplus \mathbf{q}$$

with $\mathbf{q} \cong T_{\pi(j)}N$. If $\mathbf{q}^-$ is the $(0,1)$-space for j then

$$[\mathbf{q}^-, \mathbf{q}^-] \oplus \mathbf{q}^-$$

is the nilradical of a parabolic subalgebra $\mathbf{p}$. One can then show that $G \cdot j$ is equivariantly biholomorphic to the corresponding flag manifold $G^{\mathbf{C}}/P$ as described in (4.6).

We are now in a position to complete the proof of (4.2). We have seen each canonical fibration of a flag manifold gives arise to a G-orbit in Z for some inner symmetric G-space N and that all such orbits arise in this way. But, for fixed G, there are only a finite number of biholomorphism types of flag manifold (they are in bijective correspondence with the conjugacy classes of parabolic subalgebras of $\mathbf{g}^{\mathbf{C}}$) and each flag manifold admits but a finite number of canonical fibrations. Thus Z is composed of a finite number of G-orbits all of which are closed and (4.2) follows. It is interesting to note that in this way we obtain a geometric interpretation of the purely algebraic construction of the canonical fibrations: they are just the restrictions of the projection $\pi\colon J(N) \to N$ to the various realisations of F as an orbit in Z.

Examples

To fix ideas, let us describe the flag manifolds and their canonical fibrations for two simple Lie groups of rank 3. In each case there are $7 = 2^{\mathrm{rank}\,G} - 1$ flag manifolds.

First let $G = \mathrm{SU}(4)$. There are two inner symmetric G-spaces: $\mathbf{C}P^3$ and the Grassmannian $G_2(\mathbf{C}^4)$ (of course, *qua* Riemannian symmetric space $G_3(\mathbf{C}^4)$ is the same as $\mathbf{C}P^3$). The flag manifolds are given by

$$F(r_1,\ldots,r_k;\mathbf{C}^4) = \frac{\mathrm{SU}(4)}{\mathrm{S}(\mathrm{U}(r_1) \times \cdots \times \mathrm{U}(r_k))}$$

where $r_1 + \cdots + r_k = 4$ and the complex structure is induced by the inclusion

$$F(r_1,\ldots,r_k;\mathbf{C}^4) \longrightarrow G_{r_1}(\mathbf{C}^4) \times G_{r_1+r_2}(\mathbf{C}^4) \times \cdots \times G_{r_1+\cdots+r_{k-1}}(\mathbf{C}^4).$$

The symmetric space associated to $F(r_1,\ldots,r_k;\mathbf{C}^4)$ is $G_{\sum r_{2i}}(\mathbf{C}^4)$. There is an 'antipodal map' on $G_2(\mathbf{C}^4)$ given by taking the perpendicular complement of an element and so the flag manifolds that fibre canonically over $G_2(\mathbf{C}^4)$ do so twice. They are $F(1,1,1,1;\mathbf{C}^4)$, $F(1,2,1;\mathbf{C}^4)$ and $F(2,2;\mathbf{C}^4)$. Of course, $F(2,2;\mathbf{C}^4)$ is just $G_2(\mathbf{C}^4)$ itself and the fibrations are just the identity and the antipodal map. We also note that $G_2(\mathbf{C}^4)$ is quaternionic Kähler so that the discussion in section 3 applies. In particular, recall that the 2-sphere bundle of $\mathcal{C}$ is a complex submanifold of Z: in our setting this is just one of the two realisations of $F(1,2,1;\mathbf{C}^4)$ in Z. The remaining flag manifolds $F(2,1,1;\mathbf{C}^4)$, $F(1,1,2;\mathbf{C}^4)$, $F(1,3;\mathbf{C}^4)$ and $F(3,1;\mathbf{C}^4)$ fibre canonically over $\mathbf{C}P^3$ and may be identified with the various Grassmannian bundles $G_r(T^{1,0}\mathbf{C}P^3)$ described in section 3. In conclusion, we see that, for $\mathbf{C}P^3$, Z has 4 components while, for $G_2(\mathbf{C}^4)$, it has 6.

Now let $G = \mathrm{SO}(7)$. There are again three simply connected inner symmetric G-spaces: S^6; $\widetilde{G}_3(\mathbf{R}^7)$ the Grassmannian of oriented 3-planes in $\mathbf{R}^7$ and $\widetilde{G}_2(\mathbf{R}^7)$ the 5-quadric. All of these may be viewed as Grassmannians of oriented k-planes for some k and so possess an antipodal map given by reversing the orientation of the k-planes. Thus each flag manifold has two canonical fibrations. The flag manifolds are given by

$$F_{\mathrm{iso}}(r_1,\ldots,r_k;\mathbf{R}^7) = \frac{\mathrm{SO}(7)}{\mathrm{U}(r_1) \times \cdots \times \mathrm{U}(r_k) \times \mathrm{SO}(7 - 2\sum r_i)}$$

where $r_1 + \cdots + r_k \leq 3$. To see the complex structure, we realise the space $F_{\text{iso}}(r_1, \ldots, r_k; \mathbf{R}^7)$ as a family of isotropic flags in $\mathbf{C}^7 = (\mathbf{R}^7)^{\mathbf{C}}$:

$$F_{\text{iso}}(r_1, \ldots, r_k; \mathbf{R}^7) =$$
$$\{V_1 \subset \ldots \subset V_k \subset \mathbf{C}^7 : \dim_{\mathbf{C}} V_i = r_1 + \cdots + r_i; \ V_i \text{ isotropic}\}$$

and then the complex structure is induced from the natural inclusion into $G_{r_1}(\mathbf{C}^7) \times \cdots \times G_{r_k}(\mathbf{C}^7)$. Now set $r_{k+1} = \frac{1}{2}(7 - \sum r_i)$ so that $2(r_1 + \cdots + r_{k+1}) = 7$ and then the symmetric space associated to $F_{\text{iso}}(r_1, \ldots, r_k; \mathbf{R}^7)$ is $\widetilde{G}_{\sum 2r_{2i}}(\mathbf{R}^7)$ with the canonical fibrations induced by the inclusions

$$\mathrm{U}(r_1) \times \cdots \times \mathrm{U}(r_k) \times \mathrm{SO}(2r_{k+1}) \to$$
$$\mathrm{SO}(2r_1) \times \cdots \times \mathrm{SO}(2r_k) \times \mathrm{SO}(2r_{k+1}) \to \mathrm{SO}(\textstyle\sum 2r_{2i+1}) \times \mathrm{SO}(\textstyle\sum 2r_{2i}).$$

We now see that $F_{\text{iso}}(3; \mathbf{R}^7)$ fibres canonically over S^6; $F_{\text{iso}}(1; \mathbf{R}^7)$, $F_{\text{iso}}(1, 1; \mathbf{R}^7)$ and $F_{\text{iso}}(2, 1; \mathbf{R}^7)$ over $\widetilde{G}_2(\mathbf{R}^7)$ and $F_{\text{iso}}(2; \mathbf{R}^7)$, $F_{\text{iso}}(1, 2; \mathbf{R}^7)$ and $F_{\text{iso}}(1, 1, 1; \mathbf{R}^7)$ over $\widetilde{G}_3(\mathbf{R}^7)$. We remark that $\widetilde{G}_3(\mathbf{R}^7)$ is quaternionic Kähler and that the associated sphere-bundle twistor space is one of the realisations of $F_{\text{iso}}(2; \mathbf{R}^7)$. In conclusion, for S^6, Z has two components which are just $J_+(S^6)$ and $J_-(S^6)$ while the other two symmetric spaces have six components each in Z.

There is a well-known duality between symmetric spaces of compact and non-compact type and this duality extends to the twistor theory we have been discussing. For each non-compact real form G_{R} of a complex semisimple group Lie group $G^{\mathbf{C}}$, there is a unique Riemannian symmetric space G_{R}/K of non-compact type. The corresponding involution is called the *Cartan involution* of G_{R}. The above development suggests that we restrict attention to G_{R} with inner Cartan involution.

Consider now the orbits of such a G_{R} on the various flag manifolds $F = G^{\mathbf{C}}/P$. Those orbits which are open subsets of F we call *flag domains*: in our situation, an orbit is a flag domain precisely when the stabilisers contain a compact Cartan subgroup of G_{R}. It turns out that the presence of this compact Cartan subgroup is precisely what we need to define a canonical element of $\mathbf{g}_{\mathrm{R}}$ and thus an involution of $\mathbf{g}_{\mathrm{R}}$ just as in the compact case. However the involution is not necessarily a Cartan involution (i.e. the associated symmetric space need not be Riemmanian). In case that the involution is a Cartan involution we say that our flag domain is a *canonical flag domain* and then we exponentiate the involution to get a Riemannian symmetric space of non-compact type and a canonical fibration (unique in this case) of our canonical flag domain over it. We may now repeat the analysis of the compact case and, in particular, we find that G_{R} acts transitively on connected components of $Z \subset J(G_{\mathrm{R}}/K)$. Further, each component is a canonically fibred canonical flag domain so that (4.2) holds in the non-compact setting.

It is interesting to note that these flag domains have arisen in other areas: indeed, for $G^{\mathbf{C}} = \mathrm{SO}(n, \mathbf{C})$ or $\mathrm{Sp}(n, \mathbf{C})$, they form a subset of the Griffiths period matrix domains [19] that classify Hodge structures. There are intriguing relationships between the theory of variation of Hodge structure and the theory of minimal surfaces in compact Riemannian symmetric spaces that arise from this twistor space duality. For example, both flag manifolds and flag domains carry an invariant holomorphic distribution which is transverse to the fibres of the canonical fibrations. This *super-horizontal distribution* is defined at the identity coset as the $\sqrt{-1}$-eigenspace $\mathbf{g}_1$ of the canonical element. It has the following interesting property:

Theorem 4.7 ([12]) *Let X be a flag manifold or canonical flag domain and $\pi\colon X \to N$ be a homogeneous fibration onto a Riemannian symmetric space. If $\phi\colon M \to X$ is a holomorphic map of a Kähler manifold with image tangent to the super-horizontal distribution then $\pi \circ \phi\colon M \to N$ is harmonic.*

We call maps $\phi\colon M \to X$ satisfying the hypotheses of (4.7) *super-horizontal holomorphic maps*.

For instance, let $X = \mathrm{SU}(n+1)/\mathrm{S}(\mathrm{U}(1) \times \cdots \times \mathrm{U}(1)) = F(1,\ldots,1;\mathbf{C}^{n+1})$. There are $n+1$ homogeneous fibrations $\pi_i\colon X \to \mathbf{C}P^n$, $i = 1,\ldots,n$, with π_0 holomorphic and π_n anti-holomorphic. Now a super-horizontal holomorphic map $\phi\colon S^2 \to X$ is essentially just the Frenet frame of the holomorphic map $\pi_0 \circ \phi\colon S^2 \to \mathbf{C}P^n$ while each $\pi \circ \phi$ is a harmonic map $S^2 \to \mathbf{C}P^n$. It is the content of the classification theorem for harmonic 2-spheres in $\mathbf{C}P^n$ that all such harmonic maps arise in this way. Thus in this situation all harmonic maps are produced via (4.7).

On the non-compact side of the fence, the super-horizontal distribution is precisely that which defines the infinitesimal period relation. Thus the (local lifts of) period maps are precisely the super-horizontal holomorphic maps into period domains. Much progress has recently been made by Carleson-Toledo [13] on the relationship between period maps and harmonic maps into compact quotients of symmetric spaces of non-compact type and it seems likely that harmonic maps of Kähler manifolds into such quotients of sufficiently high rank are covered by period maps as in (4.7).

Finally, let us briefly describe the situation for non-inner even-dimensional Riemannian symmetric spaces. Here it can be shown that Z is nonempty and, of course, by (4.1) each orbit in Z is an almost complex submanifold on which $\mathcal{J}$ is integrable. However the orbits are no longer flag manifolds or flag domains and G no longer acts transitively on connected components of Z.

5 Flags and the Loop Group

There is another realisation of the canonical fibrations of flag manifolds that serves to introduce a twistor space of a quite different type. For this, assume that G is of adjoint type (i.e. has trivial centre) and let ΩG denote the infinite-dimensional manifold of based loops in G: the loop group. In fact ΩG is a

Kähler manifold [23] and may be viewed as a flag manifold $\mathcal{G}^{\mathbf{C}}/\mathcal{P}$ where $\mathcal{G}^{\mathbf{C}}$ is the manifold of loops in $G^{\mathbf{C}}$ and $\mathcal{P}$ is the subgroup of those that extend holomorphically to the disc [17, 23]. We have various fibrations $\rho_\lambda \colon \Omega G \to G$ given by evaluation at $\lambda \in S^1$ and in some ways ρ_{-1} behaves like a canonical fibration making ΩG into a universal twistor space for G. For instance, it is a theorem of Uhlenbeck [28] that any harmonic map of S^2 into G is of the form $\rho_{-1} \circ \Phi$ for some 'super-horzontal' holomorphic map $\Phi \colon S^2 \to \Omega G$.

The flag manifolds of G embed in ΩG as conjugacy classes of geodesics and we find a particular embedding of this kind using the canonical element. Indeed, our assumption that G be centre-free means that $\exp 2\pi\xi = e$ for any canonical element ξ. Thus if $F = G/H = G^{\mathbf{C}}/P$ is a flag manifold with ξ the canonical element of $\mathbf{p}$, we may define a map $\Gamma \colon F \to \Omega G$ by setting

$$\Gamma(eH) = (e^{\sqrt{-1}t} \mapsto \exp t\xi)$$

and extending by equivariance. Moreover, if N is the inner symmetric space associated to F, we have a totally geodesic immersion $\gamma \colon N \to G$ defined by setting $\gamma(x)$ equal to the element of G that generates the involution at x. We now have:

Proposition 5.1 $\Gamma \colon F \to \Omega G$ *is a totally geodesic, holomorphic, isometric immersion and the following diagram commutes*

$$
\begin{array}{ccc}
F & \xrightarrow{\ \Gamma\ } & \Omega G \\
\pi_1 \downarrow & & \downarrow \rho_{-1} \\
N & \xrightarrow{\ \gamma\ } & G
\end{array}
$$

where π_1 is a canonical fibration.

Thus we have a third realisation of the canonical fibrations as the trace of ρ_{-1} on certain conjugacy classes of geodesics. The reader is invited to ponder on the relation between these three constructions.

6   Conclusion

We have seen that the construction of complex manifolds associated to a Riemannian manifold N by the above methods requires stringent conditions on the curvature of N. However, the construction can be carried through for a fairly large class of geometrically interesting Riemannian manifolds. As for applications, the twistor theory of quaternionic Kähler and hyperkähler manifolds is highly developed [21, 24] while for symmetric spaces it is only just beginning (see [12], though, for applications to minimal surfaces) and there are many unanswered questions. Let us finish by mentioning a few of these.

- We have seen that there is a good theory of flag spaces fibring over Riemannian symmetric spaces. What can be said about the case of pseudo-Riemannian symmetric spaces? Certainly, one can produce such fibrations of non-canonical flag domains.

- If we view the Riemannian symmetric spaces as 'flat' examples, are there 'curved' analogues of the above theory? The example of quaternionic Kähler manifolds suggests that there are far less twistor spaces in the non-symmetric case but there is still a satisfactory theory. A good test case for this would be the complex paraconformal manifolds of Bailey-Eastwood [3] which are 'curved' versions of the complex Grassmannians.

- Finally, an interesting but almost certainly ill-posed question is: what is the relationship between the theory discussed in this chapter and the highly developed twistor theory of space-time discussed elsewhere in this volume?

References

[1] M. F. Atiyah, D. G. Drinfeld, N. J. Hitchin, Y. I. Manin, *Construction of instantons*, Phys. Lett. **65A** (1978), 185–187.

[2] M. F. Atiyah, N. J. Hitchin & I. M. Singer, *Self-duality in four-dimensional Riemannian geometry* Proc. Roy. Soc. Lond. **A362** (1978), 425–461.

[3] T. N. Bailey & M. G. Eastwood, *Complex paraconformal manifolds— their differential geometry and twistor theory*, Preprint.

[4] M. Berger, *Sur les groupes d'holonomie des variétés à connexion affine et des variétés riemanniennes*, Bull. Soc. Math. France. **15** (1955), 279–330.

[5] A. Borel, *Kählerian coset spaces of semi-simple Lie groups*, Proc. Nat. Acad. Sci. USA. **40** (1954), 1147–151.

[6] A. Borel & F. Hirzebruch, *Characteristic classes and homogeneous spaces, I*, Amer. J. Math. **80** (1958), 459–538.

[7] R. L. Bryant, *Conformal and minimal immersions of compact surfaces into the 4-sphere*, J. Diff. Geom. **17** (1982), 455–473.

[8] R. L. Bryant, *Lie groups and twistor spaces*, Duke Math. J. **52** (1985), 223–261.

[9] D. Burns, *Harmonic mappings from $\mathbb{C}P^1$ to $\mathbb{C}P^n$*, in 'Proc. Tulane Conf.', Lecture Notes in Math. 949 (Springer, Berlin, 1982), pp. 48–56.

[10] F. E. Burstall, *Twistor methods for harmonic maps*, in 'Differential Geometry' (ed. V. L. Hansen), Lecture Notes in Math. 1263 (Springer, Berlin, 1987), pp. 55–96.

[11] F. E. Burstall, *Recent developments in twistor methods for harmonic maps*, in 'Harmonic mappings, twistors and σ-models' (ed. P. Gauduchon, World Scientific, Singapore, 1988), pp. 158–176.

[12] F. E. Burstall & J. H. Rawnsley, *Twistor theory for Riemannian symmetric spaces with applications to harmonic maps of Riemann surfaces*, (preprint) 1989.

[13] J. A. Carleson & D. Toledo, *Harmonic mappings of Kähler manifolds to locally symmetric spaces*, (Utah preprint) 1988.

[14] A. M. Din & W. J. Zakrzewski, *General classical solutions in the $\mathbb{C}P^{n-1}$ model*, Nucl. Phys. B. **174** (1980), 397–406.

[15] M. Dubois-Violette, *Structures complexes au-dessus des variétés*, In: *Mathématiques et Physique* Progress in Math. 37 (Birkhauser, Boston, 1983).

[16] J. Eells & J. C. Wood, *Harmonic maps from surfaces to complex projective spaces*, Adv. in Math. **49** (1983), 217–263.

[17] D. S. Freed, *The geometry of loop groups*, J. Diff. Geom. **28** (1988), 223–276.

[18] V. Glaser & R. Stora, *Regular solutions of the $\mathbb{C}P^n$ models and further generalisations*, CERN preprint, 1980.

[19] P. A. Griffiths, *Periods of integrals on algebraic manifolds, III*, Publ. Math. I.H.E.S **38** (1970), 125–180.

[20] N. J. Hitchin, *Monopoles and geodesics*, Comm. Math. Phys. **83** (1982), 579–602.

[21] N. J. Hitchin, A. Karlhede, U. Lindström & M. Roček, *Hyperkähler metrics and supersymmetry*, Comm. Math. Phys. **108** (1987), 535–589.

[22] N. R. O'Brian & J. H. Rawnsley, *Twistor spaces*, Ann. Glob. Anal. and Geom. **3** (1985), 29–58.

[23] A. N. Pressley & G. Segal, *Loop Groups*, Oxford Math. Monographs, 1986.

[24] S. M. Salamon, *Quaternionic Kähler manifolds*, Invent. Math. **67** (1982), 143–171.

[25] S. M. Salamon, *Riemannian Geometry and holonomy groups*, Pitman Research Notes in Math. 201, 1989.

[26] I. M. Singer & J. A. Thorpe, *The curvature of 4-dimensional Einstein spaces*, in 'Global Analysis, Papers in honor of K. Kodaira' (ed. D. C. Spenser and S. Iyanaga, Princeton University Press, New Jersey, 1969), pp. 355–365.

[27] A. J. Small, *A twistorial interpretation of the Weierstrass representation formulae*, Ph.D. thesis, University of Warwick 1989.

[28] K. Uhlenbeck, *Harmonic maps into Lie groups (classical solutions of the chiral model)*, J. Diff. Geom. **30** (1989), 1–50.

[29] J. C. Wood, *Twistor constructions for harmonic maps*, in 'Differential Geometry and Differential Equations' (eds. C. H. Gu, M. Berger & R. L. Bryant), Lecture Notes in Math. 1255 (Springer, Berlin, 1988), pp. 130–159.

[30] R. S. Ward, *On self-dual gauge fields*, Phys. Lett. **61A** (1977), 81–82.

[31] J. A. Wolf, *The action of a real semi-simple Lie group on a complex flag manifold. I: Orbit structure and holomorphic arc components*, Bull. Amer. Math. Soc. **75** (1969), 1121–1237.

Twistors, Ambitwistors, and Conformal Gravity

C.R. LeBrun

1 Introduction

In the last several years, a new understanding has emerged concerning the remarkable depth of the relationship linking twistor geometry to the Einstein vacuum equations. This is particularly striking insofar as twistor ideas are rooted in conformal geometry, whereas the Einstein equations

$$R_{ab} = \tfrac{1}{4} R g_{ab}$$

are certainly not conformally invariant. These equations nonetheless display such a powerful affinity for conformal geometry that they leave an indelible fingerprint on the conformal classes of Einstein metrics; even more surprisingly, the resulting conformally invariant conditions are exactly those which arise naturally out of complex deformation theory by means of the ambitwistor correspondence. As this correspondence also provides an appropriate setting for the study of the Yang-Mills and Dirac equations in curved space-time, it would seem to give us not merely a useful method for solving certain interesting PDEs, but also something that would seem to be of far greater significance—a geometric framework which, in an entirely non-trivial manner, 'predicts' which equations are of physical importance! For my own part, I am strongly tempted to conclude that the curious efficacy of twistor methods for the study of the PDEs of fundamental physics cannot be accidental, but rather presages the emergence of a different kind of setting for realistic physical theories.

The reader may find the following stylistic remarks helpful:

- We work throughout with complex space-times, although we will make some concluding remarks regarding real structures in section 6.

- The holomorphic tangent bundle of a complex manifold X will simply be denoted by TX, rather than by the more precise $T^{1,0}X$.

- If $E \to X$ is a holomorphic vector bundle, $0_E \subset E$ denotes the zero section and $\mathbf{P}E = (E - 0_E)/(\mathbf{C} - 0)$ denotes the associated holomorphic bundle of projective spaces.

2   The Ambitwistor Correspondence

Suppose that $\mathcal{M}$ is a complex 4-manifold, and let $\mathbf{g} \in \Gamma(\mathcal{M}, \mathcal{O}(\odot^2 T^*\mathcal{M}))$ be a holomorphic non-degenerate symmetric 2-tensor on $\mathcal{M}$. We will say that $\mathbf{g}$ is a *complex-Riemannian metric* on $\mathcal{M}$, and $(\mathcal{M}, \mathbf{g})$ will be called a *complex space-time*. We can then associate to to $(\mathcal{M}, \mathbf{g})$ a family of curves called *null geodesics* by considering those inextendible connected one-dimensional complex submanifolds $\gamma \subset \mathcal{M}$ for which any tangent vector field $v \in \Gamma(\gamma, \mathcal{O}(T\gamma))$ satisfies

$$\nabla_v v \propto v$$
$$\mathbf{g}(v, v) \equiv 0,$$

where ∇ denotes the Levi-Civita connection associated with $\mathbf{g}$. Knowing these curves determines the *conformal class* of the complex metric $\mathbf{g}$, since a vector is null iff it is tangent to some null geodesic γ; what is less apparent is that the conformal class conversely determines the set of null geodesics.

To understand this latter fact, let us notice that the projectivized cotangent bundle of $\mathcal{M}$ carries a natural contact structure, by which we mean a line-bundle-valued 1-form $\Theta \in \Gamma(\mathbf{P}T^*\mathcal{M}, \Omega^1(\mathbf{L}))$, such that

$$\Theta \wedge (d\Theta)^{\wedge 3} \neq 0;$$

indeed, Θ is the usual canonical 1-form $p_j dq^j$ defined by

$$\Theta|_{[\phi]} = \pi^*(\phi),$$

where $\pi : \mathbf{P}T^*\mathcal{M} \to \mathcal{M}$ is the canonical projection. Knowing a conformal class on $\mathcal{M}$ amounts to knowing the hypersurface

$$\mathcal{Q} = \{[\phi] \in \mathbf{P}T^*\mathcal{M} \mid \mathbf{g}^{-1}(\phi, \phi) = 0\}$$

of null covectors. Now the restriction

$$\vartheta = \Theta|_{\mathcal{Q}}$$

is not a contact structure; rather, $\xi := \ker(\vartheta \wedge d\vartheta)$ defines a rank 1 sub-bundle $\xi \subset TQ$, and this distribution ξ is then tangent to a holomorphic foliation of $\mathcal{M}$ by complex curves $\hat{\gamma} \subset Q$. Each leaf of this foliation then projects to a holomorphic curve $\gamma = \pi[\hat{\gamma}]$, and one then checks that each such curve in $\mathcal{M}$ is a null geodesic; indeed, the leaf through $[\phi_a] \in \mathbf{P}T^*\mathcal{M}$ projects to the null geodesic tangent to ϕ^a. This gives a manifestly conformally invariant construction of the null geodesics.

Let $\mathcal{N}$ denote the set of null geodesics of $(\mathcal{M}, \mathbf{g})$. Since this is the same as defining $\mathcal{N}$ as the leaf space of the above foliation, this carries a natural topology—namely, let $q : Q \to \mathcal{N}$ denote the tautological 'quotient' map (assigning, to each point of Q, the leaf through it) and equip $\mathcal{N}$ with the quotient topology. If we assume that $(\mathcal{M}, \mathbf{g})$ is geodesically convex, $\mathcal{N}$ is then Hausdorff and the foliation tangent to ξ has trivial holonomy, so that [14] $\mathcal{N}$ has a unique complex structure making q a holomorphic map of maximal rank. The complex manifold $\mathcal{N}$ is then called the *ambitwistor space* of $(\mathcal{M}, \mathbf{g})$. Letting $p : Q \to \mathcal{M}$ denote the restriction to Q of the canonical projection $\pi : \mathbf{P}T^*\mathcal{M} \to \mathcal{M}$, we have a holomorphic double fibration

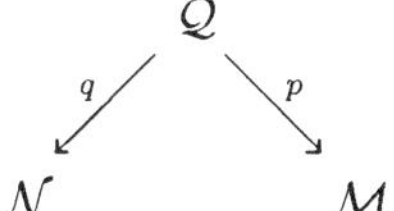

called the *ambitwistor correspondence*, interrelating $\mathcal{M}$ and $\mathcal{N}$. The line-bundle-valued 1-form ϑ descends to $\mathcal{N}$, giving it the structure of a complex contact manifold; i.e. there is a line-bundle-valued 1-form θ on $\mathcal{N}$ with $q^*\theta = \vartheta$, and $\theta \wedge (d\theta)^{\wedge 2} \neq 0$.

Every point $x \in \mathcal{M}$ gives rise to a complex submanifold $Q_x := q[p^{-1}(x)]$ of $\mathcal{N}$ isomorphic to $\mathbf{P}_1 \times \mathbf{P}_1$; we will call this submanifold the *sky* of x, since it exactly represents the set of light rays through x. Since, by construction, ϑ vanishes on any fiber of p, it follows that θ vanishes on every sky; i.e. the skies of $\mathcal{M}$ are *Legendrian submanifolds* [1] of $\mathcal{N}$. Since the restriction of $\mathbf{L} \to \mathbf{P}T^*\mathcal{M}$ to $p^{-1}(x) \cong \mathbf{P}_1 \times \mathbf{P}_1$ is the line bundle $\mathcal{O}(1,1)$ (= the divisor of the diagonal $\mathbf{P}_1 \subset \mathbf{P}_1 \times \mathbf{P}_1$), it follows that the the line-bundle $(\wedge^5 T\mathcal{N})^{1/3}$ in which θ takes its values must restrict to Q_x as $\mathcal{O}(1,1)$, and the fact that Q_x is Legendrian then implies that the normal bundle $\mathbf{N} \to Q_x$ of each sky is isomorphic to $J^1\mathcal{O}(1,1)$. In particular, there is an exact sequence

$$0 \to \Omega^1 \otimes \mathcal{O}(1,1) \to \mathbf{N} \to \mathcal{O}(1,1) \to 0,$$

where $\Omega^1 \cong \mathcal{O}(-2,0) \oplus \mathcal{O}(0,-2)$ is the cotangent bundle of $Q_x \cong \mathbf{P}_1 \times \mathbf{P}_1$, implying that

$$H^m(Q_x, \mathcal{O}(\mathbf{N})) = \begin{cases} 4 & \text{if } m = 0 \\ 0 & \text{otherwise}, \end{cases}$$

so that the 4-parameter family of compact complex submanifolds given by $\{Q_x \subset \mathcal{N} | x \in \mathcal{M}\}$ is therefore complete in the sense of Kodaira [11]. Thus the complex structure of $\mathcal{N}$ encodes the conformal geometry of $(\mathcal{M}, \mathbf{g})$, in the sense that, assuming that $\mathcal{M}$ is, say, geodesically convex[1], we can recreate $\mathcal{M}$ from $\mathcal{N}$ as a connected component of the the 2-quadrics $\mathbf{P}_1 \times \mathbf{P}_1$ contained in $\mathcal{M}$.

3 Twistors and Ambitwistors

In the case of the 4-quadric $\mathbf{Q}_4 \subset \mathbf{P}_5$, which is the natural conformal compactification of $(\mathbf{C}^4, \sum_{j=1}^4 (dz^j)^{\otimes 2})$, the corresponding ambitwistor space is given by

$$\mathbf{A} := \{([Z^\alpha], [W_\alpha]) \in \mathbf{P}_3 \times \mathbf{P}_3 \mid \sum_{\alpha=1}^4 Z^\alpha W_\alpha = 0\}.$$

This is a manifestation of a rather more general phenomenon relating the ambitwistor correspondence to Roger Penrose's classic *twistor correspondence* [19, 20].

Indeed, instead of considering null *curves* in $(\mathcal{M}, \mathbf{g})$, we might choose to look for (totally) null *surfaces*, meaning complex 2-dimensional submanifolds $\Sigma \subset \mathcal{M}$ for which $\mathbf{g}|_\Sigma \equiv 0$. Such submanifolds are necessarily totally geodesic—making them interesting, but also making them rare. In fact, there are typically no such submanifolds at all! Rare though they be, they nonetheless come in two flavors; for if we let

$$* : \textstyle\bigwedge^2 T\mathcal{M} \to \bigwedge^2 T\mathcal{M}$$

denote the *star-operator* $\frac{1}{2}e^{ab}{}_{cd}$ of $\mathbf{g}$, then a totally null surface Σ automatically has the property that $\bigwedge^2 T\Sigma$ is an eigenspace of $*$, with eigenvalue ± 1, so that we may label such surfaces as α-*surfaces* and β-*surfaces* in accordance with the sign of the eigenvalue[2]. The condition for the existence of α-surfaces is then that the Weyl curvature C of $(\mathcal{M}, \mathbf{g})$satisfy $C = *C$, where the star-operator treats C as a bundle-valued 2-form; such a space-time is then called *self-dual*. (In terms of curvature spinors [20], our condition becomes $\tilde{\Psi}_{A'B'C'D'} = 0$). For a geodesically convex self-dual space-time, we then have a 3-manifold $\mathcal{P}$ of α-surfaces, called the *twistor space* of $(\mathcal{M}, \mathbf{g})$, and a

---

[1]This actually still holds with much weaker convexity hypotheses, as will be explained in Section 6.

[2]Note that $*$ is locally determined up to a $\pm$ sign, the choice of which is called called a 'complex orientation.' Changing this 'orientation' interchanges our labels of self-duality and anti-self-duality.

double fibration

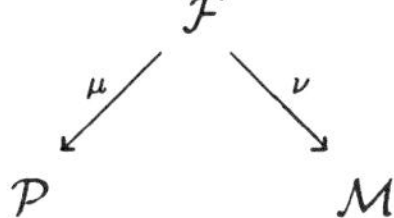

called the *twistor correspondence*. The fibers of ν are then $\mathbf{P}_1$'s, while the fibers of μ are complex surfaces.

Now, quite generally, a null geodesic is contained in at most one α-surface and at most one β-surface. In the case of a self-dual space-time $(\mathcal{M}, \mathbf{g})$, this gives rise to a holomorphic projection

$$\lambda : \mathcal{N} \to \mathcal{P}$$

by sending a null geodesic to the unique α-surface containing it. The fibers of this map are Legendrian submanifolds, and thus there is a natural induced map

$$\hat{\lambda} : \mathcal{N} \to \mathbf{P}(T^*\mathcal{P})$$

which turns out to be an inclusion. Thus

Proposition 3.1 *The ambitwistor space $\mathcal{N}$ of a self-dual space-time $(\mathcal{M}, \mathbf{g})$ is an open set in the projectivized cotangent bundle of its twistor space $\mathcal{P}$.*

Proof For more details see [16]. $\square$

A particular consequence of this is that the embedding $\mathsf{A} \hookrightarrow \mathbf{P}_3 \times \mathbf{P}_3$ occurring in the conformally flat case has an analogue in the self-dual case. Namely, if $\kappa \to \mathcal{P}$ denotes the canonical line-bundle defined by $\mathcal{O}(\kappa) := \Omega^3_{\mathcal{P}}$, the jet bundle $J^1\kappa \to \mathcal{P}$ is a rank 4 vector bundle, and we have a canonical inclusion $\Omega^1 \otimes \kappa \hookrightarrow J^1\kappa$. This gives rise to an inclusion $\mathbf{P}(T^*\mathcal{M}) \hookrightarrow \mathbf{P}(J^1\kappa)$, since $\mathbf{P}(T^*\mathcal{M}) = \mathbf{P}(\kappa \otimes T^*\mathcal{M})$. Together with Proposition (3.1), this gives us an embedding of $\mathcal{N}$ in $\mathbf{P}(J^1\kappa)$ as a complex hypersurface. In the conformally flat case, this exactly reconstructs $\mathsf{A} \hookrightarrow \mathbf{P}_3 \times \mathbf{P}_3$.

One might be tempted to ask whether this embedding has an analogue in the case of arbitrary conformal curvature. As we shall see, the answer is no—but, in the process, we will be led to the very doorstep of Einstein's equations.

4 Thickenings and Poisson Structures

Consider for a moment the general problem of realizing a given complex manifold as a complex hypersurface $\mathbf{X}$ in some complex manifold $\mathbf{Y}$. This can obviously always be done—for example, if $\mathbf{L} \to \mathbf{X}$ is any holomorphic line bundle, we may take $\mathbf{Y}$ to be the total space of $\mathbf{L}$ and let $\mathbf{X} \hookrightarrow \mathbf{Y}$ be the zero section of the line bundle. However, this construction does not typically account for all possible embeddings, even on the level of germs at $\mathbf{X}$—there is no reasonable holomorphic analogue of the tubular neighborhood theorem!

To get a better feeling for the problem, notice that another invariant of an embedding $\mathbf{X} \hookrightarrow \mathbf{Y}$ is the 'extended tangent bundle' $\hat{T}\mathbf{X} := (T\mathbf{Y})|_{\mathbf{X}}$, which is an extension

$$0 \to T\mathbf{X} \to \hat{T}\mathbf{X} \to \mathbf{L} \to 0$$

of the 'normal bundle' $\mathbf{L}$ by the tangent bundle of $\mathbf{X}$, and is therefore characterized by an element of $H^1(\mathbf{X}, \mathcal{O}(T\mathbf{X} \otimes \mathbf{L}^*))$. If $\mathbf{X} \hookrightarrow \mathbf{Y}$ were just the zero-section of a line bundle, this invariant would vanish; but it is, in fact, typically non-trivial, as it is, for example, in the case of $\mathsf{A} \hookrightarrow \mathbf{P}_3 \times \mathbf{P}_3$. The following question would therefore seem to be pertinent: Given a holomorphic vector bundle $V \to \mathbf{X}$ and an inclusion $T\mathbf{X} \hookrightarrow V$, can we find an embedding $\mathbf{X} \hookrightarrow \mathbf{Y}$ such that $\hat{T}\mathbf{X} = V$? The answer, generally speaking, is *certainly not*. In fact, there is an infinite hierarchy of obstructions that must vanish in order to merely be able to solve this problem on the level of formal power series!

To make this more precise, notice that there is a ringed space

$$\mathbf{X}^{(m)} := (\mathbf{X}, \mathcal{O}_{(m)} := (\mathcal{O}_{\mathbf{Y}}/\mathcal{I}^{m+1})|_{\mathbf{X}}),$$

called the mth *infinitesimal neighborhood* of $\mathbf{X} \subset \mathbf{Y}$, associated with the m-jet of an embedding $\mathbf{X} \hookrightarrow \mathbf{Y}$; here $\mathcal{I} \subset \mathcal{O}_{\mathbf{Y}}$ denotes the ideal of functions vanishing on $\mathbf{X}$. If $V = \hat{T}\mathbf{X}$ is the extended tangent bundle of some embedding, then we have

$$\mathcal{O}_{(1)} = \{(f, \varphi) \in \mathcal{O} \oplus \mathcal{O}(V^*) \mid \varphi|_{T\mathbf{X}} = df\}, \tag{1}$$

and that, more generally, given a holomorphic vector bundle $V \to \mathbf{X}$ and a holomorphic embedding $T\mathbf{X} \hookrightarrow V$, we can use equation (1) to define a sheaf of rings on $\mathbf{X}$ locally modeled on $\mathcal{O}[\zeta]/(\zeta^2)$. This is then an example of a *thickening*, meaning an 'abstract infinitesimal neighborhood'; to be precise, an mth order thickening $\mathbf{X}^{(m)}$ of a complex manifold $\mathbf{X}$ consists of a sheaf of rings $\mathcal{O}_{(m)}$ on $\mathbf{X}$ and a ring homomorphism $\alpha : \mathcal{O}_{(m)} \to \mathcal{O}$ which is locally isomorphic to the obvious homomorphism $\mathcal{O}[\zeta]/(\zeta^{m+1}) \to \mathcal{O}$. A problem more modest than the one above is to determine when a thickening of order m can be extended to order $m + 1$, and, if it extends, in how many different ways it does so. The answer [9] is then that the obstruction to extending a thickening $\mathbf{X}^{(m)}$, $m \geq 1$, to a thickening $\mathbf{X}^{(m+1)}$ lives in $H^2(\mathbf{X}, \mathcal{O}(\hat{T}\mathbf{X} \otimes \mathbf{L}^{*\otimes(m+1)}))$, while the freedom of extension is parameterized by $H^1(\mathbf{X}, \mathcal{O}(\hat{T}\mathbf{X} \otimes \mathbf{L}^{*\otimes(m+1)}))$.

In the case of an ambitwistor space $\mathcal{N}$, we have a complex contact structure available, and this turns out to be a crucial guide through the labyrinth of the above obstructions. Indeed, over $\mathcal{N}$ we have a standard line bundle $\mathbf{L}$ in which the contact form $\theta \in \Gamma(\mathcal{N}, \Omega^1(\mathbf{L}))$ takes its values[3],

---

[3]Thus $\mathbf{L}^{\otimes 3} \cong \kappa^*$, where κ is the canonical line bundle defined by $\mathcal{O}(\kappa) := \Omega^5_{\mathcal{N}}$.

and we will consider only thickenings with this normal bundle. Moreover, the contact structure induces on the total space $\mathcal{L}$ of this bundle a natural *Poisson structure* [22], meaning that the sheaf $\mathcal{O}_{\mathcal{L}}$ of holomorphic functions comes equipped with a Lie algebra structure

$$\{\,,\} : \mathcal{O} \times \mathcal{O} \to \mathcal{O}$$

induced by the rule

$$\{f, g\} := \tau(df, dg)$$

where the *exelissic form* $\tau \in \Gamma(\mathcal{L}, \mathcal{O}(\wedge^2 T\mathcal{L}))$ is defined as follows. The adjoint of the contact form $\theta : T\mathcal{N} \to \mathbf{L}$ defines an inclusion $\mathbf{L}^* \hookrightarrow T^*\mathcal{N}$ and thereby defines a complex symplectic form ω on the complement $\mathbf{L}^* - 0_{\mathbf{L}^*}$ of the zero section by pulling back the symplectic form of $T^*\mathcal{N}$. However, there is a canonical biholomorphism $\Phi : \mathbf{L} - 0_{\mathbf{L}} \cong \mathbf{L}^* - 0_{\mathbf{L}^*}$ which is characterized by $\langle \Phi(v), v \rangle \equiv 1$. Then

$$\tau := (\Phi^*\omega)^{-1}$$

is defined on $\mathcal{L}$ except at the zero section $\mathcal{N} = 0_{\mathbf{L}}$, and we now notice that it in fact has a holomorphic continuation to all of $\mathcal{L}$. In fact, in local coordinates for which the contact form on $\mathcal{N}$ is represented by

$$\vartheta = du + \sum_{j=1}^{2} p_j \, dq^j,$$

we have

$$\tau = t \left[(t\frac{\partial}{\partial t} + \sum p_j \frac{\partial}{\partial p_j}) \wedge \frac{\partial}{\partial u} + \sum \frac{\partial}{\partial q^j} \wedge \frac{\partial}{\partial p_j} \right],$$

where the local coordinate on t on the total space $\mathcal{L}$ of the line bundle $\mathbf{L} \to \mathbf{X}$ is the fiber coordinate induced by the trivialization of $\mathbf{L}$ for which $\theta \in \Gamma(\mathcal{N}, \Omega^1(\mathbf{L}))$ is represented by ϑ.

Now if $\mathcal{O}_m$ denotes the structure sheaf of the mth infinitesimal neighborhood of the zero section $\mathcal{N} = 0_{\mathbf{L}} \subset \mathbf{L}$, the exelissic form τ defines a map

$$\begin{aligned} \mathcal{O}_{m-1}/\mathbf{C} &\to \operatorname{Der}(\mathcal{O}_m) \\ f &\mapsto \{f, \cdot\}, \end{aligned}$$

and the image of this map is a sheaf of nilpotent Lie algebras that we can exponentiate to give a certain sheaf $\mathcal{G}_m$ of groups of automorphisms of $\mathcal{O}_m$. A thickening of $\mathcal{N}$ is then said to be of *Poisson type* if its transition functions can be taken to be elements of $\mathcal{G}_m$. Equivalently, a thickening is of Poisson type iff its equivalence class is in the image of

$$H^1(\mathcal{N}, \mathcal{G}_m) \to H^1(\mathcal{N}, \operatorname{Aut}(\mathcal{O}_m)).$$

Elements of the torsor $H^1(\mathcal{N}, \mathcal{G}_m)$ are called *Poisson thickenings*.

A particular case of this construction is that, at the level of first order thickenings, one has a special class of candidates for the extended tangent bundle, given by the image of the map

$$\tau_* : H^1(\mathcal{N}, \mathcal{O}/\mathbf{C}) \to H^1(\mathcal{N}, \mathcal{O}(T\mathcal{N} \otimes \mathbf{L}^*)).$$

In particular, when we throw in the logarithm log $: \mathcal{O}^* \to \mathcal{O}/\mathbf{C}$ we get a natural morphism

$$\tau_*(\log)_* : H^1(\mathcal{N}, \mathcal{O}_*) \to H^1(\mathcal{N}, \mathcal{O}(T\mathcal{N} \otimes \mathbf{L}^*))$$

associating an extended tangent bundle to every complex line bundle over $\mathcal{N}$. In fact, there is nothing mysterious about this morphism; if we feed it a line bundle $\mathbf{L}_+$, it spits out the extended tangent bundle

$$\hat{T}\mathcal{N} := [T\mathcal{N} \oplus (\mathbf{L}_- \otimes J^1\mathbf{L}_+)/\underline{\mathbf{C}}]/\mathbf{D},$$

where $\mathbf{L}_-$ denotes $\mathbf{L} \otimes \mathbf{L}_+^*$, $\underline{\mathbf{C}}$ denotes the trivial line sub-bundle induced by the inclusion $\mathbf{L}_- \otimes \mathbf{L}^* \otimes \mathbf{L}_+ \hookrightarrow \mathbf{L}_- \otimes T^*\mathcal{N} \otimes \mathbf{L}_+ \hookrightarrow \mathbf{L}_- \otimes J^1\mathbf{L}_+$, and $\mathbf{D} = \mathbf{D}^* \otimes \mathbf{L}$ is included diagonally into the Whitney sum. For the line bundle $\mathbf{L}_+$, we now choose the 'spinor' line bundle[4], whose fiber over a null geodesic γ with auto-parallel tangent vector field $v^{AA'}$ is defined to be the (conformally invariant) solution space of the ODE

$$\nabla_v \pi_{A'} = 0, \quad v^{AA'}\pi_{A'} = 0.$$

The resulting thickening is then called the *canonical first order Poisson thickening* of the ambitwistor space $\mathcal{N}$. This agrees with the first-order thickening that arises in the previously mentioned construction for the half-conformally-flat case, and is sufficiently natural as to have been originally discovered by completely different methods [13].

Now the obstruction theory for extending Poisson thickenings is rather simpler than that needed for ordinary thickenings, in the sense that the obstruction to extending a given m^{th} order Poisson thickening of $\mathcal{N}$ to a Poisson thickening of order $m + 1$ is an element of $H^2(\mathcal{N}, \mathcal{O}(\mathbf{L}^{*\otimes m}))$, while the freedom of extension is parameterized by $H^1(\mathcal{N}, \mathcal{O}(\mathbf{L}^{*\otimes m}))$ should this obstruction vanish. For $m \geq 1$, this obstruction theory is related to that for the associated ordinary thickenings by the natural maps

$$H^*(\mathcal{N}, \mathcal{O}(\mathbf{L}^{*\otimes m})) \to H^*(\mathbf{X}, \mathcal{O}(\hat{T}\mathbf{X} \otimes \mathbf{L}^{*\otimes(m+1)}))$$

---

[4]It is perhaps even more natural to allow ourselves to consider more general choices for the line bundle $\mathbf{L}_+$ and the associated extended tangent bundle $\hat{T}\mathcal{N}$, merely requiring that $\mathbf{L}_+$ have the 'right' Chern class–i.e. that its restriction to some sky be isomorphic to $\mathcal{O}(0, 1)$. This has the effect of coupling everything to an electromagnetic field in a manner that will be described in section 6.

induced by τ and split by θ. This has the effect that a Poisson thickening can be extended to the next order as an *ordinary* thickening iff it can be extended as a *Poisson* thickening.

Applying the Penrose transform [4] [8] to the above cohomology groups translates these obstruction groups into tensors on $\mathcal{M}$, as indicated by the following table:

m	$H^1(\mathcal{N},\mathcal{O}(\mathbf{L}^{*\otimes m}))$	$H^2(\mathcal{N},\mathcal{O}(\mathbf{L}^{*\otimes m}))$
1	$\mathcal{O}[-2]$	0
2	0	$\mathcal{O}[-2]$
3	0	$\mathrm{Ker}\ \nabla^a : \mathcal{O}_a[-2] \to \mathcal{O}[-4]$
4	0	$\mathrm{Ker}\ \nabla^a : \mathcal{O}^{\ddagger}_{(ab)}[-2] \to \mathcal{O}_b[-4]$
5	0	$\mathrm{Ker}\ \nabla^a : \mathcal{O}^{\ddagger}_{(abc)}[-2] \to \mathcal{O}^{\ddagger}_{(bc)}[-4]$

(Here the entries are spaces of tensor fields on $\mathcal{M}$; we use $\ddagger$ to denote the trace-free tensors of the type indicated by the indices, while the numbers in square brackets are conformal weights.) It turns out [17] that the freedom of extension from order 1 to order 2 exactly compensates for the obstruction to extending from order 2 to order 3; i.e. the canonical first order Poisson thickening extends uniquely to third order. The higher order obstructions then represent conformally invariant tensor fields determined by the metric. In the next section, we will analyze these obstructions, and find that they are closely related to Einstein's equations.

5 Conformal Gravity

The obstructions arising in the last chapter were first analyzed by Baston and Mason [3] on the level of linearized and second order deformations of A. This prompted them to conjecture that, quite generally, the condition for the existence of a sixth order thickening of $\mathcal{N}$ were given by the following conformally invariant field equations:

$$B_{ab} = 0 \tag{2}$$
$$E_{abc} = 0. \tag{3}$$

Here the *Bach tensor* [2] is defined by

$$B_{ab} := (\nabla^c\nabla^d + \tfrac{1}{2}R^{cd})C_{acbd},$$

where C_{abcd} denotes the Weyl curvature, while

$$E_{abc} := \tilde{\Psi}_{A'B'C'D'}\nabla^{DD'}\Psi_{ABCD} - \Psi_{ABCD}\nabla^{DD'}\tilde{\Psi}_{A'B'C'D'}$$

is the so-called *Eastwood-Dighton tensor* [7]. A complex space-time satisfying (2) will be said to be *Bach-flat*. A complex space-time satisfying both (2) and (3) will be called a solution to the *conformal Einstein equations*. The justification for the last bit of terminology is the following result [3, 12]:

Proposition 5.1 *Suppose that $(\mathcal{M}, \mathbf{g})$ is a solution of equations (2) and (3) with algebraically general Weyl curvature. Then there exists a conformal factor α such that $\hat{\mathbf{g}} := \alpha^2 \mathbf{g}$ is Einstein, meaning that the Ricci curvature of $\hat{R}$ of $\hat{\mathbf{g}}$ satisfies*

$$\hat{R}_{ab} = \tfrac{1}{4}\hat{\mathbf{g}}.$$

Now notice that the Bach and Eastwood-Dighton tensors are both symmetric trace-free tensors, and are both conformally invariant, with conformal weight -2, meaning that under

$$\mathbf{g} \mapsto \hat{\mathbf{g}} = \alpha^2 \mathbf{g}$$

we have

$$B \mapsto \hat{B} = \alpha^{-2} B \quad \text{and} \quad E \mapsto \hat{E} = \alpha^{-2} E.$$

Moreover, they are both invariant under biholomorphisms, meaning that if $\phi : \mathcal{M} \to \hat{\mathcal{M}}$ is a biholomorphism, then these tensors depend upon the metric in such a manner that

$$B(\phi^*(\mathbf{g})) = \phi^*(B(\mathbf{g}))$$
$$E(\phi^*(\mathbf{g})) = \pm\phi^*(E(\mathbf{g})),$$

where the above $\pm$ sign depends upon the choice of sign for the associated star operators $* : \bigwedge^2 \to \bigwedge^2$. Moreover, the value at $x \in \mathcal{M}$ of these tensors is a holomorphic function of the m-jet of $\mathbf{g}$ at x for some $m \in \mathbb{N}$.

This motivates the following definition:

Definition 5.2 *An admissible invariant is a symmetric trace-free tensor field $D(\mathbf{g})$ of conformal weight -2, which is invariant under biholomorphisms, and such that, for some $m \in \mathbb{N}$, the value of $D(\mathbf{g})$ at x depends holomorphically on the m-jet of $\mathbf{g}$ at x.*

The interest of this definition lies in the fact that such creatures are, in fact, quite rare:

Proposition 5.3 *Admissible invariants have the following properties:*

1. *Any admissible invariant D_a vanishes.*

2. *Any admissible invariant D_{ab} is a constant multiple of the Bach tensor.*

3. *Any admissible invariant D_{abc} is a constant multiple of the Eastwood-Dighton tensor.*

Proof Let us only consider jets of metrics of the form

$$\begin{aligned}
\mathbf{g}_{ab} &= \delta_{ab} + r_{(ab)(cd)}x^c x^d + s_{(ab)(cde)}x^c x^d + t_{(ab)(cdef)}x^c x^d x^f \\
&\quad + \cdots + u_{(ab)(cde\cdots f)}x^c x^d \cdots x^f,
\end{aligned}$$

where δ_{ab} denotes the flat metric on $\mathbf{C}^4$ and x^a are the standard complex coordinates on $\mathbf{C}^4$, since any m-jet can be put in this form by making a coordinate transformation. Thus each component of D is given by an entire holomorphic function $h(r, s, t, \cdots, u)$. But if we make the coordinate transformation $x \mapsto \alpha x$ coupled with the conformal transformation $\mathbf{g} \mapsto \alpha^{-2}\mathbf{g}$, the effect on the above normal form will be

$$
\begin{aligned}
r &\mapsto \alpha^2 r \\
s &\mapsto \alpha^3 s \\
t &\mapsto \alpha^4 t \\
&\ \ \vdots \\
u &\mapsto \alpha^m u
\end{aligned}
$$

while the effect on D will be given by

$$
D_{\underbrace{ab\cdots c}_{l}} \mapsto \alpha^{2+l} D_{\underbrace{ab\cdots c}_{l}}.
$$

Thus our functions h satisfy the mixed homogeneity condition

$$
h(\alpha^2 r, \alpha^3 s, \alpha^4 t, \ldots, \alpha^m u) = \alpha^{2+l} h(r, s, t, \ldots, u),
$$

and, being holomorphic at 0, are actually polynomials.

For $l = 1$, we conclude that D_a is linear in s and independent of the other variables. If we now further restrict our class of coordinate systems by assuming that we are working in *geodesic spray coordinates*, and assume, using the the conformal invariance [10], that $\nabla_{(a} \cdots \nabla_b R_{cd)}|_{x=0} = 0$, this says that D_a is linear in the first covariant derivative of the Weyl tensor. Using the natural exponential extension of the action of $SO(4, \mathbf{C}) = [SL(2, \mathbf{C}) \times SL(2, \mathbf{C})]/\mathbb{Z}_2$ on the tangent space at the origin, we conclude that $D_a(\mathbf{g})$ would represent an $SL(2, \mathbf{C}) \times SL(2, \mathbf{C})$-equivariant map

$$
\mathbf{C}_{(ABCD)EE'} \oplus \mathbf{C}_{(A'B'C'D')EE'} \longrightarrow \mathbf{C}_{AA'}
$$

and so must vanish, since there are no common irreducible components shared by these two representations.

For $l = 2$, we proceed similarly. This time h could be the sum of a linear function of t and a quadratic function of r. We then use conformal normal coordinates to identify r and t with the Weyl curvature and its second covariant derivative. Remembering to use the Bianchi identities for the second derivatives of the Weyl curvature, the homomorphisms of $SL(2, \mathbf{C}) \times SL(2, \mathbf{C})$-modules corresponding to the linear and quadratic pieces of $D_{(ab)}$ would be

$$
[\mathbf{C}_{(ABCD)} \oplus \mathbf{C}_{(A'B'C'D')}]^{\otimes 2} \longrightarrow \mathbf{C}_{(AB)(A'B')}
$$

and

$$
\mathbf{C}_{(AB)(A'B')} \oplus \mathbf{C}_{(ABCDE)F(E'F')} \oplus \mathbf{C}_{(A'B'C'D'E')F'(EF)} \longrightarrow \mathbf{C}_{(AB)(A'B')};
$$

but the first vanishes by Schur's lemma, while the second is determined up to scale. Thus the Bach tensor B_{ab} is, up to a constant factor, the only such invariant.

For $l = 3$, h must be bilinear in r and s. Comparison of representations tells us this time that our invariant must be of the form

$$c_1 \Psi_{ABCD} \nabla^{DD'} \tilde{\Psi}_{A'B'C'D'} + c_2 \tilde{\Psi}_{A'B'C'D'} \nabla^{DD'} \Psi_{ABCD}.$$

But one then notices that under the conformal change $\mathbf{g} \mapsto \alpha^2 \mathbf{g}$ this becomes

$$\alpha^{-2}(c_1 \Psi_{ABCD} \nabla^{DD'} \tilde{\Psi}_{A'B'C'D'} + c_2 \tilde{\Psi}_{A'B'C'D'} \nabla^{DD'} \Psi_{ABCD})$$

$$+\alpha^{-2}2(c_1 + c_2)\Upsilon^{DD'}\Psi_{ABCD}\tilde{\Psi}_{A'B'C'D'},$$

where $\Upsilon = \alpha^{-1}d\alpha$, and so is conformally invariant with weight -2 iff $c_1 = -c_2$. $\qquad\square$

In fact, the proof actually gives us a bit more than is stated. In particular, even if an admissible invariant $D_{(abc)}$ were defined only as a function of *Bach-flat* metrics, it could still only be a constant multiple of $E_{(abc)}$.

This then leads to the following result:

Theorem 5.4 *The obstructions to extending the canonical fist-order thickening of $\mathcal{N}$ to higher order are:*

1. *There is no obstruction to extending the canonical first-order thickening of $\mathcal{N}$ to fourth order.*

2. *The obstruction to extending the canonical first-order thickening of $\mathcal{N}$ to fifth order is the Bach tensor.*

3. *The obstruction to extending the canonical first-order thickening of $\mathcal{N}$ to sixth order is the Dighton-Eastwood tensor.*

Proof(sketch) There are two essential ingredients: one must verify that these obstructions are admissible invariants, and then show that they are non-zero at orders 5 and 6. The former is essentially straight-forward once the obstructions can be shown to only depend on some jet of the metric; however, nailing this down is surprisingly involved [17]. (Of course, the obstruction to *sixth* order extension is only *a priori* defined as an invariant of those metrics with fifth order extension, but, as previously noted, we could weaken our definition of admissibility for $D_{(abc)}$, allowing it to just be defined on the set of *Bach-flat* metrics, without altering the conclusion of theorem 5.4.) As for the second critical fact—that the obstructions are non-trivial at orders 5 and 6—one may appeal to the perturbative calculations of Baston-Mason [3]. $\qquad\square$

6 Concluding Remarks

Real Structures

If $(\mathcal{M}, \mathbf{g})$ is obtained from a real-analytic pseudo-Riemannian manifold (M, g) by analytic continuation, then the ambitwistor space $\mathcal{N}$ may be correspondingly endowed with an anti-holomorphic involution

$$\sigma : \mathcal{N} \xrightarrow{\overline{\mathcal{O}}} \mathcal{N}, \sigma^2 = id_{\mathcal{N}},$$

since $\mathcal{M}$ may be taken to have a 'complex conjugation' map

$$\rho : \mathcal{M} \xrightarrow{\overline{\mathcal{O}}} \mathcal{M}, \rho^2 = id_{\mathcal{M}},$$

such that

$$\rho^* \mathbf{g} = \overline{\mathbf{g}};$$

namely, for null projective covectors $[\phi] \in \mathcal{Q}$ we may define σ by

$$\sigma(q([\phi])) := q([\rho^* \phi])),$$

which is then independent of representatives and anti-holomorphic. Conversely, given an anti-holomorphic involution $\sigma : \mathcal{N} \to \mathcal{N}$, we may define the *real slice* of $(\mathcal{M}, \mathbf{g})$ to be the set

$$M := \{x \in \mathcal{M} \mid \sigma[Q_x] = Q_x\}$$

of points with real skies, which then has a natural pseudo-Riemannian conformal structure obtained by restricting the complex conformal metric $[\mathbf{g}]$.

Convexity and Reflexivity

Geodesic convexity is perhaps the simplest hypothesis concerning $(\mathcal{M}, \mathbf{g})$ that guarantees that the space of null geodesics is a Hausdorff manifold; in a certain sense this is quite satisfying, in so far as it imposes no local constraints on the differential geometry, but, on the other hand, it may seem a bit confining from the point of view of global pseudo-Riemannian geometry. We therefore define $(\mathcal{M}, \mathbf{g})$ to be *civilized* if its space of null geodesics is a Hausdorff manifold. At present, no satisfactory general global theory of such manifolds exists, but at least we can affirm that every real-analytic *Riemannian* 4-manifold has a civilized complexification. It would be interesting to show this also works for globally hyperbolic Lorentz-signature space-times.

Geodesically convex manifolds have another interesting property, though, which is more subtle than civilization. Namely, given any civilized complex space-time $(\mathcal{M}, \mathbf{g})$, we may 'extend' $\mathcal{M}$ by setting

$$\hat{\mathcal{M}} := \{S \subset \mathcal{N} \text{ a complex submanifold} \times \mathbf{P}_1, c_1(\mathbf{L})[S] = (1, 1)\};$$

this is a complex conformal 4-manifold, and we have an open conformal embedding

$$\mathcal{M} \hookrightarrow \hat{\mathcal{M}}$$
$$x \mapsto Q_x.$$

If $\mathcal{M}$ is Stein (e.g. geodesically convex), it can also be shown that the image of this map is *closed*, so that it is, in fact, a biholomorphism if $\hat{\mathcal{M}}$ is connected. While this connectivity statement is certainly true for, say, convex subsets of complexified Minkowski space, it poses, in general, an open problem, which I shall call the *reflexivity conjecture*: for $(\mathcal{M}, \mathbf{g})$geodesically convex, $\mathcal{M} = \hat{\mathcal{M}}$.

Smoothness

Some may feel uncomfortable with the fact that the approach described herein is tailored to analytic manifolds. However, one can actually do some quite interesting things with smooth Riemannian manifolds. Indeed, while $\mathcal{N}$ need not exist for such a space, an analogue of the subset $G \subset \mathcal{N}$ of complex null geodesics meeting the real slice does exist as an abstract CR 8-manifold of codimension-2 type [15, 21]; indeed, as a smooth manifold, G arises as the total space of the Grassmannian of oriented real 2-planes $\tilde{G}_2(TM) \to M$. It turns out that G is an embeddable CR manifold iff the conformal metric $[g]$ of M is real-analytic in some coordinates. As any Riemannian Einstein metric has this property [6], it would be fascinating to see how the theory of thickenings interacts with the CR embedding problem. Pseudo-Riemannian versions of G should also be explored.

Electromagnetism

The obstruction theory of §4 works equally well if we replace the 'spinor' line bundle $\mathbf{L}_+ \to \mathcal{N}$ by some arbitrary line bundle of the same Chern class. On $\mathcal{M}$, this corresponds to introducing an electromagnetic field, and the previously 'unused' obstruction at order 4 then turns out to be the current of this field. For the fifth order obstruction, the electromagnetic stress-energy tensor could, in principle, couple to the Bach tensor, as in Merkulov's conformally invariant unified field theory [18]. Whether this happens or not is still a matter of controversy.

Supersymmetry

This work seems to have deep connections with the theory of super light rays and conformal supergravity [5, 23]. It would be extremely interesting if more direct proofs could be given using supersymmetric ideas. This linkage also suggests that there may be non-trivial generalizations of these constructions to dimension 10.

Acknowledgements

This work was supported in part by NSF grant DMS 87-04401.

References

[1] V.I. Arnold, *Mathematical Methods of Classical Mechanics*, Graduate Texts in Mathematics, Springer-Verlag, 1978.

[2] R. Bach, *Zur Weylschen Relativitätstheorie und der Weylschen Erweiterung des Krummenstensorsbegriffs*, Math. Zeitschrift **9** (1921), 110–135.

[3] R.J. Baston & L.J. Mason, *Conformal Gravity, the Einstein Equations and Spaces of Complex Null Geodesics*, Class. Quantum Grav. **4** (1987), 815–826.

[4] N.P. Buchdahl, *On the Relative deRham Sequence*, Proc. Am. Math. Soc. **87** (1983), 363–366.

[5] L.L. Chau & C.S. Lim, *Geometrical Constraints and Equations of Motion in Extended Supergravity*, Phys. Rev. Lett. **56** (1986), 294–297.

[6] D. DeTurck & J. Kazdan, *Some Regularity Theorems in Riemannian Geometry*, Ann. Scient. Ec. Norm. Sup. (4° sér.) **14** (1981), 249–260.

[7] K. Dighton, *An Introduction to the Local Theory of Twistors*, Int. J. Theor. Phys. **11** (1974), 31–43.

[8] M.G. Eastwood, *The Penrose Transform for Curved Ambitwistor Space*, Qu. J. Math. Oxford (2) **39** (1988), 427–441.

[9] M.G. Eastwood & C.R. LeBrun, *Thickenings and Supersymmetric Extensions of Complex Manifolds*, Am. J. Math. **108** (1986), 1177–1192.

[10] C.R. Graham, *Formal Conformal Normal Form*, to appear.

[11] K. Kodaira, *A Theorem of Completeness of Characteristic Systems for Analytic Families of Compact Submanifolds of a Complex Manifold*, Ann. Math. (2) **75** (1962), 146–162.

[12] C. Kozameh, E.T. Newman, & K.P. Tod, *Conformal Einstein Spaces*, Gen. Rel. Grav. **17** (1985), 343–352.

[13] C.R. LeBrun, *The First Formal Neighbourhood of Ambitwistor Space for Curved Space-Time*, Lett. Math. Phys. **6** (1982), 345–354.

[14] C.R. LeBrun, *Spaces of Complex Null Geodesics in Complex-Riemannian Geometry*, Trans. Am. Math. Soc. **278** (1983), 209–231.

[15] C.R. LeBrun, *Twistor CR Manifolds and Three Dimensional Conformal Structures*, Trans. Am. Math. Soc. **284** (1984), 601–616.

[16] C.R. LeBrun, *Thickenings and Gauge Fields*, Class. Quantum Gravity **3** (1986), 1039–1059.

[17] C.R. LeBrun, *Thickenings and Conformal Gravity*, preprint, 1989.

[18] S.A. Merkulov, *A Conformally Invariant Theory of Gravitation and Electromagnetism*, Class. Quantum Gravity **1** (1984), 349–355.

[19] R. Penrose, *Non-linear Gravitons and Curved Twistor Theory*, Gen. Rel. Grav. **7** (1976), 31–52.

[20] R. Penrose & W. Rindler, *Spinors and Space-time, vol. 2*, C.U.P., 1986.

[21] H. Rossi, *LeBrun's Nonrealizability Theorem in Higher Dimensions*, Duke Math. J. **52** (1985), 457–474.

[22] A. Weinstein, *The Local Structure of Poisson Manifolds*, J. Diff. Geom. **18** (1983), 523–557.

[23] E. Witten, *Twistor-like Transform in Ten Dimensions*, Nuc. Phys. **B266** (1986), 245–264.

The Penrose Transform

M. G. Eastwood

1 Introduction

This article is a survey of developments in the Penrose transform since [8]. Recall that in [8] the transform was precisely the homomorphism

$$\mathcal{P} : H^1(V, \mathcal{O}(-n-2)) \to \Gamma(U, \mathcal{Z}_n)$$

for

- U = an open subset of $\mathbf{M}$ (= compactified complexified Minkowski space)

- V = the corresponding open subset of $\mathbf{P}$ (= projective twistor space)

- $\mathcal{O}(-n-2)$ = the sheaf of germs of holomorphic functions homogeneous of degree $-n-2$

- $\mathcal{Z}_n$ = the sheaf of germs of holomorphic solutions of the zero-rest-mass free field equations of helicity $\frac{n}{2}$.

The transform was shown to be an isomorphism under mild topological conditions on U which hold, in particular, for the important special case of $U = \mathbf{M}^+$ and $V = \mathbf{P}^+$ (see [14] for standard twistor notation). Thus, one obtains a twistor description of positive frequency massless fields. The method of proof in [8] was to set up a general spectral sequence which specialized to give the different results for different values of n. In particular, one can see how the transform automatically falls into three cases:

$n \geq 1$	The massless fields turn out to be right-handed and are given directly by the transform
$n = 0$	The Penrose transform gives rise to solutions of the wave equation. Notice that this is a second order equation as opposed to the first order equations which arise for $n \neq 0$.
$n \leq -1$	The massless fields turn out to be left-handed and are given indirectly by the transform as potentials modulo gauge-freedom (the case $n = -1$ is degenerate).

Thus, one should view [8] as providing a cohomological 'machine' in the form of a spectral sequence which turns cohomology into differential equations. Applications of this machine are given in various articles in this collection. In particular, the four dimensional case (as above) is discussed in the contribution by Bailey and Singer.

Since [8] the Penrose transform has undergone a considerable amount of generalization and refinement. It now lies at the heart of twistor theory – many twistor constructions can be viewed as special cases. There are two main areas of development:

1. The generalization to other *homogeneous* correspondences (for example, between general Grassmannians, the twistor case being the case of $\mathsf{M} = \mathrm{Gr}_2(\mathbb{C}^4)$ versus $\mathsf{P} = \mathrm{Gr}_1(\mathbb{C}^4)$). This will generally be called the *flat* theory.

2. The construction of the Penrose transform for the 'curved' versions of this flat theory.

As one might expect, the main ingredients for part (1) come from representation theory—the high degree of symmetry is exploited. There is a close analogy with the theory of the Radon transform [13], also studied through representation theory in providing 'special functions' and consequent expansions. Part (2) is an exercise in differential geometry guided by the results of part (1). Of these two developments, part A is rather more complete and is the subject of [4]. There is feedback from this into representation theory. In particular, there is the construction of certain unitary representations (discussed in 'The Twistor Transform', this volume) and of invariant differential operators (discussed in 'Invariant Operators', this volume). These invariant operators usually have curved theory analogues—they are the operators which arise in the curved Penrose transform. Part (2) is well advanced and very much a matter of current concern [2, 11, 1]. However, it is far from complete and there is much scope for further investigation.

Rather than attempt a complete survey I shall confine myself to a few illustrative examples. A more complete account would require much more representation theory and would end up following [4].

2   The Non-linear Graviton

Though not usually regarded as such, this construction is an example of the curved Penrose transform. More usually, the construction is given entirely in terms of complex differential geometry [21, 27, 17] but this entails some slightly awkward argument (in proving that a certain natural connection is torsion-free). The approach below is via LeBrun's 'Einstein bundle' (see also his contribution to this volume).

Suppose M is a four dimensional complex conformal manifold. The non-linear graviton construction due to Penrose [21] is a local construction so we shall also assume that M is geodesically convex (for some metric in the conformal class) as in [18]. In particular, we can choose a spin structure on M. Let F denote the projective primed spin bundle over M with local coördinates $(x^a, \pi_{A'})$ as usual ($\pi_{A'}$ is homogeneous). Choose a metric in the conformal class and lift the corresponding metric connection horizontally to a differential operator on spinor fields on F. If one rescales the metric then the operator changes. For example, if g_{ab} is replaced by $\hat{g}_{ab} = \Omega^2 g_{ab}$, then

$$\widehat{\nabla}_{AA'}\phi_B = \nabla_{AA'}\phi_B - \Upsilon_{BA'}\phi_A + w\Upsilon_{AA'}\phi_B + \pi_{A'}\Upsilon_{AB'}\frac{\partial \phi_B}{\partial \pi_{B'}}$$

for ϕ_B a spinor field on M of conformal weight w. As usual $\Upsilon_a = \Omega^{-1}\nabla_a\Omega$. On functions of weight w we obtain

$$\widehat{\nabla}_{AA'}\phi = \nabla_{AA'}\phi + w\Upsilon_{AA'}\phi + \pi_{A'}\Upsilon_{AB'}\frac{\partial \phi}{\partial \pi_{B'}}.$$

In particular, if $w = 0$ and we set $\nabla_A = \pi^{A'}\nabla_{AA'}$ then

$$\widehat{\nabla}_A\phi = \nabla_A\phi.$$

In other words, ∇_A is a conformally invariant operator on ordinary functions and may thus be viewed as an invariant distribution on F. It is well known that this distribution is integrable if and only if the self-dual part $\widetilde{\Psi}_{A'B'C'D'}$ of the Weyl curvature vanishes. Specifically, one can check Frobenius' criterion as follows. Given spinors λ^A, μ^A, and $\pi^{A'}$ at a point $x^a \in M$, horizontally lift the vectors $\lambda^A\pi^{A'}$ and $\mu^A\pi^{A'}$ to the point $(x^a, \pi_{A'}) \in F$, extend arbitrarily to vector fields in the distribution and ask that their Lie bracket be again in the distribution. The vertical component of the commutator is precisely the curvature:

$$\lambda^A\pi^{A'}\mu^B\pi^{B'}\Delta_{ab}\pi_{C'} = \lambda_B\mu^B\pi^{A'}\pi^{B'}\Box_{A'B'}\pi_{C'} = \lambda_B\mu^B\pi^{A'}\pi^{B'}\pi^{D'}\widetilde{\Psi}_{A'B'C'D'}$$

where notation and formulae are taken from [22]. This must vanish whence $\widetilde{\Psi}_{A'B'C'D'} = 0$. Conversely, if $\widetilde{\Psi}_{A'B'C'D'} = 0$ then choose to extend $\pi^{A'}$ to a spinor field near x^a so that $\nabla_b\pi^{A'}$ vanishes at x^a. Extend λ^A and μ^A arbitrarily. The computation of the commutator may then be carried out on M. We find that at x^a

$$\lambda^A\pi^{A'}\nabla_{AA'}\left(\mu^B\pi^{B'}\right) - \mu^A\pi^{A'}\nabla_{AA'}\left(\lambda^B\pi^{B'}\right)$$
$$= \left(\lambda^A\pi^{A'}\nabla_{AA'}\mu^B - \mu^A\pi^{A'}\nabla_{AA'}\lambda^B\right)\pi^{B'}$$

which is again in the distribution and Frobenius is satisfied.

If $\tilde{\Psi}_{A'B'C'D'} = 0$ then M is said to be right conformally flat or conformally anti-self-dual. In this case, therefore, we may integrate the distribution on F to give a new space P as the space of leaves. We obtain the familiar double fibration:

$$(1)$$

The space P is known as the twistor space of M and the submanifolds $\nu(\mu^{-1}(Z))$ of M for $Z \in P$ are known as α-surfaces in M. Each point $x \in M$ gives rise to a 'line' $L_x = \mu(\nu^{-1}(x))$ in P whose points correspond to the α-surfaces through x as further explained in [21]. This generalizes the flat correspondence as discussed in [8, 14, 23] where $M = \mathsf{M}$ and $P = \mathsf{P}$. Intrinsically L_x is a Riemann sphere. The space-time M with its conformal structure may be recovered from P (as shown in [21, 18]) – nearby points $x, y \in M$ are null separated if and only if L_x and L_y intersect. The infinitesimal statement comes down to the form of the normal bundle to L_x as $\mathcal{O}(1) \oplus \mathcal{O}(1)$, a form which is stable under small deformations (again see Penrose's original discussion [21]).

This is the first stage of the non-linear graviton construction yielding a twistor space for a conformally anti-self-dual spacetime. It is a remarkable though extremely natural geometric construction. What is less clear is how to encode a genuine metric on M into some sort of structure on P. Ideally one would like an Einstein metric on M (as in [21]) or, more generally, Einstein with cosmological constant (as given by Richard Ward [27]).

By using LeBrun's Einstein bundle, this extra structure arises as an example of the Penrose transform. To construct the Einstein bundle one can proceed as follows. Suppose ϕ is a function of conformal weight 1 (sometimes called a conformal density of weight 1). Then, for any choice of metric in the conformal class, $\nabla_A \phi$ has weight 0 and transforms under rescaling as

$$\widehat{\nabla}_A \phi = \nabla_A \phi + \Upsilon_A \phi.$$

Applying ∇_A again gives $\nabla_A \nabla_B \phi$ which transforms as

$$\begin{aligned}
\widehat{\nabla}_A \widehat{\nabla}_B \phi &= \nabla_A(\nabla_B \phi + \Upsilon_B \phi) - \Upsilon_B(\nabla_A \phi + \Upsilon_A \phi) \\
&= \nabla_A \nabla_B \phi + (\nabla_A \Upsilon_B - \Upsilon_B \Upsilon_A)\,\phi.
\end{aligned}$$

Although $\nabla_A \nabla_B \phi$ is not conformally invariant, it may be modified by means of a curvature term to produce an invariant differential operator (see 'Invariant Operators', this volume). Specifically, with the usual conventions [22], let $-2\Phi_{ab}$ denote the trace free Ricci tensor associated to the metric g_{ab} in the conformal class. Define Φ_{AB} on F by $\Phi_{AB} = \pi^{A'}\pi^{B'}\Phi_{ABA'B'}$. Then, under rescaling g_{ab} by Ω^2, we obtain the transformation

$$\Phi_{AB} \mapsto \hat{\Phi}_{AB} = \Phi_{AB} - \nabla_A \Upsilon_B + \Upsilon_A \Upsilon_B.$$

Combining this with the calculation above yields

$$D_{AB} = \nabla_A \nabla_B + \Phi_{AB},$$

a conformally invariant operator acting on functions of weight 1. By construction, ∇_A acts along the fibres of μ. Hence, so does D_{AB} and one can ask for ϕ of conformal weight 1 which are annihilated by D_{AB} obtaining thereby a vector bundle E on P. Explicitly, the fibre of E at Z is the finite-dimensional vector space

$$\{\phi \text{ of conformal weight 1 defined on } \mu^{-1}Z \text{ such that } D_{AB}\phi = 0\}.$$

That this is finite-dimensional follows immediately from the differential equation since the second derivatives are determined by the function. In fact, the dimension of this solution space is three exactly as in the flat case—the best way to see this is via an observation due to Bailey that α-surfaces inherit from the conformal structure on M a flat projective structure and moreover that D_{AB} is a projectively invariant differential operator. An analogous bundle E was introduced by LeBrun [19] (see also his contribution to this volume) in the context of ambitwistors and we retain his terminology—E is called the *Einstein bundle* on P. The reason for this terminology will soon become clear.

Theorem 2.1 *There is an isomorphism (the Penrose transform)*

$$\Gamma(P, E) \cong \left\{ \begin{array}{c} \phi \text{ of conformal weight 1 on } M \text{ satisfying} \\ \nabla_{(A}^{(A'} \nabla_{B)}^{B')} \phi + \Phi_{AB}^{A'B'} \phi = 0 \end{array} \right\}.$$

Proof By the definition of E it follows that

$$\Gamma(P, E) \cong \left\{ \begin{array}{c} \phi \text{ of conformal weight 1 on } F \text{ satisfying} \\ \nabla_A \nabla_B \phi + \Phi_{AB}\phi = 0 \end{array} \right\}.$$

Now, the crucial observation is that the fibres of $\nu : F \to M$ are compact (being Riemann spheres) and so $\phi(x^a, \pi_{A'})$ is really a function of x^a alone. The equation on F now reads

$$\pi^{A'} \pi^{B'} \nabla_{AA'} \nabla_{BB'} \phi + \pi^{A'} \pi^{B'} \Phi_{ABA'B'} \phi = 0$$

and the π's can be removed to yield the theorem. Notice that the conformal invariance of the equation on M is assured by this procedure. $\qquad\square$

Corollary 2.2 *There is an isomorphism between nowhere vanishing sections of E over P and Einstein metrics in the conformal class on M. Here, Einstein means with cosmological constant i.e. vanishing trace-free Ricci curvature.*

Proof The 'nowhere vanishing' stipulation implies (by elementary theory of differential equations) that we may choose ϕ to be nowhere vanishing. If we use ϕ^{-2} to rescale the metric then $\hat{\phi} \equiv 1$ and our differential equation now reads $\hat{\Phi}_{ab} = 0$. In other words, the trace-free Ricci curvature of the rescaled metric vanishes as advertized. This argument is evidently reversible. $\square$

LeBrun's argument in the ambitwistor case [19] is exactly parallel. An extra feature in the twistor case, however, is that there is a simple intrinsic description of E on P. This is what gives the non-linear graviton in its usual form.

Theorem 2.3 *There is a canonical isomorphism*

$$E = \Omega^1(2)$$

where Ω^1 is the bundle of 1-forms and $\mathcal{O}(2)$ is a line bundle on P to be explained in the proof.

Proof The tangent bundle of P may be described on M in terms of a connecting vector field subject to a Jacobi equation. Writing this up on F shows that the tangent bundle corresponds to solutions of the differential equation (cf. the twistor equation; a detailed argument can be found in [20])

$$\nabla_{(A}\omega_{B)} = 0$$

where $\omega_B \in \mathcal{O}_B(1)[1]$ i.e. ω_B is homogeneous of degree 1 in $\pi_{A'}$ and of conformal weight 1. This method of describing natural bundles on P gives many more. In particular, $\mathcal{O}(k)$ on P is defined by the differential equation

$$\nabla_A\psi = 0$$

on F where ψ is homogeneous of degree k in $\pi_{A'}$. The statement of the theorem is now expressible in terms of differential equations on F. We are required to exhibit a perfect pairing between solutions of the equations

$$\nabla_{(A}\omega_{B)} = 0 \qquad \text{where} \quad \omega_B \in \mathcal{O}_B(-1)[1]$$
$$\nabla_A\nabla_B\phi + \Phi_{AB}\phi = 0 \quad \text{where} \quad \phi \in \mathcal{O}[1]$$

along each fibre of $\mu : F \to P$. To do this consider the function

$$f = 2\omega^A\nabla_A\phi - \phi\nabla_A\omega^A \in \mathcal{O}.$$

Notice that f is conformally invariant:

$$
\begin{aligned}
\widehat{f} &= 2\omega^A \widehat{\nabla}_A \phi - \phi \widehat{\nabla}_A \omega^A \\
&= 2\omega^A \nabla_A \phi + 2\omega^A \Upsilon_A \phi - \phi \nabla_A \omega^A - 2\phi \Upsilon_A \omega^A \\
&= 2\omega^A \nabla_A \phi - \phi \nabla_A \omega^A = f.
\end{aligned}
$$

Also, f is constant along the fibres of μ:

$$
\begin{aligned}
\nabla_B f &= 2\nabla_B(\omega^A \nabla_A \phi) - (\nabla_B \phi)\nabla_A \omega^A - \phi \nabla_B \nabla_A \omega^A \\
&= (\nabla_C \omega^C)\nabla_B \phi + 2\omega^A \nabla_B \nabla_A \phi - (\nabla_B \phi)\nabla_A \omega^A - \phi \nabla_B \nabla_A \omega^A \\
&= 2\omega^A \nabla_B \nabla_A \phi - \phi \nabla_B \nabla_A \omega^A
\end{aligned}
$$

where we have used that $\nabla_{(A}\omega_{B)} = 0$ to obtain the second equality. Further use of this equation and the commutator formula of [22] gives

$$
\begin{aligned}
\nabla_B \nabla_A \omega^A &= \nabla_B \nabla_A \omega^A - \nabla_A \nabla_B \omega^A + \nabla_A \nabla_B \omega^A \\
&= \varepsilon_{BA} \Phi^A_C \omega^C + \tfrac{1}{2} \nabla_A \delta^A_B \nabla_C \omega^C \\
&= -\Phi_{AB}\omega^B + \tfrac{1}{2}\nabla_B \nabla_C \omega^C
\end{aligned}
$$

whence $\nabla_B \nabla_A \omega^A = -2\Phi_{AB}\omega^A$ and now

$$
\nabla_B f = 2\omega^A \left(\nabla_A \nabla_B \phi + \Phi_{AB}\phi \right) = 0
$$

as required. This pairing is non-degenerate and the result follows. $\qquad\square$

Remark The identification of the tangent bundle given in the proof above easily shows that the canonical bundle $\kappa = \Omega^3$ on P is naturally isomorphic to $\mathcal{O}(-4)$ (just as in the flat case where P is complex projective space). Thus, the Einstein bundle can be given a completely intrinsic description on P as $\Omega^1 \otimes \kappa^{-\frac{1}{2}}$.

Combining the results above shows that an Einstein metric in the given conformal class on M corresponds precisely to a nowhere vanishing 1-form τ on P homogeneous of degree 2. Note, therefore, that the existence of such a 1-form is a strong condition—there are obstructions [16, 3] to finding an Einstein metric in a given conformal class. The remainder of the non-linear graviton construction is standard. Specifically, one now considers $\tau \wedge d\tau$. This is a well-defined section of $\Omega^3(4)$ but, as in the remark above, this is naturally trivialized say by ρ so that one can write $\tau \wedge d\tau = 2\Lambda\rho$ for some function Λ. Another application of the Penrose transform or use of the many lines L_x in P shows that holomorphic functions in P are necessarily constant. With the usual normalizations [22], Λ is the cosmological constant. If $\Lambda = 0$ as in Penrose's original formulation [21], then τ defines an integrable distribution $\sigma : P \to \mathbb{CP}_1$ and a relative 2-form $\mu \in \Omega^2_\sigma(2)$ is determined by $\tau \wedge \mu = \rho$. This is the original form of the construction: the foliation $\sigma : P \to \mathbb{CP}_1$ and the homogeneous symplectic structure μ clearly determine τ.

This method extends [1] to give a proof that Salamon's twistor construction for quaternionic Kähler and hyperKähler manifolds [24] is invertible. The replacement for the right-flat conformal manifolds in dimension four is that of quaternionic or torsion-free QCF as discussed in [1].

Even in four dimensions the Penrose transform gives much more. A detailed investigation of the transform for the ambitwistor correspondence is given in [11, 12] (see also LeBrun's article in this volume) whilst the twistor case (when M is right conformally flat) is a straightforward analogue thereof as explained in [2, 1] (see also the article in this volume by Bailey and Singer).

3   The Penrose Transform for Vector Bundles

This example is an illustration of part A. The correspondence is the usual flat twistor correspondence

$$(2)$$

as in [8] but the cohomology is taken with coefficients in a homogeneous vector bundle rather than just a line bundle $\mathcal{O}(-n-2)$. This is a very natural generalization from the point of view of representation theory. The general case of a correspondence between Grassmannians is described in [9] whilst [10] is an investigation of what happens for the usual twistor correspondence. I shall discuss just two examples. The general case requires some new notation (suggested by Buchdahl and used in [9]) in order to identify particular homogeneous vector bundles but these two examples can be presented in traditional spinor/twistor notation [14, 8, 22, 23]. The two notations on $\mathbf{F}$ are related by

$$(a|b|c, d) = \mathcal{O}_{\underbrace{(AB...D)}_{d-c}}(-a)[d - b]$$

where round brackets refer to homogeneity in $\pi_{A'}$ and square brackets refer to conformal weight. In [8] 'primed' and 'unprimed' conformal weights were distinguished. It is usually better to identify them and this is always assumed in what follows. This amounts to fixing a twistor volume form – in other words reducing the symmetry group from $\mathrm{GL}(4,\mathbf{C})$ to the arguably more natural $\mathrm{SL}(4,\mathbf{C})$. The two examples to be explained below are the cohomologies

$$H^*(V, \Theta(-6)) \quad \text{and} \quad H^*(V, \Omega^1(2))$$

where V corresponds to a convex open subset $U \subset \mathbf{M}$, whilst Θ and Ω^1 denote the tangent and cotangent bundles respectively. One aspect of the Penrose transform which has not changed a great deal since [8] is the topological discussion of pull-back along the fibres of μ. There are two important contributions—one is the theorem of Buchdahl [6] which gives precise conditions under which the topological part of the Penrose transform is valid and the other is the investigation due to Singer [26] of what happens when the conditions are violated (for example when dealing with sourced fields—see also the article by Bailey and Singer in this collection). The conclusion of this topological discussion is that if U is sufficiently topologically simple (certainly if U is convex) then there is an isomorphism

$$H^r(V, \mathcal{O}(B)) \cong H^r(\nu^{-1}(U), \mu^{-1}\mathcal{O}(B))$$

for any vector bundle B on V where $\mu^{-1}\mathcal{O}(B)$ is the sheaf of germs of sections of $\mu^* B$ constant along the fibres of μ. The improved method concerns the interpretation of the intermediate cohomology $H^r(\nu^{-1}(U), \mu^{-1}\mathcal{O}(B))$ down on $U \subset \mathbf{M}$. In this discussion the case $r = 0$ is especially simple (as in the non-linear graviton example).

First consider the case $B = \Theta(-6)$. There is a resolution

$$0 \to \mu^{-1}\Theta(-6) \to \mathcal{O}_A(-5)[1] \xrightarrow{\nabla_B} \mathcal{O}_{(AB)}(-4) \xrightarrow{\nabla^A \nabla^B} \mathcal{O}(-2)[-4] \to 0$$

where the operator $\nabla_B = \pi^{B'} \nabla_{BB'}$ gives the Jacobi equation as for the non-linear graviton example. Recall that in [8] resolutions along the fibres of μ were given by the relative deRham sequence. The resolution above, however, is considerably more efficient. It may be derived from the deRham resolution as in [10] but has since been identified as a Bernstein-Gelfand-Gelfand resolution—see [4] for details. As in [8], this resolution gives rise to a spectral sequence on U. To identify the terms in this spectral sequence we need the direct images (computed as in [8] but more generally [9, 4] via the Bott-Borel-Weil theorem)

$$\nu^1_* \mathcal{O}_A(-5)[1] = \mathcal{O}_{A(A'B'C')}$$
$$\nu^1_* \mathcal{O}_{(AB)}(-4) = \mathcal{O}_{(AB)(A'B')}[-1]$$
$$\nu^1_* \mathcal{O}(-2)[-4] = \mathcal{O}[-5]$$

and now the spectral sequence gives

$$\begin{array}{c}
q \\
\Big\uparrow \\
\Gamma(U, \mathcal{O}_{A(A'B'C')}) \xrightarrow{\nabla^{C'}_B} \Gamma(U, \mathcal{O}_{(AB)(A'B')}[-1]) \xrightarrow{\nabla^{AA'}\nabla^{BB'}} \Gamma(U, \mathcal{O}[-5]) \\
\Big\vert \\
0 \rule{2cm}{0.4pt} 0 \rule{2cm}{0.4pt} 0 \longrightarrow \\
\hfill p
\end{array}$$

so that, in particular, the Penrose transform gives

$$H^1(V, \Theta(-6)) \cong \left\{ \begin{array}{c} \text{Unweighted holomorphic spinor fields} \\ \psi_{AA'B'C'} = \psi_{A(A'B'C')} \text{ on } U \text{ satisfying} \\ \nabla^{A'}_{(A}\psi_{B)A'B'C'} = 0 \end{array} \right\}.$$

These differential equations which the transform throws up are automatically conformally invariant and are closely allied to the usual massless field equations.

Now consider the case $B = \Omega^1(2)$. There is a Bernstein-Gelfand-Gelfand resolution

$$0 \to \mu^{-1}\Omega^1(2) \to \mathcal{O}[1] \xrightarrow{\nabla_A \nabla_B} \mathcal{O}_{(AB)}(2)[-1] \xrightarrow{\nabla^B} \mathcal{O}_A(3)[-3] \to 0$$

in which one can recognize the Einstein bundle of the non-linear graviton example. The only non-vanishing direct images are

$$\nu_* \mathcal{O}[1] = \mathcal{O}[1]$$
$$\nu_* \mathcal{O}_{(AB)}(2)[-1] = \mathcal{O}_{(AB)(A'B')}[1]$$
$$\nu_* \mathcal{O}_A(2)[-3] = \mathcal{O}_{A(A'B'C')}$$

and the E_1-level of the spectral sequence is

$$\begin{array}{c}
q \\
\uparrow \\[6pt]
0 \qquad\qquad\qquad 0 \qquad\qquad\qquad 0 \\[10pt]
\Gamma(U,\mathcal{O}[1]) \xrightarrow{\nabla_{AA'}\nabla_{BB'}} \Gamma(U,\mathcal{O}_{(AB)(A'B')}[-1]) \xrightarrow{\nabla_B^{C'}} \Gamma(U,\mathcal{O}_{A(A'B'C')}) \longrightarrow p
\end{array}$$

giving

$$H^0(V,\Omega^1(2)) \cong \{\phi \in \Gamma(U,\mathcal{O}[1]) \text{ s.t. } \nabla_{(A}^{(A'}\nabla_{B)}^{B')}\phi = 0\},$$

as we already know for the Einstein bundle, and

$$H^1(V,\Omega^1(2)) \cong \frac{\{\xi_{ABA'B} \in \Gamma(U,\mathcal{O}_{(AB)(A'B')}[1]) \text{ s.t. } \nabla_{(A'}^A \xi_{B'C')AB} = 0\}}{\{\xi_{AB}^{A'B'} = \nabla_{(A}^{(A'}\nabla_{B)}^{B')}\phi \text{ for some } \phi \in \Gamma(U,\mathcal{O}[1])\}}$$

which is very much akin to the potential/gauge description of left-handed massless fields. Indeed, there is a Bernstein-Gelfand-Gelfand resolution on **M** (extending the generalized deRham sequence of Buchdahl [5]) which reads

$$\mathcal{O}[1] \xrightarrow{\nabla_{AA'}\nabla_{BB'}} \mathcal{O}_{(AB)(A'B')}[1] \begin{array}{c} \xrightarrow{\nabla_{C'}^B} \mathcal{O}_{A(A'B'C')} \xrightarrow{\nabla_C^{B'}} \\ \searrow_{\nabla_C^{B'}} \quad \mathcal{O}_{(ABC)A'} \quad \nearrow_{-\nabla_{C'}^B} \end{array} \mathcal{O}_{(AC)(A'C')}[-1] \xrightarrow{\nabla^{AA'}\nabla^{CC'}} \mathcal{O}[-5] \to 0$$

as a resolution of $\mathbb{C}_{[\alpha\beta]}$ (the skew dual bi-twistors) whence it follows immediately that

$$H^1(V, \Omega^1(2)) \cong \{\phi_{ABCA'} \in \Gamma(U, \mathcal{O}_{(ABC)A'}) \text{ s.t. } \nabla^A_{(A'}\phi_{B')ABC} = 0\}$$

where the field $\phi_{ABCA'}$ is obtain from the potential $\xi_{ABA'B'}$ by

$$\phi_{ABCA'} = \nabla^{B'}_{(A}\xi_{BC)A'B'}.$$

As for the usual case of massless fields there is a curved Penrose transform obtained by replacing the operators above by their curved analogues (e.g. $\nabla_A \nabla_B$ acquires Φ_{AB} as a correction term as in the Einstein bundle) but just as for massless fields the Bernstein-Gelfand-Gelfand sequence on M fails to be a complex, the Weyl curvature giving rise to Buchdahl conditions whence potentials/gauge are no longer equivalent to fields.

An important aspect of the Penrose transform is the twistor transform it induces (see for example the article thereon in this collection). By combining the two examples above a twistor transform emerges:

Theorem 3.1 *There is an $SU(2,2)$-equivariant isomorphism*

$$\mathcal{T} : H^1(\mathbb{P}^+, \Omega^1(2)) \cong H^1(\mathbb{P}^{*-}, \Theta(-6))$$

known as the twistor transform.

Proof Both are isomorphic to

$$\{\phi_{ABCA'} \in \Gamma(U, \mathcal{O}_{(ABC)A'}) \text{ s.t. } \nabla^A_{(A'}\phi_{B')ABC} = 0\}$$

as in our calculation above. $\qquad\qquad\qquad\qquad\qquad\qquad\qquad\square$

The isomorphism is a sort of cohomological integral intertwining operator and when combined with the natural Hermitian pairing between

$$H^1(\overline{\mathbb{P}^+}, \Omega^1(2)) \quad \text{and} \quad H^1(\overline{\mathbb{P}^{*-}}, \Theta(-6)),$$

gives rise to an Hermitian form on $H^1(\overline{\mathbb{P}^+}, \Omega^1(2))$ and hence to a unitary representation of $SU(2,2)$. More details are in 'The Twistor Transform', this volume.

4 Higher Dimensional Conformal Manifolds

In dimensions other than four one cannot have a conformally half-flat manifold since the Weyl curvature corresponds to an *irreducible* representation of $SO(n)$. Thus, the natural generalization of twistor theory applies only to homogeneous spaces. Unfortunately, it is difficult even to describe what happens without the language of representation theory. The same comment applies the more so to proofs where the Bernstein-Gelfand-Gelfand resolution and the Bott-Borel-Weil theorem are the principal ingredients. In fact, the results are extremely straightforward once this machinery is under control but the resulting technology is the subject matter of [4] and it would be pointless to repeat anything so general in this review. Instead, I shall just state what happens in a couple of sample cases in six dimensions. In fact, just as in four dimensions (where one exploits the isomorphism $SO(6) \cong SL(4)$), one can proceed in six dimensions by special techniques (e.g. [15]). The methods of [4], however, are completely general.

In any dimension, the compactified complexified Minkowski space is simply the non-singular quadric in $\mathbf{CP}_{n+1}$, evidently a homogeneous space for $SO(n+2)$. For the rest of this example let $\mathbf{M}$ denote the quadric of dimension six. The 3-planes in $\mathbf{M}$ fall into two distinct families. Choose one of these families and call them α-planes. Write $\mathbf{P}$ for the space of α-planes. Evidently, one has a correspondence

$$\tag{3}$$

Let κ denote the canonical bundle on $\mathbf{P}$ (thus, $\kappa = \Omega^6$ since $\mathbf{P}$ has dimension 6). One can resolve $\mu^{-1}\kappa$ on $\mathbf{F}$ by the relative deRham sequence (which agrees with the Bernstein-Gelfand-Gelfand resolution in this case)

$$0 \to \mu^{-1}\kappa \to \Omega^0_\mu(\kappa) \to \Omega^1_\mu(\kappa) \to \Omega^2_\mu(\kappa) \to \Omega^3_\mu(\kappa) \to 0.$$

The direct images under ν are easily computed (given the technology of [4]) and it turns out that the non-zero such are

$$\nu^3_*\Omega^0_\mu(\kappa) = \Omega^3_- \quad \nu^3_*\Omega^1_\mu(\kappa) = \Omega^4$$
$$\nu^3_*\Omega^2_\mu(\kappa) = \Omega^5 \quad \nu^3_*\Omega^3_\mu(\kappa) = \Omega^6$$

where Ω^3_- denotes the sheaf of anti-self-dual 3-forms. The Penrose transform thus yields an isomorphism

$$\mathcal{P} : H^3(V, \kappa) \cong \{\omega \in \Gamma(U, \Omega^3_-) \text{ s.t. } d\omega = 0\}$$

for $U \subset \mathbf{M}$ and $V = \mu(\nu^{-1}(U))$ subject to the usual topological restrictions. This space may reasonably be called the anti-self-dual Maxwell fields on U.

Now consider $H^r(V, \mathcal{O})$. To apply the Penrose transform one can utilize the resolution

$$0 \to \mu^{-1}\mathcal{O} \to \Omega^0_\mu \to \Omega^1_\mu \to \Omega^2_\mu \to \Omega^3_\mu \to 0$$

and it turns out that the non-zero direct images are

$$\nu_*\Omega^0_\mu = \Omega^0 \quad \nu_*\Omega^1_\mu = \Omega^1$$
$$\nu_*\Omega^2_\mu = \Omega^2 \quad \nu_*\Omega^3_\mu = \Omega^3_+.$$

Thus, as an example we obtain

$$\mathcal{P} : H^2(V, \mathcal{O}) \cong \frac{\{\Phi \in \Gamma(U, \Omega^2) \text{ s.t. } d_+\Phi = 0\}}{\{\Phi = d\theta \text{ for some } \theta \in \Gamma(U, \Omega^1)\}}$$

which is a potential/gauge description of again the anti-self-dual Maxwell fields.

Thus, there are two different ways of describing fields of the same helicity on twistor space. This is in contrast to the classical case (in four dimensions) where cohomology on twistor space gives fields of any helicity but just once. In other words the twistor transform in dimension six reads:

Theorem 4.1 *For open sets $V \subset \mathbf{P}$ as above, there is a canonical isomorphism*

$$\mathcal{T} : H^2(V, \mathcal{O}) \cong H^3(V, \kappa).$$

This phenomenon alternates in even dimensions—starting with the flat conformal manifold of dimension $2n$, the twistor transform acts between cohomology on twistor space and its dual if n is even whereas if n is odd, the transform gives cohomology back on twistor space. This pattern evidently follows the action of the longest element of the corresponding Weyl group. In odd dimensions the twistor space and its dual coincide so the twistor transform has no choice.

5 A Computation of K-types

In this final example the Penrose transform is used to re-interpret some cohomology considered by Schmid [25] in his investigation of the discrete series. This example was worked out with Ed Dunne during his visit to Adelaide in 1987 (see also [7]).

Let Φ be an Hermitian form on $\mathbf{C}^3$ of type $(+ - -)$. Let $\mathbf{F}$ denote the corresponding flag manifold:

$$\mathbf{F} = \{L_1 \subset L_2 \text{ s.t. } L_j \text{ is a linear subspace of } \mathbf{C}^3 \text{ of dimension } j\}.$$

The form Φ distinguishes an open subset thereof:

$$\mathbf{F}^{-+-} = \{L_1 \subset L_2 \text{ s.t. } \Phi|_{L_1} \text{ is negative definite and } \Phi|_{L_2} \text{ has type } (+-)\}$$

(whereas $\mathbf{F}^{+--} = \{L_1 \subset L_2 \text{ s.t. } \Phi|_{L_1} \text{ is positive definite and } \Phi|_{L_2} \text{ has type } (+-)\}$ etc.). This is a homogeneous space SU(1,2)$/T$ where $T = $ S(U(1) $\times$

U(1) × U(1)) is the maximal torus. It inherits from $\mathbf{F}$ an SU(1,2)-invariant complex structure.

There are some natural complex line-bundles on $\mathbf{F}$ and hence on $\mathbf{F}^{-+-}$ which may be described as follows. Let $\mathbf{P}$ denote complex projective 2-space. The form Φ on $\mathbb{C}^3$ partitions $\mathbf{P}$ into $\mathbf{P}^+$, $\mathbf{P}^0$, and $\mathbf{P}^-$ according to the sign of $\Phi(Z^\alpha, Z^\alpha)$ where Z^α denote homogeneous coordinates. Planes in $\mathbb{C}^3$ may be represented as points in $\mathbf{P}^*$, the dual projective space, and then

$$\mathbf{F} = \{(Z^\alpha, W_\alpha) \in \mathbf{P} \times \mathbf{P}^* \text{ s.t. } Z^\alpha W_\alpha = 0\}$$

where $Z^\alpha W_\alpha$ is the natural dual pairing. Let $\mathcal{O}(k)$ denote the usual line bundle on $\mathbf{P}$ corresponding to functions homogeneous of degree k. We may form the line bundles $\mathcal{O}(k, \ell)$ on $\mathbf{P} \times \mathbf{P}^*$ whose local sections may be identified with functions with the homogeneity

$$f(\lambda Z^\alpha, \mu W_\alpha) = \lambda^k \mu^\ell f(Z^\alpha, W_\alpha)$$

and finally restrict to $\mathbf{F}$. The cohomology we shall investigate is

$$H^1(\mathbf{F}^{-+-}, \mathcal{O}(-4, -3)).$$

Notice that one could also inquire as to the structure of

$$H^1(\mathbf{F}^{+--}, \mathcal{O}(-4, -3)) \text{ and } H^1(\mathbf{F}^{--+}, \mathcal{O}(-4, -3)),$$

$\mathbf{F}^{+--}$ and $\mathbf{F}^{--+}$ being the other two open orbits for the action of SU(1,2) on $\mathbf{F}$. Whilst these may also be treated (rather easily) with the Penrose transform, they are less interesting since [25], as representations of SU(1,2), they give the holomorphic and antiholomorphic discrete series as opposed to $H^1(\mathbf{F}^{-+-}, \mathcal{O}(-4, -3))$ which corresponds to discrete series representations which are of mixed type. Schmid [25] investigates this cohomology through arguments of pseudoconvexity in order to establish its K-types under the action of SU(1,2). In what follows we shall obtain the same conclusion by means of a Penrose transform.

The transform is derived from the following geometry. For each $P^\alpha \in \mathbf{P}^+$ and $Q_\alpha \in \mathbf{P}^{*+}$ one obtains a line in $\mathbf{F}^{-+-}$ as in the following picture

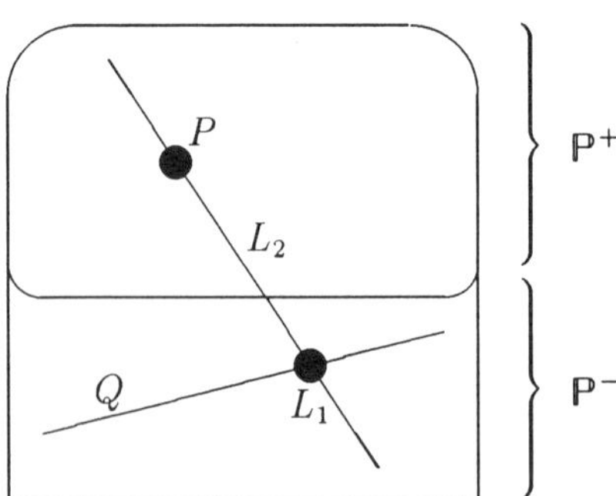

where $Q = \{Z^\alpha \in \mathbf{P} \text{ s.t. } Z^\alpha Q_\alpha = 0\}$. Algebraically this is

$$\{(Z^\alpha, W_\alpha) \text{ s.t. } Z^\alpha Q_\alpha = 0, \quad W_\alpha P^\alpha = 0, \quad Z^\alpha W_\alpha = 0\}.$$

Restricting a cohomology class in $H^1(\mathbf{F}^{-+-}, \mathcal{O}(-4, -3))$ to this line and allowing (P^α, Q_α) to vary gives a field on $\mathbf{P}^+ \times \mathbf{P}^{*+}$ as in the classical Penrose transform. The abstract form of the transform requires a little abstract representation theory. Instead, the remainder of the discussion will be phrased in terms of the traditional integral formulae. Write

$$P^\alpha = (1, x_A) \qquad Q_\alpha = (1, y^A)$$
$$Z^\alpha = (\omega, \pi_A) \qquad W_\alpha = (\eta, \xi^A).$$

Then, the line as above has equations

$$\begin{aligned}
\omega + y^A \pi_A &= 0 \\
\eta + \xi^A x_A &= 0 \\
\omega \eta + \xi^A \pi_A &= 0
\end{aligned}$$

the general solution of which is

$$(Z^\alpha, W_\alpha) = (-y^A \pi_A, \pi_A, x^A \pi_A, y^B \pi_B x^A + \pi^A)$$

as π_A varies. The Penrose transform gives the field

$$\phi_{ABCDE}(x_F, y^F) = \oint \pi_A \pi_B \pi_C \pi_D \pi_E f(-y^F \pi_F, \pi_F, x^F \pi_F, y^G \pi_G x^F + \pi^F) \pi^H d\pi_H$$

noting that the integrand has been arranged to have total homogeneity zero. The field equations are

$$\frac{\partial \phi_{ABCDE}}{\partial x_A} = 0 = \frac{\partial \phi_{ABCDE}}{\partial y_A}$$

as can be verified by differentiating under the integral sign. The usual arguments show that this is an isomorphism:

$$H^1(\mathbf{F}^{-+-}, \mathcal{O}(-4, -3)) \cong \left\{ \begin{array}{l} \phi_{ABCDE} \in \Gamma(\mathbf{P}^+ \times \mathbf{P}^{*+}, \mathcal{O}_{(ABCDE)}) \text{ s.t.} \\ \frac{\partial \phi_{ABCDE}}{\partial x_A} = 0 = \frac{\partial \phi_{ABCDE}}{\partial y_A} \end{array} \right\}.$$

Strictly speaking there are some conformal weights of which one can, with more care, keep track. The great advantage of having the cohomology realized in this way is that $\mathbf{P}^+ \times \mathbf{P}^{*+}$ is a product of balls and one can therefore easily expand in power series. Thus, the general solution of the field equations is

$$\alpha_{ABCDE} + \begin{array}{c} \beta_{ABCDEF} x^F \\[6pt] \gamma_{ABCDEF} y^F \end{array} + \begin{array}{c} \zeta_{ABCDEFG} x^F x^G \\[6pt] \eta_{ACDEFG} x^F y^G \\[6pt] \theta_{ABCDEFG} y^F y^G \end{array} + \ldots$$

where all the coefficients are symmetric in their indices. The maximal compact subgroup

$$K = \mathrm{S}(\mathrm{U}(1) \times \mathrm{U}(2)) \subset \mathrm{SU}(1,2)$$

acts on $\mathbf{P}^+ \times \mathbf{P}^{*+}$ by fixing the centre about which these power series are taken. The form of the power series immediately yields the decomposition into K-types. A little more care over conformal weights gives the U(1) action whilst the SU(2) action is manifest. The resulting cone of K-types is exactly as predicted by Blattner's formula [25].

References

[1] T. N. Bailey and M. G. Eastwood. Complex paraconformal manifolds— their differential geometry and twistor theory. Preprint.

[2] R. J. Baston. *The Algebraic Construction of Invariant Differential Operators*. D.Phil. thesis. Oxford University 1985.

[3] R. J. Baston and L. J. Mason. Conformal gravity, the Einstein equations, and spaces of complex null geodesics. Class. Quantum Grav. **4** (1987), 815–826.

[4] R. J. Baston and M. G. Eastwood, *The Penrose Transform: its Interaction with Representation Theory*. Oxford University Press 1989.

[5] N. P. Buchdahl. A generalized deRham sequence. Twistor Newsletter **10** (1980), 11–13.

[6] N. P. Buchdahl. On the relative deRham sequence. Proc. A.M.S. **87** (1983), 363–366.

[7] E. G. Dunne and M. G. Eastwood. A twistor transform for the discrete series: the case of SU(1, 1). Twistor Newsletter **26** (1988), 26–30.

[8] M. G. Eastwood, R. Penrose, and R. O. Wells, Jr.. Cohomology and massless fields. Commun. Math. Phys. **78** (1981), 305–351.

[9] M. G. Eastwood. The generalized Penrose-Ward transform. Math. Proc. Camb. Phil. Soc. **97** (1985), 165–187.

[10] M. G. Eastwood. A duality for homogeneous bundles on twistor space. Jour. L.M.S. **31** (1985), 349–356.

[11] M. G. Eastwood. The Penrose transform for curved ambitwistor space. Quart. J. Math. **39** (1988), 427–441.

[12] A. R. Gover. Conformally invariant operators of standard type. Quart. J. Math. Oxford (2) **40** (1989), 197–207.

[13] S. Helgason. *The Radon Transform*. Prog. Math. vol. 5. Birkhäuser 1980.

[14] S. A. Hugget and K. P. Tod. *An Introduction to Twistor Theory*. L.M.S. Stud. Text. vol. 4. Cambridge University Press 1985.

[15] L. P. Hughston. Applications of SO(8) spinors. In: *Gravitation and Geometry* (eds. W. Rindler and A. Trautman. Bibliopolis 1987) pp. 243–277.

[16] C. Kozameh, E. T. Newman, and K. P. Tod. Conformal Einstein spaces. Gen. Rel. Grav. **17** (1985), 343–352.

[17] C. R. LeBrun. $\mathcal{H}$-space with a cosmological constant. Proc. R. Soc. Lond. **A380** (1982), 171–185.

[18] C. R. LeBrun. Spaces of complex null geodesics in complex-Riemannian geometry. Trans. A.M.S. **278** (1983), 209–231.

[19] C. R. LeBrun. Ambi-twistors and Einstein's equations. Class. Quantum Grav. **2** (1985), 555–563.

[20] C. R. LeBrun. Thickenings and gauge fields. Class. Quantum Grav. **3** (1986), 1039–1059.

[21] R. Penrose. Non-linear gravitons and curved twistor theory. Gen. Rel. Grav. **7** (1976), 31–52.

[22] R. Penrose and W. Rindler, *Spinors and space-time, Vol. 1*. Cambridge University Press 1984.

[23] R. Penrose and W. Rindler, *Spinors and space-time, Vol. 2*. Cambridge University Press 1986.

[24] S.M. Salamon. Quaternionic Kähler manifolds. Invent. Math. **67** (1982), 143–171.

[25] W. Schmid. Homogeneous complex manifolds and representations of semisimple Lie groups. Proc. Nat. Acad. Sci. U.S.A. **59** (1968), 56–59.

[26] M. A. Singer. *A General Theory of Global Twistor Descriptions*. D.Phil. thesis. Oxford University 1987.

[27] R. S. Ward. Self-dual space-times with cosmological constant. Commun. Math. Phys. **78** (1980), 1–17.

Notation for the Penrose Transform

E.G. Dunne

Several articles in this volume (and also the book by Baston and Eastwood [1]) use a convenient notation for representations, flag varieties and parabolic subgroups of complex Lie groups It is based on the Dynkin diagram notation for simple Lie algebras. In this note the basic rules are illustrated by examples—the reader is referred to [1] for a complete description. If you are unfamiliar with the theory of roots and weights, I recommend [2], whose general notation will be used in this note.

Let us start with the Dynkin diagram for a given complex simple Lie algebra, $\mathbf{g}$. For example:

$$
\bullet\!\!-\!\!-\!\!-\!\!-\!\!\bullet\!\!\Longrightarrow\!\!\bullet \quad \text{for} \quad \mathbf{so}(7,\mathbb{C})
$$
$$
\text{or}
$$
$$
\bullet\!\!-\!\!-\!\!-\!\!-\!\!\bullet\!\!-\!\!-\!\!-\!\!-\!\!\bullet \quad \text{for} \quad \mathbf{sl}(4,\mathbb{C})
$$

When used for groups, the notation indicates the simply connected complex Lie group which corresponds to the algebra.

The conjugacy classes of parabolic subalgebras are determined by subsets of the simple roots. Thus, I will assume that a choice of simple roots has been fixed, call it Ψ. Now, the parabolic that corresponds to $\mathcal{S}_{\mathbf{p}} = \{\alpha_{i_1}, \alpha_{i_2}, \ldots, \alpha_{i_r}\} \subseteq \{simple\ roots\}$, is the subalgebra of $\mathbf{g}$ generated by all the positive root spaces and the negative root spaces for the elements of $\mathcal{S}_p$. Note that this is usually bigger than the subspace of $\mathbf{g}$ *spanned* by these root spaces, as is clear from the example of the 'parabolic' $\mathbf{p} = \mathbf{g}$. The Dynkin diagram for the parabolic is written by leaving dots at $\alpha_{i_1}, \alpha_{i_2}, \ldots, \alpha_{i_r}$ in the diagram and replacing the remaining dots by crosses. Thus, the Borel for the given choice of simple roots has all $\times$'s and no $\bullet$'s. This is the example to use to remember whether $\times$ indicates the inclusion or the exclusion of the corresponding simple root. Another example is the 'parabolic' which is the whole algebra. The diagram for this has dots at every simple root, which coincides with the usual notation. Some less trivial examples are:

$$
\times\!\!-\!\!\bullet\!\!-\!\!\bullet \quad \longleftrightarrow \quad
\begin{pmatrix}
* & * & * & * \\
0 & * & * & * \\
0 & * & * & * \\
0 & * & * & *
\end{pmatrix}
$$

$$
\times\!\!-\!\!\times\!\!-\!\!\bullet \quad \longleftrightarrow \quad
\begin{pmatrix}
* & * & * & * \\
0 & * & * & * \\
0 & 0 & * & * \\
0 & 0 & * & *
\end{pmatrix}
$$

$$
\times\!\!-\!\!\bullet\!\!-\!\!\times \quad \longleftrightarrow \quad
\begin{pmatrix}
* & * & * & * \\
0 & * & * & * \\
0 & * & * & * \\
0 & 0 & 0 & *
\end{pmatrix}
$$

$$
\times\!\!-\!\!\times\!\!-\!\!\times \quad \longleftrightarrow \quad
\begin{pmatrix}
* & * & * & * \\
0 & * & * & * \\
0 & 0 & * & * \\
0 & 0 & 0 & *
\end{pmatrix}
$$

It should now be clear how the scheme works, at least in the case of $\mathbf{sl}(n, \mathbf{C})$, whose parabolics are the easiest to represent by matrices.

Examples for the orthogonal groups require the introduction of some notation. Consider the case of $\mathbf{g} = \mathbf{so}(7, \mathbf{C})$. Let $\mathbf{t}$ be the Cartan subalgebra:

$$
\mathbf{t} =
\left(
\begin{array}{cc|cc|cc|c}
0 & \theta_1 & 0 & 0 & 0 & 0 & 0 \\
-\theta_1 & 0 & 0 & 0 & 0 & 0 & 0 \\
\hline
0 & 0 & 0 & \theta_2 & 0 & 0 & 0 \\
0 & 0 & -\theta_2 & 0 & 0 & 0 & 0 \\
\hline
0 & 0 & 0 & 0 & 0 & \theta_3 & 0 \\
0 & 0 & 0 & 0 & -\theta_3 & 0 & 0 \\
\hline
0 & 0 & 0 & 0 & 0 & 0 & 0
\end{array}
\right)
$$

Let $t(\theta_1, \theta_2, \theta_3)$ indicate the matrix above. The weight $\lambda = (\lambda_1, \lambda_2, \lambda_3) \in \mathbf{t}^*$ is the linear functional on $\mathbf{t}$ such that

$$
\begin{aligned}
\lambda(\, t(\theta_1, \theta_2, \theta_3)\,) &= \lambda(\theta_1, \theta_2, \theta_3) \\
&= i \cdot \{\lambda_1 \theta_1 + \lambda_2 \theta_2 + \lambda_3 \theta_3\}.
\end{aligned}
$$

Notice the factor of $i = \sqrt{-1}$ in the definition. Then the roots for $\mathbf{t}$ in $\mathbf{g}$ are:

$$
\begin{array}{lll}
\pm\alpha_{12} = \pm(1,-1,0), & \pm\alpha_{13} = \pm(1,0,-1), & \pm\alpha_{23} = \pm(0,1,-1), \\
\pm\beta_{12} = \pm(1,1,0), & \pm\beta_{13} = \pm(1,0,1), & \pm\beta_{23} = \pm(0,1,1), \\
\pm\gamma_1 = \pm(1,0,0), & \pm\gamma_2 = \pm(0,1,0), & \pm\gamma_3 = \pm(0,0,1)
\end{array}
$$

The standard choice of simple roots is $\Psi = \{\alpha_{12}, \alpha_{23}, \gamma_3\}$. The Killing form on $\mathbf{t}^*$ is just the standard inner product on the co-ordinates, $\theta_1, \theta_2, \theta_3$. From this, it is easy to see

$$
\begin{aligned}
(\alpha_{12}, \alpha_{23}) &= -1 \\
(\alpha_{12}, \gamma_3) &= 0 \\
(\alpha_{23}, \gamma_3) &= -1 \\[1em]
(\alpha_{12}, \alpha_{12}) &= 2 \\
(\alpha_{23}, \alpha_{23}) &= 2 \\
(\gamma_3, \gamma_3) &= 1
\end{aligned}
$$

which coincides with the diagram by choosing:

$$
\overset{\alpha_{12}}{\bullet} \rule[0.4ex]{3em}{0.4pt} \overset{\alpha_{23}}{\bullet} \xrightarrow{\hspace{2em}} \overset{\gamma_3}{\bullet}
$$

so α_{13} corresponds to the first node and so on. The root spaces for the orthogonal algebras are a bit tedious to describe. Here are a few:

$$
\underbrace{\begin{pmatrix}
0 & 0 & z & -iz & 0 & 0 & 0 \\
0 & 0 & iz & z & 0 & 0 & 0 \\
-z & -iz & 0 & 0 & 0 & 0 & 0 \\
iz & -z & 0 & 0 & 0 & 0 & 0 \\
0 & 0 & 0 & 0 & 0 & 0 & 0 \\
0 & 0 & 0 & 0 & 0 & 0 & 0 \\
0 & 0 & 0 & 0 & 0 & 0 & 0
\end{pmatrix}}_{\mathbf{g}^{\alpha_{12}}}
\qquad
\underbrace{\begin{pmatrix}
0 & 0 & z & iz & 0 & 0 & 0 \\
0 & 0 & -iz & z & 0 & 0 & 0 \\
-z & iz & 0 & 0 & 0 & 0 & 0 \\
-iz & -z & 0 & 0 & 0 & 0 & 0 \\
0 & 0 & 0 & 0 & 0 & 0 & 0 \\
0 & 0 & 0 & 0 & 0 & 0 & 0 \\
0 & 0 & 0 & 0 & 0 & 0 & 0
\end{pmatrix}}_{\mathbf{g}^{-\alpha_{12}}}
$$

For the β's:

$$
\underbrace{\begin{pmatrix}
0 & 0 & z & iz & 0 & 0 & 0 \\
0 & 0 & iz & -z & 0 & 0 & 0 \\
-z & -iz & 0 & 0 & 0 & 0 & 0 \\
-iz & z & 0 & 0 & 0 & 0 & 0 \\
0 & 0 & 0 & 0 & 0 & 0 & 0 \\
0 & 0 & 0 & 0 & 0 & 0 & 0 \\
0 & 0 & 0 & 0 & 0 & 0 & 0
\end{pmatrix}}_{\mathbf{g}^{\beta_{12}}}
\qquad
\underbrace{\begin{pmatrix}
0 & 0 & -z & iz & 0 & 0 & 0 \\
0 & 0 & -iz & -z & 0 & 0 & 0 \\
-z & iz & 0 & 0 & 0 & 0 & 0 \\
iz & z & 0 & 0 & 0 & 0 & 0 \\
0 & 0 & 0 & 0 & 0 & 0 & 0 \\
0 & 0 & 0 & 0 & 0 & 0 & 0 \\
0 & 0 & 0 & 0 & 0 & 0 & 0
\end{pmatrix}}_{\mathbf{g}^{-\beta_{12}}}
$$

For the γ's:

$$
\underbrace{\begin{pmatrix}
0 & 0 & 0 & 0 & 0 & 0 & z \\
0 & 0 & 0 & 0 & 0 & 0 & iz \\
0 & 0 & 0 & 0 & 0 & 0 & 0 \\
0 & 0 & 0 & 0 & 0 & 0 & 0 \\
0 & 0 & 0 & 0 & 0 & 0 & 0 \\
0 & 0 & 0 & 0 & 0 & 0 & 0 \\
-z & -iz & 0 & 0 & 0 & 0 & 0
\end{pmatrix}}_{\mathbf{g}^{\gamma_1}}
\qquad
\underbrace{\begin{pmatrix}
0 & 0 & 0 & 0 & 0 & 0 & z \\
0 & 0 & 0 & 0 & 0 & 0 & -iz \\
0 & 0 & 0 & 0 & 0 & 0 & 0 \\
0 & 0 & 0 & 0 & 0 & 0 & 0 \\
0 & 0 & 0 & 0 & 0 & 0 & 0 \\
0 & 0 & 0 & 0 & 0 & 0 & 0 \\
-z & iz & 0 & 0 & 0 & 0 & 0
\end{pmatrix}}_{\mathbf{g}^{-\gamma_1}}
$$

and so on.

When a Dynkin diagram is used to represent a flag variety, it is the variety G/P where G is the Lie group corresponding to $\mathbf{g}$ and P is the subgroup of G with Lie algebra $\mathbf{p}$ corresponding to the altered Dynkin diagram. In the examples from the previous paragraph we have:

$$\times\!\!-\!\!-\!\!-\!\!\bullet\!\!-\!\!-\!\!-\!\!\bullet \quad \longleftrightarrow \quad \mathbf{CP}^3$$

$$\times\!\!-\!\!-\!\!-\!\!\times\!\!-\!\!-\!\!-\!\!\bullet \quad \longleftrightarrow \quad \mathbf{F}_{12}$$

$$\times\!\!-\!\!-\!\!-\!\!\bullet\!\!-\!\!-\!\!-\!\!\times \quad \longleftrightarrow \quad \mathbf{F}_{13}$$

$$\times\!\!-\!\!-\!\!-\!\!\times\!\!-\!\!-\!\!-\!\!\times \quad \longleftrightarrow \quad \mathbf{F}_{123}$$

Here $\mathbf{F}_{12}$ is the partial flag variety of lines inside planes in $\mathbf{C}^4$. $\mathbf{F}_{13}$ is the variety of lines inside 3-dimensional subspaces in $\mathbf{C}^4$. The full flag variety of lines inside planes inside 3-dimensional subspaces in $\mathbf{C}^4$ is $\mathbf{F}_{123}$. For $\mathbf{so}(7,\mathbf{C})$,

$$\times\!\!-\!\!-\!\!-\!\!\bullet\!\!\Rightarrow\!\!\bullet \quad \longleftrightarrow \quad \mathbf{C}S^5$$

$$\bullet\!\!-\!\!-\!\!-\!\!\bullet\!\!\Rightarrow\!\!\times \quad \longleftrightarrow \quad \mathbf{Z}^7$$

where $\mathbf{C}S^5$ is the quadric in $\mathbf{CP}^6$ defined by $Q(x) = x_0^2 + x_1^2 + \ldots + x_6^2 = 0$. It is a five dimensional complex projective variety which is a complexification of the five dimensional sphere. The space $\mathbf{Z}^7$ is the projective space of pure spinors for $SO(7,\mathbf{C})$. It is a six dimensional complex projective variety.

For representations, the notation has adopted a convention which works well when using the Bott-Borel-Weil theorem. This is due to the way representations correspond to vector bundles on the generalized flag varieties G/P. (At other times the notation is opposite from what one might expect.) For representations of the full algebra, each node in the diagram represents the fundamental weight, λ_i, corresponding to the root, α_i, at that note. The fundamental weights are those defined by:

$$2\frac{\kappa(\lambda_i, \alpha_j)}{\kappa(\alpha_j, \alpha_j)} = \delta_{ij}.$$

Thus they form an integral basis for the weight lattice. That is any integral weight is of the form

$$\lambda = m_1\lambda_1 + \ldots + m_n\lambda_n.$$

The diagram notation for this is to put an m_i over the ith node. So far so good. This gives a notation for weights. For Bott's theorem using lowest weights for representations avoids some of the shifting by ρ which would otherwise occur. But since space is rather limited in these diagrams, there isn't room for the minus signs that would be required for the anti-dominant weights which parametrize irreducible representations. Thus the convention which was adopted is that

$$\overset{a}{\bullet}\!\!-\!\!\overset{b}{\bullet}\!\!-\!\!\overset{c}{\bullet}$$

indicates the representation of $\mathbf{sl}(4, \mathbf{C})$ with lowest weight $-\,(a\lambda_1 + b\lambda_2 + c\lambda_3)$, assuming that a, b and c are non-negative integers. Forgetting the minus sign which is implied in the notation can lead to serious mistakes.

For a parabolic subgroup, $\times\!\!-\!\!\bullet\!\!-\!\!\bullet$, say, with Levi factor $L = ZL_0$, with Z central and L_0 semi-simple, its irreducible representations are characters of Z tensored with irreducible representations of L_0. Thus, they are denoted:

$$\overset{p}{\times}\!\!-\!\!\overset{q}{\bullet}\!\!-\!\!\overset{r}{\bullet}$$

where $q, r \geq 0$ and p is an arbitrary integer. Letting

$$h_{12} = \begin{pmatrix} 1 & 0 & 0 & 0 \\ 0 & -1 & 0 & 0 \\ 0 & 0 & 0 & 0 \\ 0 & 0 & 0 & 0 \end{pmatrix} \qquad h_{23} = \begin{pmatrix} 0 & 0 & 0 & 0 \\ 0 & 1 & 0 & 0 \\ 0 & 0 & -1 & 0 \\ 0 & 0 & 0 & 0 \end{pmatrix}$$

$$h_{34} = \begin{pmatrix} 0 & 0 & 0 & 0 \\ 0 & 0 & 0 & 0 \\ 0 & 0 & 1 & 0 \\ 0 & 0 & 0 & -1 \end{pmatrix}$$

the lowest weight of this representation is

$$h_{12} \mapsto -p \qquad\qquad h_{23} \mapsto -q$$
$$h_{34} \mapsto -r.$$

Notice that this convention means that the one-dimensional representations of a Borel subalgebra (or of a parabolic, for that matter) are described by the negative of how the centre acts. Thus, $\overset{a}{\times}\!\!-\!\!\overset{b}{\times}\!\!-\!\!\overset{c}{\times}$ indicates the character of $\times\!\!-\!\!\times\!\!-\!\!\times$ where h_{12} acts by $-a$, etc.

A homogeneous vector bundle on a generalized flag variety, G/P, is determined by the representation of P on the fibre at the identity. Thus, the irreducible vector bundles are parametrized by the irreducible representations of P. The notation $\overset{p}{\times}\!\!-\!\!\overset{q}{\bullet}\!\!-\!\!\overset{r}{\bullet}$ as a vector bundle on $\mathbf{CP}^3$ indicates the homogeneous bundle induced by the representation of $P = \times\!\!-\!\!\bullet\!\!-\!\!\bullet$ with lowest weight $-\overset{p}{\times}\!\!-\!\!\overset{q}{\bullet}\!\!-\!\!\overset{r}{\bullet}$.

References

[1] R. J. Baston and M. G. Eastwood, *The Penrose Transform: its interaction with representation theory.* Oxford University Press, Oxford, 1989.

[2] J. E. Humphreys, *Introduction to Lie Algebras and Representation Theory*, Graduate Texts in Mathematics, **9**. Springer-Verlag, New York, 1980.

The Twistor Transform

E.G. Dunne   M.G. Eastwood

1 Introduction and Motivation

There are two ways of describing a free Maxwell field on Minkowski space:

1. As a 2-form, F, subject to $dF = 0$ and $d * F = 0$ where $*$ is the Hodge $*$-operator.

2. As an equivalence class of 1-forms, Φ subject to $d * d\Phi = 0$ where two 1-forms are said to be equivalent if their difference is of the form df for some function f.

Locally, these two descriptions are the same, the relationship between them being given by $F = d\Phi$. The 1-form Φ is called a *potential* for the *field F*. Globally, this mapping $\Phi \mapsto F$ is neither injective nor surjective. There is a topological obstruction to finding a Φ for a given F and a similar topological obstruction to finding an f such that $df = \Phi$ when $d\Phi = 0$. This global inequivalence is of great interest since it requires one to determine experimentally which, if any, is the physically relevant description. The rather unexpected answer is 'none of the above.' Nowadays, it is well known that some combination of these two is required. The equivalence class of potentials Φ may be precisely re-interpreted as a connection on a trivial line bundle and allowing non-trivial line bundles with connection allows arbitrary fields F (as the curvature of this connection). The Aharonov-Bohm effect [2] provides a physical justification for this reformulation. The resulting physical theory and generalizations (connections on vector bundles) are known as gauge theories (see, for instance, [1, 8, 45]). In twistor theory this leads to Richard Ward's *twisted photon* [47] and the *Ward correspondence* [3, 47] for self-dual Yang-Mills fields.

The direction for this article and the original *twistor transform* come from situations where the two descriptions are equivalent or, at least, nearly so. In particular, let us study Maxwell fields on $M = $ compactified Minkowski space [35]. M is homeomorphic to $S^1 \times S^3$. Now, $H^2(M, \mathbf{R}) = 0$ so we may conclude that:

$$d : \{potentials\} \longrightarrow \{fields\}$$

is surjective. If we temporarily weaken our notion of equivalence so that equivalent potentials are now those which give rise to the same field, then we obtain an isomorphism:

$$d : \left\{ \begin{array}{c} \textit{equivalence classes} \\ \textit{of potentials} \end{array} \right\} \xrightarrow{\;\cong\;} \{\textit{fields}\}.$$

Now the group of global orientation preserving conformal transformations acts on M, sending potentials to potentials and fields to fields. Since $*$ is conformally invariant on 2-forms in 4 dimensions, d is an intertwining operator between these two representations of this group. As is usual in twistor theory [34], this group (or rather its 4-fold cover) is identified as $SU(2,2)$. As is usual in representation theory [26, 46, 48], we complexify these representations, i.e. all fields and potentials from now on will be complex valued. It is generally desirable to have *unitary representations* if possible. So the question arises as to whether this Maxwell representation of $SU(2,2)$ is unitarizable; can we construct an invariant inner product (a positive definite Hermitian symmetric form)? A standard manœuvre in such a construction is to look for invariant Hermitian pairings between two different realizations of the same representation. This is an important rôle of the classical *integral intertwining operators* as in Knapp and Stein [29, 30]. We shall have more to say about this later.

Suppose we try to pair a potential Φ with a field G. It is natural to combine them as $\Phi \wedge \overline{G}$ and then integrate this 3-form over an S^3 in $M = S^1 \times S^3$:

$$\langle \Phi, G \rangle = \int_{S^3} \Phi \wedge \overline{G}.$$

This only depends on Φ up to equivalence for, if $d\Phi = d\Psi$ then, on S^3, there is a smooth function f such that $df = \Phi - \Psi$. Whence $d(f\overline{G}) = \Phi \wedge \overline{G} - \Psi \wedge \overline{G}$ and

$$\int \Phi \wedge \overline{G} = \int \Psi \wedge \overline{G}$$

by Stokes' theorem. Thus, we obtain an Hermitian form on Maxwell fields by:

$$\langle F, G \rangle = \int \Phi \wedge \overline{G}$$

where Φ is any potential for F. Also, note that it is symmetric: if Φ is a potential for F and Γ is a potential for G, then:

$$\begin{aligned}
\overline{\langle F, G \rangle} &= \int \overline{\Phi} \wedge G = \int \overline{\Phi} \wedge d\Gamma \\
&= \int d\overline{\Phi} \wedge \Gamma - d(\overline{\Phi} \wedge \Gamma) \\
&= \int \overline{F} \wedge \Gamma = \int \Gamma \wedge \overline{F} \\
&= \langle G, F \rangle.
\end{aligned}$$

So $\langle \ , \ \rangle$ is Hermitian symmetric.

We now need to check invariance of this form. This clearly boils down to the question of whether our definition is independent of the choice of cycle, S^3. If $\Phi \wedge \overline{G}$ were closed, then Stokes' theorem would guarantee independence. But it's not closed:

$$d(\Phi \wedge \overline{G}) = F \wedge \overline{G}.$$

The escape from this predicament is to insist that F is self-dual;

$$*F = iF,$$

noting that closure of such F then guarantees that it satisfies Maxwell's equations. If G is also self-dual, then $\overline{G}$ is anti-self-dual. So $F \wedge \overline{G} = 0$ and $\langle \ , \ \rangle$ is independent of the choice of S^3, as required.

The only remaining question is whether $\langle \ , \ \rangle$ is positive definite. It is not. However, the self-dual Maxwell fields do not comprise an irreducible representation of $\mathrm{SU}(2,2)$, but rather split into:

$$\{ \textit{positive frequency self-dual Maxwell fields on } M \ \}$$
$$\oplus$$
$$\{ \textit{negative frequency self-dual Maxwell fields on } M \ \}$$

each of which *is* irreducible. Here, *positive frequency* (resp. *negative frequency*) means that in the Fourier decomposition only terms in $e^{-ik_a x^a}$ occur for which k_a is future pointing (resp. past pointing). A detailed discussion may be found in [4] and [14]. One can check (for example, by Fourier analysis) that a suitable orientation for S^3 ensures that $\langle \ , \ \rangle$ is positive definite on the positive frequency fields and negative definite on negative frequency fields. This is the classical construction of what is usually called the *scalar product*, [18, 28]. We emphasize that a crucial ingredient is the intertwining operator

$$d : \left\{ \begin{array}{c} \textit{equivalence classes} \\ \textit{of potentials} \end{array} \right\} \to \{ \textit{fields} \} \, .$$

Let us try to re-interpret this construction from the twistor point of view. This was accomplished via explicit integration in [36] and via cohomology in [14]. It is now well-known (cf. [15] and 'The Penrose Transform', in this collection) that the positive frequency self-dual Maxwell fields have an especially natural interpretation on twistor space via the Penrose transform. We may as well start with real analytic fields on M in which case [4]:

$$\left\{ \begin{array}{l} \textit{real analytic positive frequency} \\ \textit{self-dual Maxwell fields on } M \end{array} \right\} \cong \left\{ \begin{array}{l} \textit{holomorphic self-dual} \\ \textit{Maxwell fields on } |\mathbf{M}^+| \end{array} \right\}$$

and now the Penrose transform gives

$$\mathcal{P} : H^1(|\mathbf{P}^+|, \Omega^3) \xrightarrow{\cong} \left\{ \begin{array}{l} \textit{holomorphic self-dual} \\ \textit{Maxwell fields on } |\mathbf{M}^+| \end{array} \right\} \, .$$

Here, we employ standard twistor notation [23] whereby $\mathbf{M}^+$ and $\mathbf{P}^+$ are suitable open orbits of $SU(2,2)$ on $\mathbf{M} = Gr_2(\mathbf{C}^4)$ and $\mathbf{P} = \mathbf{CP}_3 = \mathbf{P}(\mathbf{C}^4)$ respectively. Dual (projective) twistor space is $\mathbf{P}^* = \mathbf{P}((\mathbf{C}^4)^*)$. The sheaves of holomorphic functions and of holomorphic 3-forms are denoted $\mathcal{O}$ and Ω^3, respectively. Also, the absolute value bars are used throughout to indicate the closure of a set, an overbar being reserved to indicate complex conjugation. The Penrose transform is a completely natural construction and, in particular, this isomorphism with cohomology is $SU(2,2)$-equivariant. It is clear that the twistor description of this representation is somewhat cleaner than the massless field version—the differential equations have disappeared and the $SU(2,2)$ action is automatic. It is now natural to ask how the scalar product arises in this twistor setting.

Motivated by the classical construction it is well to look for a suitable intertwining operator. The cohomology $H^1(|\mathbf{P}^+|, \Omega^3)$ gives rise directly to fields on $\mathbf{M}^+$ whereas the 'potential modulo gauge' description of the same fields finds interpretation on dual twistor space as $H^1(|\mathbf{P}^{*-}|, \mathcal{O})$ (see [15]). Notice that $|\mathbf{M}^+|$ is simply connected (indeed contractable) and has a Stein neighbourhood basis so that the original field description and the 'potential modulo gauge' description agree. Thus, the intertwining operator which provides the crucial ingredient in the classical construction is interpreted in the twistor description as an intertwining operator (which, for convenience, we write going the other way)

$$\mathcal{T} : H^1(|\mathbf{P}^+|, \Omega^3) \xrightarrow{\cong} H^1(|\mathbf{P}^{*-}|, \mathcal{O}) \tag{1}$$

known as a *twistor transform*. Given this operator, it is now very easy to describe the scalar product in a way which is manifestly $SU(2,2)$-invariant. If $\Gamma \in H^1(|\mathbf{P}^{*-}|, \mathcal{O})$ then $\overline{\Gamma} \in H^1(|\mathbf{P}^-|, \mathcal{O})$ and so we need a pairing

$$H^1(|\mathbf{P}^+|, \Omega^3) \otimes H^1(|\mathbf{P}^-|, \mathcal{O}) \to \mathbf{C}.$$

Such a pairing can be obtained by taking representative forms σ on $|\mathbf{P}^+|$ of type $(3,1)$ and τ on $|\mathbf{P}^-|$ of type $(0,1)$ and integrating the 5-form $\sigma \wedge \tau$ over $\mathbf{P}^0$. This is equivalent to taking the Mayer-Vietoris connecting homomorphism

$$H^1(|\mathbf{P}^+|, \Omega^3) \otimes H^1(|\mathbf{P}^-|, \mathcal{O}) \xrightarrow{\wedge} H^2(\mathbf{P}^0, \Omega^3) \xrightarrow{\delta} H^3(\mathbf{P}, \Omega^3) = \mathbf{C}$$

and in [14] was dubbed the *dot product*. Thus, for $F, G \in H^1(|\mathbf{P}^+|, \Omega^3)$ we define

$$\langle F, G \rangle = F.\overline{\mathcal{T}G} \tag{2}$$

One can check, as in [14], that this agrees with the classical construction on massless fields.

The twistor construction has many advantages, the main one being that it may be generalized. Many important representations of a reductive Lie group occur naturally as cohomology on a homogeneous space. We would like to know when these representations are unitarizable. To this end it is a good plan to avoid space-time arguments and work just with cohomology. Thus, we want to understand and generalize as much as possible the twistor transform (1)

$$\mathcal{T} : H^1(|\mathbf{P}^+|, \Omega^3) \xrightarrow{\cong} H^1(|\mathbf{P}^{*-}|, \mathcal{O}).$$

From these twistor transforms it is straightforward to write down a Hermitian form and check that it is symmetric. Finally, one can investigate positivity by checking it on K-types. This has been done, for example, for the ladder representations of $\mathrm{SU}(p, q)$ in [11, 16].

An investigation of the twistor transform can take many directions. For a complete understanding one should really have a *formula* for $\mathcal{T}$. This was accomplished over twenty years ago [36] using the language of twistor diagrams (see various articles on *twistor propagators* in [25]) and this is still an intriguing formulation resembling the classical integral intertwining operators. These propagators are expressed in terms of cohomology in [11, 14]. Another direct approach to the twistor transform is as a version of the Penrose transform and was given by Singer in [44].

However, one can investigate the twistor transform simply as an abstract isomorphism, using whatever methods are available. It is this aspect which we shall review in the main part of this article. The principal such method is via the Penrose transform as in our example of Maxwell fields above. Thus, one can view the twistor transform as a combination of two Penrose transforms. (It is worth bearing in mind the real analogy where the Fourier transform on $\mathbf{R}^n$ is expressible as a combination of the one dimensional Fourier transform and the Radon transform [21, 22].) This is a straightforward investigation now that the machinery of the Penrose transform is well-understood [6]. Indeed, [6] contains a preliminary such investigation. It also contains a comparison of the twistor construction of the discrete series for $\mathrm{SU}(1, 1)$ with the classical construction. This comparison provides further motivation for the twistor approach.

Another interesting method is to compare the cohomologies as *Harish-Chandra modules*. Basically, the Harish-Chandra module of a representation of a reductive Lie group G with maximal compact subgroup K is the set consisting of the vectors v in the representation such that $\{g.v \mid g \in G\}$ spans a finite dimensional subspace. The Harish-Chandra module is a representation of the Lie algebra $\mathbf{g}$ of G and of the group K such that the actions are compatible. It turns out that for many purposes, in particular for classification problems, it is sufficient to work with Harish-Chandra modules. Moreover, it is frequently simpler to work on the level of these K-finite vectors, since representations of compact groups and complex Lie algebras involve less analysis and topology than those of non-compact groups. The reader should consult [46] for a full description of Harish-Chandra modules.

Now, an isomorphism on the level of K-types indicates the existence of a more geometric isomorphism—a twistor transform. This is explained in [9] for the case of $G = \mathrm{SU}(1,2)$.

It seems that the twistor construction is particularly suitable in regard to discrete series representations and the analytic continuation thereof. This philosophy reflects Schmid's original investigation [38, 39] though subsequent work has been largely on L^2-cohomology which is quite different from the approach which we are taking here. Constructions on ordinary Dolbeault cohomology avoid the careful analysis needed for L^2-cohomology. Moreover, the approach for unitarizing representations in the analytic continuation of the holomorphic discrete series is essentially the same as for the discrete series itself. This is already true for $\mathrm{SU}(1,1)$ as explained in [6, 16]. As such, this goes beyond the L^2 constructions which require significant extra work [31, 32, 37, 49]. What is lost in the twistor approach to the discrete series, however, is the action of the Weyl group as contained in the Langlands formulae [43]. Also, when using twistors, the positive definiteness of the inner product is no longer obvious, as it is in the L^2 theory.

Finally, we should mention that there are reasons for studying the twistor transform other than for the purpose of constructing unitary representations. We have already said that the twistor transform is closely analogous to the Fourier transform—the construction of invariant inner products reflects the fact that the Fourier transform preserves L^2. However, the Fourier transform has many more applications, such as in solving differential equations where the transform takes linear differential operators with constant coefficients into polynomial multiplication. The twistor transform also has this property, often referred to as *twistor quantization* [14, 24, 36]. It may therefore be used to solve differential equations, as explained in [13], where Murray's transform [33] is obtained through such an application of the twistor transform.

Section 2 summarizes the twistor transform in the context of $\mathrm{SU}(p,q)$. This generalizes the above discussion in two ways. Firstly, there is the straightforward geometric extension of the twistor correspondence. Secondly, there is the possibility of using the general homogeneous line bundle $\mathcal{O}(\ell)$ as coefficients. For $\mathrm{SU}(2,2)$ one obtains the massless field representations and, in general, a family known as the ladder representations. In Section 3 we include two further extensions of the basic case. The first of these is to $\mathrm{SO}(2,6)$ instead of $\mathrm{SO}(2,4) \overset{1:2}{\cong} \mathrm{SU}(2,2)$ (leaving the case $\mathrm{SO}(2,m)$ to the imagination). The second extension is to stick with $\mathrm{SU}(2,2)$ but to use a general homogeneous vector bundle as coefficients rather than just a line bundle. As we shall see, the twistor transform is available in both cases though it's detailed properties await further investigation. In Section 4 we provide evidence for a rather general twistor transform appertaining to the construction of the general discrete series. This evidence is in the form of a calculation of K-types which proves the existence of a twistor transform on the level of formal power series expansions. We feel that this is strong support for our general programme, namely that of intertwining representations by means of the twistor transform and hence proving unitarity by geometric means.

2 Ladder Representations of $SU(p,q)$

The first generalized twistor transform for representations of groups other than the conformal group was the construction [11] of the ladder representations of SU(p,q). These are representations which lie in the analytic continuation of the holomorphic discrete series. See [17, 27] for a classification of unitary highest weight modules. The ladder representations are those for which the parameter is, in a sense, the most singular. They were first shown to be unitary in [28]. Rawnsley, Wolf and Schmid, in [37], develop a general harmonic theory for indefinite metrics which enables them to extend the usual L^2-cohomology construction of the discrete series to a construction of certain singular unitary representations, which include most of the ladder representations. The full set of ladder representations for SU(p,q) was constructed using L^2-cohomology in [31, 32] where a Penrose transform plays an important rôle. In the Rawnsley-Schmid-Wolf result, the choice of complex structure on the underlying homogeneous space is linked to the parameter for the representation, thus losing any possible action of the Weyl group on the parameters, as in the discrete series [40, 42]. The approach in [11] avoids the technical difficulties of L^2-cohomology by using a twistor transform. Because there are only two open SU(p,q) orbits in the relevant complex flag variety, this is the simplest extension of the original construction of [36].

Thus, we assume fixed a complex vector space, $\mathbf{T}$, of dimension $(N+1)$ with a Hermitian form Φ with signature (p,q), $p + q = N + 1$. We shall also assume, for simplicity, that $2 \le p \le q$ (if $p = 1$ (or even if $p = 0$), the general procedure is still valid but the conclusions are subject to certain small modifications as in [11]). Let $G = \mathrm{SU}(p,q)$ be the subgroup of $\mathrm{SL}(N+1, \mathbb{C})$ which preserves Φ. The projective space $\mathbf{P} = \mathbb{C}\mathbf{P}_N$ splits into three orbits, $\mathbf{P}^+, \mathbf{P}^-$ and $\mathbf{P}^0$, under the action of G. Here $\mathbf{P}^+$ (respectively, $\mathbf{P}^-$) is the space of lines $x \subset \mathbf{T}$ such that $\Phi|_x$ is positive (respectively, negative) definite. The space $\mathbf{P}^0$ (sometimes denoted by PN) consists of those lines on which Φ is null. Thus, $\mathbf{P}^0$ is a real hypersurface in $\mathbf{P}$. Note that $\mathbf{P}^+$ and $\mathbf{P}^-$ are open subsets, hence complex manifolds. The fact that these are the only open G-orbits makes the construction of the pairing simpler. We let $\mathbf{M}$ denote the Grassmannian of p-dimensional subspaces of $\mathbf{T}$. There are $\min(p,q)+1$ open G-orbits in $\mathbf{M}$ corresponding to the possible ways in which Φ can restrict as a non-degenerate form on a space of dimension p. We are interested in the orbit $\mathbf{M}^+$ on whose points Φ restricts to be positive definite. We note this orbit is naturally a noncompact Hermitian symmetric space. In particular, $\mathbf{M}^+$ is Stein.

As shown originally in [11] and developed in full generality in [6], the composition of two Penrose transforms leads to the isomorphism:

$$\mathcal{T} \ : \ H^{p-1}(\mathbf{P}^+, \mathcal{O}(-n-p)) \xrightarrow{\cong} H^{q-1}(\mathbf{P}^{*-}, \mathcal{O}(n-q)) \qquad (3)$$

known as the *twistor transform* for $SU(p,q)$. The sheaf $\mathcal{O}(-k)$ corresponds to the kth power of the tautological line bundle on $\mathbf{P}$. Thus, when $p = q = n = 2$, $\mathcal{O}(-4) = \Omega^3$ and this coincides with (1). The proof that this is an isomorphism is similar to the proof for the isomorphisms (1) and (7). For example, when $n \geq 1$, one shows that the left-hand side is isomorphic to fields on $\mathbf{M}^+$ satisfying a differential equation analogous to the zero rest mass equations of helicity $n/2$ on Minkowski space. Similarly, the right-hand side is naturally isomorphic to potentials modulo gauge for the same fields on $\mathbf{M}^{*-}$. But $\mathbf{M}^{*-}$ is the same as $\mathbf{M}^+$, whence the isomorphism. Again, one can extend the isomorphism to the closures of $\mathbf{P}^+$ and $\mathbf{P}^-$ yielding:

Theorem 2.1 ([11]) *There is an* $SU(p,q)$*-equivariant map*

$$\mathcal{T} \ : \ H^{p-1}(|\mathbf{P}^+|, \mathcal{O}(-n-p)) \longrightarrow H^{q-1}(|\mathbf{P}^{*-}|, \mathcal{O}(n-q)).$$

which is an isomorphism.

The Hermitian form defining $SU(p,q)$ establishes a conjugate linear isomorphism between $\mathbf{T}$ and $\mathbf{T}^*$, which we denote $z \mapsto \bar{z}$. Then, for a holomorphic function f defined on an open subset of $\mathbf{T}$, we can define $\overline{f}$, a holomorphic function on the corresponding open subset of $\mathbf{T}^*$, by $\overline{f(z)} = \bar{f}(\bar{z})$. Repeating the process takes holomorphic functions on subsets of $\mathbf{T}^*$ to holomorphic functions on the corresponding subsets of $\mathbf{T}$. Moreover, if $\Psi \in H^{q-1}(|\mathbf{P}^{*-}|, \mathcal{O}(n-q))$, then $\bar{\Psi} \in H^{q-1}(|\mathbf{P}^-|, \mathcal{O}(n-q))$. This gives us a conjugate linear isomorphism between these two spaces. Combining this with the twistor transform gives a conjugate linear isomorphism:

$$H^{p-1}(|\mathbf{P}^+|, \mathcal{O}(-n-p)) \ \bar{\cong} \ H^{q-1}(|\mathbf{P}^-|, \mathcal{O}(n-q)).$$

For $\phi, \psi \in H^{p-1}(|\mathbf{P}^+|, \mathcal{O}(-n-p))$ we have $\delta(\phi \cup \overline{\mathcal{T}\psi}) \in H^{p+q-1}(\mathbf{P}, \Omega^{p+q-1})$ where δ is the connecting map in the Mayer-Vietoris sequence. Since the group $H^{p+q-1}(\mathbf{P}, \Omega^{p+q-1})$ is naturally isomorphic to $\mathbf{C}$, we have an invariant inner product on $H^{p-1}(|\mathbf{P}^+|, \mathcal{O}(-n-p))$, which we denote by $\langle \phi, \psi \rangle$. The principal results in [16] can be summarized as:

Theorem 2.2 *The inner product* $\langle \ , \ \rangle$ *on* $H^{p-1}(|\mathbf{P}^+|, \mathcal{O}(-n-p))$ *is positive definite. The subspace* $H^{p-1}(|\mathbf{P}^+|, \mathcal{O}(-n-p))$ *contains natural cohomology classes known as 'elementary states' which are dense in* $\mathcal{H}$*, the resulting Hilbert space. Though it is not their definition, these are the K-finite vectors for the representation of* $SU(p,q)$ *on* $\mathcal{H}$*.*

3 Two Examples

Given the diverse audience of this volume, we feel it will be helpful to have
two very explicit examples at hand. The first involves a real form of $SO(8,\mathbb{C})$,
a group familiar to the physicists because of triality. The second is an extension of the standard twistor transform to the case of certain vector bundles
on $\mathbb{CP}_3$. This example, probably of more interest to mathematicians, demon-
strates why it is necessary to use some of the general techniques (such as the
Bernstein-Gelfand-Gelfand resolution), rather than the methods which work
for standard twistor theory in four dimensions.

Example A

In this example we describe a twistor transform for the group $SO(2,6)$, a real
form of $SO(8,\mathbb{C})$. Let $\mathbf{S}^+$ be the reduced spinor representation of $SO(2,6)$,
which corresponds to the diagram

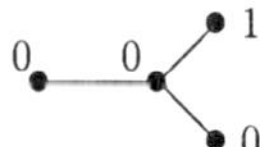

in the notation of [6]. (See [50] in this volume for an explanation of this use
of the Dynkin diagram notation in this context.) We shall be considering
line bundles over $SO(2,6)$ orbits in the space $\mathbb{P}$, which is the $SO(8,\mathbb{C})$ orbit
of a highest weight vector in the projectivized space $\mathbf{P}(\mathbf{S}^+)$. In the Dynkin
diagram notation, it is the variety:

$$\mathbb{P} = \quad\bullet\!\!\!-\!\!\!<\overset{\times}{}\quad.$$

As shown in [20], $\mathbf{S}^+$ has a natural quaternionic structure so that $\mathbf{S}^+ \cong \mathbf{H}^4$
($\mathbf{H}$ = the quaternions). Also, it carries a natural $\mathbf{H}$-Hermitian skew form ϵ
preserved by $Spin(2,6)$. For a suitable basis ϵ has the form:

$$\epsilon(x,y) = \bar{x}_1 j y_1 + \ldots + \bar{x}_4 j y_4 \tag{4}$$

where j is the usual unit quaterion. Here we are following the convention of
[20] whereby the action of quaternions on a quaternionic vector space is on
the *right*. We write elements $x \in \mathbf{H}^4$ as a sum of complex vectors: $x = \xi + j\eta$.
Then, ϵ takes the form:

$$\begin{aligned}
\epsilon(x,x) &= (\bar{\eta}_1\xi_1 + \bar{\eta}_2\xi_2 + \bar{\eta}_3\xi_3 + \bar{\eta}_4\xi_4 - \bar{\xi}_1\eta_1 - \bar{\xi}_2\eta_2 - \bar{\xi}_3\eta_3 - \bar{\xi}_4\eta_4) \\
&\quad + j(\xi_1^2 + \xi_2^2 + \xi_3^2 + \xi_4^2 + \eta_1^2 + \eta_2^2 + \eta_3^2 + \eta_4^2).
\end{aligned}$$

So we can write $\epsilon = i\epsilon_1 + j\epsilon_2$ where ϵ_1 is a $\mathbb{C}$-Hermitian symmetric form on
$\mathbb{C}^8 = \mathbf{H}^4$ with signature $(4,4)$ and ϵ_2 is a $\mathbb{C}$-symmetric form on $\mathbb{C}^8$. Up to finite

covering, therefore, $SO(2,6)$ is isomorphic to $SU(4,4) \cap SO(8, \mathbb{C})$ (both are double-covered by $\mathrm{Spin}(2,6)$). This is by no means obvious—it is one of the consequences of triality. Let us consider the projective space $\mathbf{P}(\mathbf{S}^+) \cong \mathbf{CP}_7$. Define:

$$\begin{aligned}
\mathbb{P}^+ &= \{[z] \in \mathbf{CP}_7 \mid \epsilon(z,z) = ir \text{ for } r > 0\} \\
\mathbb{P}^0 &= \{[z] \in \mathbf{CP}_7 \mid \epsilon(z,z) = 0\} \\
\mathbb{P}^- &= \{[z] \in \mathbf{CP}_7 \mid \epsilon(z,z) = ir \text{ for } r < 0\}
\end{aligned}$$

Then $\mathbb{P} = \mathbb{P}^+ \cup \mathbb{P}^0 \cup \mathbb{P}^-$ is the quadric in $\mathbf{CP}_7$ defined by $Q(z) = \epsilon_2(z,z) = 0$. By working on the projective space, the action of $SO(2,6)$ is well defined. From the irreducibility of $\mathbf{S}^+$ as a projective representation of $SO(2,6)$ we can deduce that each of $\mathbb{P}^+, \mathbb{P}^0, \mathbb{P}^-$ is a single orbit for $SO(2,6)$. Moreover, the orbits $\mathbb{P}^+$ and $\mathbb{P}^-$ are open, complex submanifolds of $\mathbb{P}$. The orbit $\mathbb{P}^0$ is the common boundary of the open orbits, hence has real dimension eleven. It is naturally a CR-manifold. This is a close analogue of the situation for standard twistors where $\mathbf{PT} = \mathbf{P}^+ \cup \mathbf{P}^0 \cup \mathbf{P}^-$ and each piece is a single $SO(2,4)$ (i.e. $SU(2,2)$) orbit.

There is the operation on $\mathbf{H}^4$ which is multiplication on the right by $j \in \mathbf{H}$. This provides us with a conjugation J on $\mathbb{P}$ as follows. Since ϵ is $\mathbf{H}$-Hermitian skew:

$$\epsilon(xj, yj) = -j\epsilon(x,y)j.$$

Thus, for $[x] \in \mathbb{P}^+$, $\epsilon(x,x) = ir$ with $r > 0$, so

$$\begin{aligned}
\epsilon(xj, xj) &= -j(ir)j = ijrj \\
&= irj^2 = -ir.
\end{aligned}$$

That is, $J(x) = xj$ is in $\mathbb{P}^-$. Similarly, $J : \mathbb{P}^- \to \mathbb{P}^+$ and $J : \mathbb{P}^0 \to \mathbb{P}^0$.

By analogy with the $SU(2,2)$ case, we anticipate a twistor transform for line bundles on $\mathbb{P}$ which provides a $\mathrm{Spin}(2,6)$-invariant inner product on the corresponding cohomology spaces. In particular, since the maximal compact subvarieties in $\mathbb{P}^+$ are copies of $\mathbf{CP}_3$, having complex dimension three, we expect to get non-trivial cohomology only for $H^3(\mathbb{P}^\pm, \mathcal{L})$ when $\mathcal{L}$ is a sufficiently negative homogeneous line bundle on $\mathbb{P}$. When $\mathcal{L}$ is sufficiently positive we get only $H^2(\mathbb{P}^\pm, \mathcal{L}) \neq 0$.

To be specific, the generalized Penrose transform as in [6] establishes the isomorphism for 'right-handed fields'

$$H^3\!\left(\mathbb{P}^+,\ \overset{0}{\bullet}\!\!-\!\!\overset{0}{\bullet}\!\!<^{\times\,-k-3}_{\bullet\,0}\right) \cong \ker\left\{ \overset{-k-1\ 0}{\underset{\times}{\bullet}}\!\!<^{\bullet\,0}_{\bullet\,k\text{-}1} \xrightarrow{\ -k-2\ 0\ } \underset{\times}{\bullet}\!\!<^{\bullet\,1}_{\bullet\,k\text{-}2} \right\} \tag{5}$$

when $k \geq 3$. Using similar arguments, we see that there is also a Penrose transform which yields these same fields as potential modulo gauge:

$$H^2\left(\mathbf{IP}^-,\ \begin{smallmatrix}0 & 0 & \times^{k-3}\\ & & \bullet\,0\end{smallmatrix}\right) \cong \frac{\ker\left\{\begin{smallmatrix}-3 & 0 & \bullet^{k-2}\\ \times & \bullet & \longrightarrow & -4 & 0 & \bullet^{k-1}\\ & \bullet\,1 & & \times & \bullet & \bullet\,0\end{smallmatrix}\right\}}{\operatorname{im}\left\{\begin{smallmatrix}-2 & 1 & \bullet^{k-3}\\ \times & \bullet & \longrightarrow & -3 & 0 & \bullet^{k-2}\\ & \bullet\,0 & & \times & \bullet & \bullet\,1\end{smallmatrix}\right\}}$$

$$\cong\ \ker\left\{\begin{smallmatrix}-k-1 & 0 & \bullet^{0}\\ \times & \bullet & \longrightarrow & -k-2 & 0 & \bullet^{1}\\ & \bullet\,{k-1} & & \times & \bullet & \bullet\,{k-2}\end{smallmatrix}\right\}. \tag{6}$$

The second isomorphism comes from the Bernstein–Gelfand–Gelfand resolution. Combining (5) and (6) yields the isomorphism known as the *twistor transform*:

$$\mathcal{T}\ :\ H^3\left(\mathbf{IP}^+,\ \begin{smallmatrix}0 & 0 & \times^{-k-3}\\ & & \bullet\,0\end{smallmatrix}\right) \xrightarrow{\ \cong\ } H^2\left(\mathbf{IP}^+,\ \begin{smallmatrix}0 & 0 & \times^{k-3}\\ & & \bullet\,0\end{smallmatrix}\right). \tag{7}$$

Also, J induces a conjugate linear isomorphism:

$$H^2\left(\mathbf{IP}^+,\ \begin{smallmatrix}0 & 0 & \times^{k-3}\\ & & \bullet\,0\end{smallmatrix}\right) \longrightarrow H^2\left(\mathbf{IP}^-,\ \begin{smallmatrix}0 & 0 & \times^{k-3}\\ & & \bullet\,0\end{smallmatrix}\right)$$

which we denote by J again. As in [6], standard techniques allow us to work with the *closures* $|\mathbf{IP}^+|$ and $|\mathbf{IP}^-|$ yielding:

$$H^3\left(|\mathbf{IP}^+|,\ \begin{smallmatrix}0 & 0 & \times^{-k-3}\\ & & \bullet\,0\end{smallmatrix}\right) \xrightarrow{\ \mathcal{T}\ } H^2\left(|\mathbf{IP}^+|,\ \begin{smallmatrix}0 & 0 & \times^{k-3}\\ & & \bullet\,0\end{smallmatrix}\right)$$

$$\xrightarrow{\ J\ } H^2\left(|\mathbf{IP}^-|,\ \begin{smallmatrix}0 & 0 & \times^{k-3}\\ & & \bullet\,0\end{smallmatrix}\right).$$

We can now put together all the pieces. Let

$$\phi,\psi \in H^3\left(|\mathbf{IP}^+|,\ \begin{smallmatrix}0 & 0 & \times^{-k-3}\\ & & \bullet\,0\end{smallmatrix}\right).$$

Then $J\mathcal{T}\psi \in H^2\left(|\mathbf{IP}^-|,\ \begin{smallmatrix}0 & 0 & \times^{k-3}\\ & & \bullet\,0\end{smallmatrix}\right)$. Using the cup product we have

$$\phi \cup (J\mathcal{T}\psi) \in H^5\left(|\mathbf{IP}^+| \cap |\mathbf{IP}^-|,\ \begin{smallmatrix}0 & 0 & \times^{-6}\\ & & \bullet\,0\end{smallmatrix}\right).$$

The connecting map, δ, for the Mayer-Vietoris sequence is:

$$\delta \;:\; H^5(|\mathbf{IP}^+|\cap|\mathbf{IP}^-|,\; \substack{0\;\;\;0\;\;\;\times\text{-}6\\ \bullet\!\!-\!\!\bullet\!\!<\!\!\bullet\,0}\;) \to H^6(|\mathbf{IP}^+|\cup|\mathbf{IP}^-|,\; \substack{0\;\;\;0\;\;\;\times\text{-}6\\ \bullet\!\!-\!\!\bullet\!\!<\!\!\bullet\,0}\;)$$

but the term on the right-hand side is just $H^6(\mathbf{IP},\Omega^6) = \mathbb{C}$. Thus, we have demonstrated

Theorem 3.1 *The pairing on* $H^3(|\mathbf{IP}^+|,\; \substack{0\;\;\;0\;\;\;\times\text{-}k\text{-}3\\ \bullet\!\!-\!\!\bullet\!\!<\!\!\bullet\,0}\;)$ *defined by* $\langle\phi,\psi\rangle =$

$\delta(\phi\cup(J\mathcal{T}\psi))$ *is a* $\mathrm{Spin}(2,6)$*-invariant (Hermitian) product.*

This pairing is analogous to the inner product on the cohomology groups which parameterize massless fields. Although we shall not address the issue of positive definiteness here, the natural approach would be that of [16], where definiteness is checked on the K-finite vectors and a density argument is used to complete the proof.

Example B

This is a case of a twistor transform for *vector* bundles. We shall discuss it for the standard (i.e. $\mathrm{SL}(4,\mathbb{C})$) twistor correspondence. The machine established in [6] applies here to define a Penrose transform. The associated fields on Minkowski space are no longer the familiar massless fields. However, the fundamental principles still hold in this more general setting. The fields on subsets of $\mathbf{M}^*$ can be represented by cohomology classes on $\mathbf{P}^{*-}$ using potential modulo gauge. It is essential to use the Bernstein-Gelfand-Gelfand resolution (cf. [7] or [6]) to see the equality.

Let us now apply the basic techniques of [6] to this situation. There are the two isomorphisms produced by the Penrose transform:

$$H^1(\mathbf{P}^+,\; \substack{\text{-}k\;\;\;\;b\;\;\;\;\;c\\ \times\!\!-\!\!\bullet\!\!-\!\!\bullet}\;) \cong \ker\left\{\mathbf{M}^+;\; \substack{k\text{-}2\;\;\;b\text{-}k+1\;\;\;c\\ \bullet\!\!-\!\!\times\!\!-\!\!\bullet}\;\longrightarrow\; \substack{k\text{-}b\text{-}3\;\;\;\text{-}k\;\;\;b+c+1\\ \bullet\!\!-\!\!\times\!\!-\!\!\bullet}\right\}$$

and

$$H^1(\mathbf{P}^{*-},\; \substack{p\;\;\;\;q\;\;\;\;n\\ \bullet\!\!-\!\!\bullet\!\!-\!\!\times}\;)$$
$$\cong\; \ker\left\{\mathbf{M}^{*-};\; \substack{p+q+n+2\;\;\text{-}q\text{-}n\text{-}3\;\;q\\ \bullet\!\!-\!\!\times\!\!-\!\!\bullet}\;\longrightarrow\; \substack{q+n+1\;\;\text{-}p\text{-}q\text{-}n\text{-}4\;\;p+q+1\\ \bullet\!\!-\!\!\times\!\!-\!\!\bullet}\right\}$$

where $k - 4 \geq b + c$ and $n \geq 0$. The second isomorphism, which is the case of left-handed fields for $\mathbf{M}^*$, is via potentials modulo gauge. Now, $\mathbf{M}^+$ is canonically isomorphic to $\mathbf{M}^{*-}$, so we can compare the parameters on the right-hand sides. Some simple algebra leads to:

$$p = b, \quad n = k - b - c - 4, \quad q = c.$$

Thus we may deduce an isomorphism for the left-hand sides:

$$\mathcal{T}\ :\ H^1(\mathbf{P}^+,\ \overset{-k}{\times}\!\!-\!\!-\!\!\overset{b}{\bullet}\!\!-\!\!-\!\!\overset{c}{\bullet}\ )\ \cong\ H^1(\mathbf{P}^{*-},\ \overset{b}{\bullet}\!\!-\!\!-\!\!\overset{c}{\bullet}\!\!-\!\!-\!\!\overset{k-b-c-4}{\times}\ ). \tag{8}$$

This can also be phrased simply without using the Dynkin diagram notation. Namely, let E be the vector bundle on $\mathbf{P}$ corresponding to the diagram $\overset{-k}{\times}\!\!-\!\!-\!\!\overset{b}{\bullet}\!\!-\!\!-\!\!\overset{c}{\bullet}$. Also use E to denote the bundle $\overset{c}{\bullet}\!\!-\!\!-\!\!\overset{b}{\bullet}\!\!-\!\!-\!\!\overset{-k}{\times}$ on $\mathbf{P}^*$. Then the twistor transform (8) becomes

$$H^1(\mathbf{P}^+, E) \cong H^1(\mathbf{P}^{*-}, E^* \otimes \Omega^3).$$

There is also the anti-linear isomorphism:

$$H^1(\mathbf{P}^{*-}, E^* \otimes \Omega^3) \ \bar{\cong}\ H^1(\mathbf{P}^-, E^* \otimes \Omega^3)$$

induced by conjugation. Thus for $\phi, \psi \in H^1(|\mathbf{P}^+|,\ \overset{-k}{\times}\!\!-\!\!-\!\!\overset{b}{\bullet}\!\!-\!\!-\!\!\overset{c}{\bullet}\ )$ we may pair them via:

$$H^2(|\mathbf{P}^0|, \Omega^3) \ni \phi \cup \overline{(\mathcal{T}\psi)} \ \overset{\delta}{\longrightarrow}\ H^3(\mathbf{P}, \Omega^3) \ =\ \mathbb{C}$$

where δ is the Mayer-Vietoris connecting homomorphism. Thus, we have an $SU(2,2)$-invariant pairing on $H^1(\mathbf{P}^+, E)$ defined by $\langle \phi, \psi \rangle = \delta(\phi \cup \overline{(\mathcal{T}\psi)})$.

We should note that the bounds which we have placed on the parameters k, b and c force our representations to be related to the holomorphic discrete series. As is shown in [12] the isomorphism holds for any vector bundle E such that $H^1(\mathbf{P}, E) = 0$ and $H^2(\mathbf{P}, E) = 0$. However, the pairing is not necessarily positive definite when E is not a line bundle; there are restrictions on which of these modules can be unitary. These can be found in [17, 27].

4 K-types and the Twistor Transform

This is another direction of research aimed at constructing the discrete series representations on Dolbeault cohomology as in [38] with their unitary structure induced from a twistor transform. A careful analysis [9] of the case of $SU(1,2)$ shows that this is an entirely reasonable aim. We shall sketch the general plan; the details are the subject of [10].

As an example, consider the familiar case of $G = SU(2,2)$ acting on the open orbit, $\mathbf{M}^+$. The *holomorphic* discrete series representations are those which are realizable as holomorphic sections of certain homogeneous vector bundles over $\mathbf{M}^+$. A typical such example is the representation on L^2 holomorphic 4-forms, where L^2 is defined with respect to the manifestly invariant inner product

$$\langle \omega, \eta \rangle = \int_{\mathbf{M}^+} \omega \wedge \bar{\eta}.$$

Sitting over $\mathbf{M}$ is $\mathbf{G}$, the space of full flags in $\mathbb{C}^4$. The pre-image of $\mathbf{M}^+$ under the natural projection is $\mathbf{G}^{++--}$, meaning that the Hermitian form has the indicated signatures (namely $+$, $++$, $++-$, and $++--$) when restricted to each part of the flag. This is one of the six open orbits of SU(2, 2) on $\mathbf{G}$:

$$\mathbf{G}^{++--}$$

$$\mathbf{G}^{+-+-}$$

$$\mathbf{G}^{-++-} \qquad\qquad \mathbf{G}^{+--+}$$

$$\mathbf{G}^{-+-+}$$

$$\mathbf{G}^{--++}$$

As is well known [38] each orbit corresponds to a different type of discrete series. This correspondence is effected by taking cohomology of the appropriate line bundles on the various orbits. In the simplest case, the holomorphic discrete series, we note that

$$H^2(\mathbf{G}^{++--}, \Omega^6) \cong \Gamma(\mathbf{M}^+, \Omega^4).$$

This isomorphism is a consequence of a calculation of direct images which may be undertaken using the machinery of [6]. It is natural to ask that SU(2, 2) be represented on the corresponding L^2-cohomology.

From the example of standard twistor theory and Examples A and B, it is clear that a twistor construction should end up pairing $H^2(\mathbf{G}^{++--}, \Omega^6)$ with $H^3(\mathbf{G} - \mathbf{G}^{++--}, \mathcal{O})$. In order to arrive at such a pairing one would first need a twistor transform that is, an SU(2, 2)-equivariant isomorphism:

$$\mathcal{T} : H^2(|\mathbf{G}^{--++}|, \Omega^6) \xrightarrow{\cong} H^3(\mathbf{G} - \mathbf{G}^{++--}, \mathcal{O}).$$

In the best of all possible worlds we would find a transform for each of the six orbits and in this way obtain the complete discrete series.

What *is* true is that such an isomorphism holds on the level of K-types. This can be proved directly by comparing the computation of Schmid's thesis [38] with that of Kempf as in [5]. Specifically one has:

Theorem 4.1 *Suppose G_0 is a real, semisimple Lie group having a compact Cartan subgroup T_0. Let K_0 be a maximal compact subgroup of G_0 containing T_0. Let Q be a maximal compact subvariety in a fixed open G_0 orbit X_0 in the flag variety $X = G/B$ and $\mathcal{L}$ be a homogeneous line bundle with parameter in the antidominant chamber for B. Then*

$$\mathcal{T} : H^s(Q^\infty, \mathcal{L}) \cong H^d_{[\tilde{Q}]}(X, \tilde{\mathcal{L}})$$

is an isomorphism of Harish-Chandra modules. Here $s = \dim_{\mathbb{C}}(Q)$, $d = \operatorname{codim}_{\mathbb{C}}(Q)$, $\tilde{\mathcal{L}}$ is the image of $\mathcal{L}$ under the affine action of the Weyl group of G, and $\tilde{Q}$ is the compact subvariety corresponding to Q under the natural conjugation on X induced by the conjugation on G which defines G_0.

Here $H^s(Q^\infty, *)$ indicates the expansion of cohomology classes in infinite formal neighbourhoods of Q and $H^d_{[\tilde{Q}]}(X, *)$ denotes relative cohomology *in the algebraic category*. The proof of the theorem is based on a theorem about the cohomology on K-orbits which we state below. We can now appeal to a theorem of Schmid [41], which states that the discrete series representations are determined (among all admissible representations) by their K-types, to conclude that the twistor transform $\mathcal{T}$ in Theorem 4.1 is an isomorphism of Harish-Chandra modules. We are currently making attempts to derive the full isomorphism directly from the geometry.

In the setting of Theorem 4.1, the compact K-orbits in X are naturally parameterzed by the *right* cosets $W_K \backslash W_G$ of the Weyl group for G. Any such coset contains a canonical element w which is distinguished by requiring that $w(\rho)$ is dominant for K, where ρ is half the sum of the positive roots for G. There is a natural involution on the compact orbits:

$$W_K . w \longmapsto W_K . w w_0$$

where w_0 is the longest element of W_G. This is given on the canonical representatives by

$$w \longmapsto w_1 w w_0$$

where w_1 is the longest element in W_K. In the case of the discrete series, this involution corresponds to the map sending a complex K_0-orbit to its conjugate. Now each compact K-orbit is a homogeneous space of the form $K/(K \cap B)$. In making this identification, it is best to chose the basepoint to correspond to the canonical representative. Let us now fix one such compact orbit Q with basepoint corresponding to $w \in W_G$. Let $\mathcal{L}$ be a homogeneous line bundle on X. It is given by a weight λ by specifying that $\mathcal{L}$ is associated to the irreducible representation of B where the Cartan subalgebra acts by *minus* the weight. This is in keeping with the system which makes the Borel-Weil-Bott theorem easiest to state. Then $\mathcal{L}|_Q$ is given by the weight $w\lambda$, following the same scheme.

$\tilde{Q}$ is the closed K-orbit which corresponds to $\tilde{w} = w_1 w w_0 \in W_G$. So $\tilde{w}$ is the image of w under the natural involution. $\tilde{\mathcal{L}}$ is the line bundle on X given by the weight $w_0\lambda - 2\rho$. Thus $\tilde{\mathcal{L}}|_{\tilde{Q}}$ is given by $w_1 w w_0(w_0\lambda - 2\rho) = w_1 w \lambda + 2 w_1 w \rho$. We denote the normal and conormal bundles of Q and $\tilde{Q}$ by $\mathcal{N}, \tilde{\mathcal{N}}$ and $\mathcal{N}^*, \tilde{\mathcal{N}}^*$. Theorem 4 then follows from

Theorem 4.2 $H^s(Q, \mathcal{L} \otimes \mathcal{N}^{*(k)}) \cong H^0(\tilde{Q}, \tilde{\mathcal{L}} \otimes \tilde{\mathcal{N}}^{(k)} \otimes \det \tilde{\mathcal{N}})$ *where the superscript* (k) *indicates the* k^{th} *symmetric power.*

which is itself a straightforward application of the Bott-Borel-Weil therorem.

The twistor transform

$$\mathcal{T} : H^s(Q^\infty, \mathcal{L}) \overset{\cong}{\longrightarrow} H^{d-1}(X - \tilde{Q}, \tilde{\mathcal{L}})$$

arises by using the isomorphism [19] (in the algebraic category)

$$H^{d-1}(X - \tilde{Q}, \tilde{\mathcal{L}}) \cong H^d_{[\tilde{Q}]}(X, \tilde{\mathcal{L}}).$$

We now get a pairing as follows. If $\Psi \in H^{d-1}(X - \tilde{Q}, \tilde{\mathcal{L}})$, then the conjugation on X defines $\overline{\Psi} \in H^d(X - Q, \overline{\tilde{\mathcal{L}}})$ where $\overline{\tilde{\mathcal{L}}}$ is the line bundle on X given by weight $-w_0(w_0\lambda - 2\rho) = -\lambda - 2\rho$. The weight -2ρ gives rise to the canonical bundle Ω^n on X. Thus, $\overline{\tilde{\mathcal{L}}} = \mathcal{L}^* \otimes \Omega^n$ and we may use the connecting homomorphism δ of the Mayer-Vietoris sequence to form:

$$\begin{aligned}
\langle \phi, \psi \rangle &= \delta(\phi \cup \overline{T\psi}) \\
&\in H^n(X, \Omega^n) \\
&= \mathbb{C}
\end{aligned}$$

One now needs to check that the pairing provides a positive definite inner product. As in Example A, this should be done along the lines of [16].

Acknowledgement

This work was supported by a grant from the Australian Research Council.

References

[1] E. S. Abers and B. W. Lee, Gauge theories, *Phys. Rep.* **9C** (1973), 1-141.

[2] Y. Aharonov and D. Bohm, Significance of electromagnetic potentials in the quantum theory, *Phys. Rev.* **115** (1959), 485-491.

[3] M. F. Atiyah and R. S. Ward, Instantons and algebraic geometry, *Comm. Math. Phys.* **55** (1977), 111-124.

[4] T. N. Bailey, L. Ehrenpreis and R. O. Wells, Jr., Weak solutions of the massless field equations, *Proc. Roy. Soc. London* **A384** (1982), 403-425.

[5] R. J. Baston, Local cohomology, elementary states and evaluation, *Twistor Newsletter* **22** (1986), 8-13.

[6] R. J. Baston and M. G. Eastwood, *The Penrose Transform: its interaction with representation theory*, Oxford University Press, 1989.

[7] I. N. Bernstein, I. M. Gelfand and S. I. Gelfand, Differential operators on the base affine space and a study of g-modules, in *Lie Groups and Their Representations*, ed. I. M. Gelfand. Adam Hilgar, 1975, pp. 21-64.

[8] H. J. Bernstein and A. V. Phillips, Fiber bundles and quantum theory, *Scientific American* **245** (1981), 94-109.

[9] E. G. Dunne and M. G. Eastwood, A twistor transform for the discrete series: the case of SU(1,2), *Twistor Newsletter* **26** (1988), 26-30.

[10] E. G. Dunne and M. G. Eastwood, A twistor transform for the discrete series, in preparation.

[11] M. G. Eastwood, The generalized twistor transform and unitary representations of $SU(p,q)$, preprint, Oxford University 1983.

[12] M. G. Eastwood, A duality for homogeneous bundles on twistor space, *J. Lond. Math. Soc.* **31**, (1985), 349-356.

[13] M. G. Eastwood, On Michael Murray's twistor correspondence, *Twistor Newsletter* **19** (1985), 24-25.

[14] M. G. Eastwood and M. L. Ginsberg, Duality in twistor theory, *Duke Math. J.* **48** (1981), 177-196.

[15] M. G. Eastwood, R. Penrose and R. O. Wells, Jr., Cohomology and massless fields, *Comm. Math. Phys.* **78** (1981), 305-351.

[16] M. G. Eastwood and A. M. Pilato, On the density of twistor elementary states, preprint, Adelaide University 1988.

[17] T. J. Enright, R. Howe and N. R. Wallach, Classification of unitary highest weight modules, in *Representation Theory of Reductive Groups: Proceedings of the University of Utah Conference, 1982*, Progress in Mathematics **40**, Birkäuser, 1983.

[18] M. Fierz, Über die relativistische Theorie kräftefreier Teilchen mit beliebigem Spin, *Helv. Phys. Acta* **12** (1939), 3-37.

[19] R. Hartshorne, *Residues and Duality*, Lecture Notes in Mathematics **20**, Springer-Verlag, 1966.

[20] F. R. Harvey, *Spinors and Calibrations*, Perspectives in Mathematics, Academic Press, 1990.

[21] S. Helgason, *The Radon Transform*, Progress in Mathematics **5**, Birkhäuser, 1980.

[22] S. Helgason, *Groups and Geometric Analysis*, Academic Press, 1984.

[23] S. A. Huggett and K. P. Tod, *An Introduction to Twistor Theory*, London Mathematical Society Student Texts, 4, Cambridge University Press, 1985.

[24] L. P. Hughston, *Twistors and Particles*, Lecture Notes in Physics **97**, Springer-Verlag, 1979.

[25] L. P. Hughston and R. S. Ward, (eds.) *Advances in Twistor Theory*, Research Notes in Mathematics **37**, Pitman, 1979.

[26] J. E. Humphreys, *Introduction to Lie Algebras and Representation Theory*, Graduate Texts in Mathematics **9**, Springer-Verlag, 1972.

[27] H. P. Jakobsen, Hermitian symmetric spaces and their unitary highest weight modules, *J. Func. An.* **52** (1983), 385-412.

[28] H. P. Jakobsen and M. Vergne, Wave and Dirac operators, and representations of the conformal group, *J. Func. An.* **24** (1977), 52-106.

[29] A. W. Knapp, *Representation Theory of Semisimple Groups*, Princeton University Press, 1986.

[30] A. W. Knapp and E. M. Stein, Intertwining operators for semisimple groups, *Ann. of Math.* **93** (1971), 489-578; Intertwining operators for semisimple groups II, *Invent. Math.* **60** (1980), 9-84.

[31] L. A. Mantini, An integral transform in L^2-cohomology for the ladder representations of $U(p,q)$, *J. Funct. Anal.* **60** (1985), 211-242.

[32] L. A. Mantini, An L^2-cohomology construction of negative spin mass zero equations for $U(p,q)$, *J. Math. Anal. Appl.* **136** (1988), 419-449.

[33] M. Murray, A twistor correspondence for homogeneous polynomial differential operators, Math. Ann. **272** (1985), 99-115.

[34] R. Penrose, Twistor algebra, *J. Math. Phys.* **8** (1967), 345-366.

[35] R. Penrose, Structure of space-time, in *Battelle Rencontres: 1967, Lectures in Mathematics and Physics*, ed. C. M. DeWitt and J. A. Wheeler, Benjamin, 1968.

[36] R. Penrose and M. A. H. MacCallum, Twistor theory: an approach to the quantization of fields and space-time, *Phys. Repts.* **6C** (1972), 241-315.

[37] J. Rawnsley, W. Schmid and J. A. Wolf, Singular unitary representations and indefinite harmonic theory, *J. Func. An.* **51** (1983), 1-114.

[38] W. Schmid, *Homogeneous complex manifolds and representations of semisimple Lie groups*, Ph.D. Thesis, University of California, Berkeley, 1967.

[39] W. Schmid, Homogeneous complex manifolds and representations of semisimple Lie groups, *Proc. Nat. Acad. Sci. USA* **59** (1968), 56-59.

[40] W. Schmid, On a conjecture of Langlands, *Ann. of Math.* **93** (1971), 1-42.

[41] W. Schmid, Some properties of square-integrable representations of semisimple Lie groups, *Ann. of Math.* **102** (1975), 535-564.

[42] W. Schmid, L^2-cohomology and the discrete series, *Ann. of Math.* **103** (1976), 375-394.

[43] W. Schmid, Representations of semisimple Lie groups, Proceedings of the International Congress of Mathematics, Helsinki (1978), 195-207.

[44] M. A. Singer, Duality in twistor theory without Minkowski space, *Math. Proc. Camb. Phil. Soc.* **98** (1985), 591-600.

[45] C. H. Taubes, Physical and mathematical applications of gauge theories, *A.M.S. Notices* **33** (1986), 707-715.

[46] D. A. Vogan, *Representations of Real Reductive Lie Groups*, Progress in Mathematics **15**, Birkhäuser, 1981.

[47] R. Ward, The twisted photon: massless fields as bundles, Advances in Twistor Theory, 132-135.

[48] H. Weyl, *The Classical Groups, Second Edition* , Princeton University Press, 1946.

[49] R. Zierau, Geometric construction of certain highest weight modules, *Proc. A.M.S.* **95** (1985), 631-635.

[50] Notation for the Penrose transform, (this volume).

Invariant Operators

R.J. Baston   M.G. Eastwood

1   Introduction

Twistor theory, in both its flat and curved forms, [49] is principally a *conformally invariant* theory. This means that the associated differential geometry is somewhat weaker than the usual Riemannian geometry—the notion of distance is sacrificed whilst that of angle is preserved. This has certain advantages and disadvantages. To its credit, the weaker structure allows for greater generality—in two dimensions, for example, the notion of a Riemann surface is probably more useful than that of Riemannian geometry. Also, one may take a two step approach to problems in Riemannian geometry via conformal geometry. For instance one has the non-linear graviton construction [46] and the construction outlined in LeBrun's article in this volume whereby one first deals with a conformal manifold and then imposes additional structure to obtain Einstein metrics. A disadvantage is that conformal geometry is generally a good deal harder to work with; for instance, one no longer has the distinguished Levi Civita connection, its curvature, and the associated invariants. Instead one has to make do with the local twistor or Cartan conformal connection. This is not defined on the tangent bundle of the manifold but rather on an auxiliary bundle—the local twistor bundle. So at first sight it is not nearly as useful as the Levi Civita connection. It turns out, however, that one can use this connection and its curvature to obtain a large collection of tensor and differential invariants of a conformal structure which is more intriguing than its Riemannian counterpart and potentially more useful [17, 18, 29]. All of this boils down to differences in the algebraic properties of the orthogonal and conformal groups and, in particular, to the fact that the latter has a one dimensional center whilst the former is semisimple.

In this article we aim to sketch a little of the theory which has been developed to investigate conformal invariants. It is based on simple linear algebraic considerations for flat conformal manifolds yet it yields results on fully curved conformal manifolds. Thus, for many invariant linear differential operators on a conformally flat space the theory produces an invariant operator on curved conformal manifolds with the same symbol by adding in lower order curvature correction terms. These correction terms involve only Ricci curvature and its derivatives. Indeed, there is a basic class of invariant operators which are obtained from the exterior differential on forms by helicity raising and lowering and, for these, only trace-free Ricci curvature correction terms appear. Details of this theory may be found in [26, 6, 5, 32, 33, 34, 39].

We should admit here that in an earlier draft of this article we believed that *all* flat invariant operators had curved analogues. In dimension 4 this is what [26] purports to prove. There is, however, an error recently discovered by Graham and Mason. We are especially grateful to Robin Graham for many communications on the subject culminating in his finding a counterexample to the general existence of curved analogues (theorem 4.1 below). The error in [26] is not the disaster it might first appear. Its elimination simply eliminates a particular family of flat invariant operators which need no longer have curved analogues. In other words, the general method of [26] is correct but its implementation fails in one particular family. This still leaves most of the flat operators with curved analogues. We conjecture that this result is sharp, i.e. the operators which cannot be reached by the method of [26] have no curved analogues. There are good reasons, based on [30], as to why this conjecture should hold (the reasons for arousing suspicions of [26] in the first place). In odd dimensions all invariant operators have curved analogues. A precise explanation of all these matters appears later in this article (see §4).

The Penrose transform [4, 9, 24, 25] is itself a conformal construction. Anything the transform produces will be conformally invariant. Of course, the main output of the transform is a collection of differential operators and these must be invariant. It is remarkable that (in the flat case) *every* such operator arises from an instance of the transform. This follows from the classification of such operators in terms of homomorphisms of Verma modules [26] which are all known in the conformal case (see, for example, [11] in the *regular* case and subsection 3.3 below). On the other hand it means that one can hope that the transform itself will lead to all invariant operators for other homogeneous spaces; there is some evidence that this is true (see examples in [9, 23]) and such considerations give a second major contribution of twistor theory to representation theory (after the construction of unitary representations—see 'The Twistor Transform' in this volume). Also, *curved* versions of the Penrose transform exist [2, 5, 21] and these produce curved invariant operators directly. This leads to the best means of computing such operators known at present (see [33] and §4.2).

We should point out that the study of conformally invariant operators has a long history. The conformal invariance of Maxwell's equations and the (flat) Laplacian was known to Bateman [10] at the start of the century; the conformally invariant Laplacian $\Delta + R/6$ on a curved space goes back almost as far. Other systematic investigations have been carried out in recent years [13, 14, 15, 37, 45, 50, 57]. The topic was very popular in the 1920's especially in the works of É. Cartan, S. Sasaki, T.Y. Thomas, and O. Veblen. A summary and extensive references can be found in Schouten [51, esp. VI §§5–7].

2 Definitions and Examples

A conformal manifold M is a smooth oriented manifold with a metric defined up to scale. That is, there is a distinguished ray subbundle of the bundle $\otimes^2 T^*M$ which is everywhere positive definite in the sense that any non-zero element gives a positive definite quadratic form[1]. If one chooses local coordinates x^a and y^a compatible with the given orientation, then on the overlap the Jacobian $\det(\partial y^a / \partial x^b)$ is positive and so has an nth root. Here and throughout this article, n is the dimension of M. Using these n^{th} roots as transition functions gives a line bundle whose dual we shall denote by L. A density of weight w on M is a smooth section of L^w. Evidently a density of weight $-n$ is a volume form. A choice of metric g_{ab} in the conformal class gives rise to an associated volume form and so to a trivialization of L. This allows one to regard densities as functions. Then, if g_{ab} is replaced by $\hat{g}_{ab} = \Omega^2 g_{ab}$ for some smooth function $\Omega > 0$, the associated volume form scales by Ω^n. Hence, a density of weight w may be viewed as a function f which is replaced by $\hat{f} = \Omega^w f$ upon rescaling the metric and we shall usually adopt this viewpoint below[2]. Densities in this context are sometimes called *conformal densities*. The construction of conformally invariant objects (e.g. tensors or operators) is often accomplished by giving a definition in terms of a choice of metric in the conformal class. One must then check that as the metric is rescaled the object is unchanged.

Example 2.1 The simplest example of a conformally invariant tensor is the totally trace-free part C_{abcd} of the Riemann curvature R_{abcd}. This is known as the Weyl tensor. Our convention for the Riemann curvature is:

$$R_{abc}{}^d X^c = (\nabla_a \nabla_b - \nabla_b \nabla_a) X^d.$$

To see that C_{abcd} is invariant we first compute (see e.g. [47]) that under a rescaling of the metric the Levi Civita connection changes according to

$$\hat{\nabla}_a X^b = \nabla_a X^b + \Upsilon_a X^b - \Upsilon^b X_a + \Upsilon_c X^c \delta_a^b$$

on (unweighted) vectors and

$$\hat{\nabla}_a X_b = \nabla_a X_b - \Upsilon_a X_b - \Upsilon_b X_a + \Upsilon^c X_c g_{ab} \tag{1}$$

---

[1]Most of our discussion works equally well for metrics of arbitrary signature as well as for complex Riemannian manifolds (e.g. [42])—the usual setting for twistor theory. The obvious changes must be made: in the Lorentzian case one must replace the Laplacian by the wave operator and so on. We shall leave the necessary modifications to the interested reader.

[2]This agrees with [47, 48] but is at variance with the conventions of [3, 5, 21, 26, 32, 33] where densities are always regarded as sections of a power of L. The effect is that factors of Ω^w appear explicitly in our formulae.

on forms where $\Upsilon_a \equiv \Omega^{-1}\nabla_a\Omega$. A short calculation now shows that the Riemann tensor varies by

$$\hat{R}_{abcd} = \Omega^2(R_{abcd} + \Xi_{ac}g_{bd} - \Xi_{bc}g_{ad} + \Xi_{bd}g_{ac} - \Xi_{ad}g_{bc}) \tag{2}$$

where

$$\Xi_{ab} \equiv \nabla_a\Upsilon_b - \Upsilon_a\Upsilon_b + \tfrac{1}{2}\Upsilon_c\Upsilon^c g_{ab}.$$

In particular, the variation is entirely through traces and C_{abcd} is therefore invariant.

The Weyl curvature (except in dimension 4) is an $SO(n)$-irreducible part of the full Riemann tensor. Equation (2) suggests that the remaining part $R_{abcd} - C_{abcd}$ is best written in terms of a symmetric tensor P_{ab} defined by

$$C_{abcd} = R_{abcd} + P_{ac}g_{bd} - P_{bc}g_{ad} + P_{bd}g_{ac} - P_{ad}g_{bc} \tag{3}$$

whence equation (2) may be rewritten as

$$\hat{P}_{ab} = P_{ab} - \nabla_a\Upsilon_b + \Upsilon_a\Upsilon_b - \tfrac{1}{2}\Upsilon_c\Upsilon^c g_{ab}. \tag{4}$$

This tensor occurs time and again in conformal geometry. Its above definition and its particularly simple transformation law indicate why this should be so. Tracing equation (3) over the indices b and d shows that

$$P_{ac} = \frac{1}{2-n}\left(R_{ac} + \frac{R}{2(1-n)}g_{ac}\right)$$

where $R_{ac} \equiv R_{abc}{}^b$ is the usual Ricci curvature[3].

Example 2.2 A second invariant is the conformally invariant Laplacian (or wave operator, depending on signature) acting on a density f of weight $\frac{2-n}{2}$ (recall dim $M = $ n):

$$f \longmapsto \nabla^a\nabla_a f + \tfrac{2-n}{2}Pf$$

where P is the trace of P_{ab} and ∇_a the Levi Civita connection both taken with respect to a choice of metric in the conformal class[4].

Using equation (4), we see that

$$\hat{P} = \Omega^{-2}\left(P - \nabla^a\Upsilon_a + \tfrac{2-n}{2}\Upsilon^a\Upsilon_a\right)$$

and, using equation (1),

$$\hat{\nabla}_a\hat{f} = \Omega^{\frac{2-n}{2}}\left(\nabla_a f + \tfrac{2-n}{2}\Upsilon_a f\right)$$
$$\Rightarrow \quad \hat{\nabla}^a\hat{\nabla}_a\hat{f} = \Omega^{\frac{-2-n}{2}}\left(\nabla^a\nabla_a f + \tfrac{2-n}{2}(\nabla^a\Upsilon_a - \tfrac{2-n}{2}\Upsilon^a\Upsilon_a)f\right).$$

---

[3]We follow [47] in our sign convention for the Ricci tensor which is the reverse of the usual in Riemannian geometry.

[4]The scalar curvature $R \equiv R_{ab}{}^{ab}$ is equal to $2(1-n)P$ and so this operator can be rewritten in the more familiar form $\nabla^a\nabla_a f + \frac{n-2}{4(n-1)}Rf$. It is often known as the Yamabe operator [43].

Combining these observations gives:

$$\hat{\nabla}^a\hat{\nabla}_a\hat{f} + \tfrac{2-n}{2}\hat{P}\hat{f} = \hat{\nabla}^a\hat{\nabla}_a\hat{f} + \tfrac{2-n}{2}\hat{P}\Omega^{\frac{2-n}{2}}f$$
$$= \Omega^{\frac{-2-n}{2}}\left(\nabla^a\nabla_a f + \tfrac{2-n}{2}(P - \Omega^2\hat{P})f\right) + \tfrac{2-n}{2}\hat{P}\Omega^{\frac{2-n}{2}}f$$
$$= \Omega^{\frac{-2-n}{2}}\left(\nabla^a\nabla_a f + \tfrac{2-n}{2}Pf\right)$$

Thus we obtain an invariant operator by adding an appropriate multiple of the scalar curvature to the Laplacian, provided we regard the result as a density of weight $(-n-2)/2$. The invariance of the Laplacian in the flat case can be proved more efficiently by other means (indicated in §3 below). This rather fortunate cancellation in the curved case is typical and suggests the existence of some underlying principal allowing one to pass from flat invariant operators to curved ones.

The four dimensional case $\nabla^a\nabla_a - P = \nabla^a\nabla_a + R/6$ acting on conformal densities of weight -1 is especially important physically and the two dimensional case is just

$$\frac{1}{4}\frac{\partial^2}{\partial z\partial\bar{z}}$$

acting on functions; here the metric is $dzd\bar{z} = dx^2 + dy^2$ for $z = x + iy$.

Example 2.3 Physically, some of the most interesting conformally invariant operators arise on *spinor* fields. Notice that the concept of a spinor itself is conformally invariant. In four dimensions, for a given metric g_{ab} in the conformal class we may always find skew spinors ϵ_{AB} and $\epsilon_{A'B'}$ so that

$$g_{ab} = \epsilon_{AB}\epsilon_{A'B}$$

and these rescale equally:

$$\epsilon_{AB} \mapsto \hat{\epsilon}_{AB} = \Omega\epsilon_{AB} \quad \text{and} \quad \epsilon_{A'B'} \mapsto \hat{\epsilon}_{A'B'} = \Omega\epsilon_{A'B'}.$$

(Those unfamiliar with this two-spinor calculus should consult [47].)

Then, one has a sequence of conformally invariant massless field equations beginning with the Dirac-Weyl neutrino equation

$$\nabla^A_{A'}\phi_A = 0.$$

Here ϕ_A is a spinor-valued density of weight -1, so that under rescaling one replaces ϕ_A by $\hat{\phi}_A = \Omega^{-1}\phi_A$. On unweighted spinors one checks that the Levi Civita connection varies by

$$\hat{\nabla}_{AA'}\tau_B = \nabla_{AA'}\tau_B - \Upsilon_{BA'}\tau_A$$

and so

$$\hat{\nabla}^A_{A'}\hat{\phi}_A = \hat{\epsilon}^{BA}\hat{\nabla}_{AA'}\hat{\phi}_B$$
$$= \Omega^{-1}\nabla^A_{A'}(\Omega^{-1}\phi_A) + \Omega^{-2}\Upsilon^A_{A'}\phi_A$$
$$= \Omega^{-2}\nabla^A_{A'}\phi_A$$

whence the equation is conformally invariant if we regard $\nabla^A_{A'}\phi_A$ as a spinor-valued density of weight -2.

A similar calculation shows that the equations

$$\nabla^A_{A'}\phi_{AB...C} = 0 \quad \text{and} \quad \nabla^{A'}_A\phi_{A'B'...C'} = 0$$

acting on arbitrary symmetric spinor fields of conformal weight -1 are similarly invariant. Notice the essential role played by conformally weighted densities in all these calculations. Notice also that on any manifold whose tangent bundle can be factored as a tensor product

$$TM \cong S \otimes S' \tag{5}$$

we could hope that these calculations would produce invariant operators, provided we could make sense of ∇ and ϵ, ϵ'. This is indeed the case—the details appear in [2] where the structure (5) is called *paraconformal*.

Conformally invariant connections

The most basic invariant operator is the *local twistor connection* mentioned above. This is easily constructed in any dimension as follows (see [3] for more details). Firstly we specify the bundle[5] on which it should act. Sections of this bundle can be represented as triples

$$Z = (\sigma, \mu_a, \rho)$$

where σ, ρ are densities of weight 1 and -1 respectively and μ_a is a one-form of conformal weight 1. Thus, σ is a section of L, ρ of L^*, and μ_a of $T^*M \otimes L$. Such a representation depends on a choice of metric within the conformal class, however, and when we rescale the metric the representation varies:

$$\begin{pmatrix} \hat{\sigma} \\ \hat{\mu}_a \\ \hat{\rho} \end{pmatrix} = \begin{pmatrix} \Omega\sigma \\ \Omega(\mu_a + \Upsilon_a\sigma) \\ \Omega^{-1}(\rho + \Upsilon^a\mu_a + \frac{1}{2}\Upsilon^a\Upsilon_a\sigma) \end{pmatrix}.$$

One can easily check that this definition is consistent and hence defines a vector bundle (of rank $n + 2$) which we shall call the *local twistor bundle* and denote by $\mathcal{T}$. As shown in [3], it can be identified with the subbundle of the second jet bundle of L determined by the equation

$$\text{trace-free part of } (\nabla_a\nabla_b + P_{ab})\sigma = 0 \tag{6}$$

using P_{ab} defined by equation (3) above.

---

[5]In four dimensions, this is $\bigwedge^2$ of the usual local twistor bundle which is based on spinors [48]. The form we give generalizes most easily to arbitrary dimensions. See [7] for the generalization of the spinor form.

$\mathcal{T}$ was first introduced by Cartan [20] and Thomas [53] who also observed that it carried an invariant connection. It is given by the formula

$$\nabla_a \begin{pmatrix} \sigma \\ \mu_b \\ \rho \end{pmatrix} = \begin{pmatrix} \nabla_a \sigma - \mu_a \\ \nabla_a \mu_b + P_{ab}\sigma - g_{ab}\rho \\ \nabla_a \rho + P_{ab}\mu^b \end{pmatrix}$$

and one can verify directly that this is well-defined (i.e. independent of choice of g_{ab}). Alternatively, one can derive[6] this formula from equation (6), a conformally invariant equation as one can readily check. The procedure is to rewrite (6) as a pair of equations

$$\nabla_a \sigma - \mu_a = 0 \tag{7}$$

$$\nabla_a \mu_b + P_{ab}\sigma - g_{ab}\rho = 0. \tag{8}$$

Differentiating the second once more, substituting from the first, contracting and using the Bianchi identity we obtain

$$\nabla_a \rho + P_{ab}\mu^b = 0. \tag{9}$$

In general, of course, these three equations (7,8,9) are overdetermined and have no non-trivial solutions. Along any curve within M, however, we may regard them as a system of first order ordinary differential equations and use these to propagate Z along the curve. This gives the notion of parallel transport as above.

3   The Flat Theory

Conformal invariance can mean at least two things. On a general conformal manifold it refers to a construction which depends only on a given conformal structure, and not on any particular choices, such as a metric within the conformal class. We can be more precise by saying that a local conformal invariant (tensor or linear differential operator) should depend polynomially on finitely many derivatives of the structure at any point.

In the conformally flat case, however, we can proceed differently. For there is a *flat model* provided by the usual round sphere S^n, regarded as a conformal manifold. The global conformal transformations are induced by an action of $SO(1, n + 1)$ (by regarding the sphere as the space of null lines in the light cone of the origin in $\mathbf{R}^{1,n+1}$). Thus, one should regard S^n as a

---

[6]This argument parallels the four dimensional case [48] where one can start with the *twistor equation*. This was Penrose's original construction of *local twistor transport* only later identified [31] with Cartan's connection (a confusion which dates back as far as 1925 [53]). Since the twistor equation is first order, not involving curvature, it is perhaps more basic. In particular, the equation is directly deduced from the conformal structure. The calculation in [7] generalizes this to arbitrary dimensions.

homogeneous space for $SO(1, n+1)$ and ask for differential operators which are invariant under the action of this group. It is reasonable to call these the *conformally invariant operators* and, indeed, for $n \geq 3$ operators invariant in this sense precisely coincide with operators invariant under rescaling, by dint of Liouville's theorem [26]. Hence, we should first classify the invariant operators on S^n. We also need to give a description of tensor and related bundles on S^n to see how $SO(1, n+1)$ should act on sections of these, so that we can make sense of invariant operators on tensors, spinors, etc..

Firstly, we must view S^n as a homogeneous space. Clearly, any null line through the origin in $\mathbf{R}^{1,n+1}$ can be mapped to any other under $SO(1, n+1)$. So the action of $SO(1, n+1)$ on S^n is transitive. The stabilizer P of a particular line is easy to describe. Suppose we take the bilinear form preserved by $SO(1, n+1)$ to be

$$
n\left\{\begin{pmatrix}
0 & 0 & 1 \\
0 & \mathbf{1} & 0 \\
1 & 0 & 0
\end{pmatrix}\right.
$$

and let us choose to stabilize the line spanned by the null vector

$$
\begin{pmatrix}
1 \\ 0 \\ \vdots \\ 0
\end{pmatrix}.
$$

Clearly, this line is stabilized by precisely those matrices of the form

$$
\begin{pmatrix}
* & * & * \\
0 & A & * \\
0 & 0 & *
\end{pmatrix}
$$

where $A \in SO(n)$. Hence, P is all such matrices. The whole group of conformal motions is then generated by P and the Abelian subgroup

$$
Q = \left\{\begin{pmatrix}
1 & 0 & 0 \\
y & I & 0 \\
-\tfrac{1}{2}y^t y & -y^t & 1
\end{pmatrix}\right\} = \left\{\exp\begin{pmatrix}
0 & 0 & 0 \\
y & 0 & 0 \\
0 & -y^t & 0
\end{pmatrix}\right\}.
$$

We can use this subgroup to identify $\mathbf{R}^n \hookrightarrow S^n$:

$$
\mathbf{R}^n \ni \begin{pmatrix}
y_1 \\ y_2 \\ \vdots \\ y_n
\end{pmatrix} = y \mapsto \exp\begin{pmatrix}
0 & 0 & 0 \\
y & 0 & 0 \\
0 & -y^t & 0
\end{pmatrix}\begin{pmatrix}
1 \\ 0 \\ \vdots \\ 0
\end{pmatrix} = \begin{pmatrix}
1 \\ y \\ -\tfrac{1}{2}y^t y
\end{pmatrix}.
$$

This provides the usual flat coördinates corresponding to stereographic projection. The action of Q corresponds to translation in $\mathbf{R}^n$. In particular, the Q-invariant linear differential operators on functions can be naturally identified with the polynomial ring

$$\mathbf{C}[\partial^1, \partial^2, \ldots, \partial^n]$$

where $\partial^i = \partial/\partial y_i$. Our task is now to identify those operators which are also invariant under P for these will be the fully $SO(1, n+1)$-invariant ones. Since P and Q do not commute, P does not preserve the Q-invariant operators. Nevertheless, P fixes the origin in $\mathbf{R}^n$ and so an operator in $\mathbf{C}[\partial^1, \partial^2, \ldots, \partial^n]$ is invariant if and only if the result of applying it to a function and evaluating at the origin is independent of conjugation by elements of P. The action of P gives rise to an action of the corresponding Lie algebra $\mathbf{p}$ by differential operators and a notion of infinitesimal invariance which is equivalent to genuine P-invariance. Specifically, for $D \in \mathbf{C}[\partial^1, \partial^2, \ldots, \partial^n]$ to be invariant it is necessary and sufficient that

$$[X, D]|_0 = 0$$

for any differential operator X arising from $\mathbf{p}$. These operators X are easily calculated and may be described as follows. An element of $\mathbf{p}$ may be written as:

$$\begin{pmatrix} \lambda & -x^t & 0 \\ 0 & A & x \\ 0 & 0 & -\lambda \end{pmatrix} \qquad \text{where } A^t = -A.$$

The various pieces give rise to:

$$\lambda \;\mapsto\; \lambda y_i \partial^i$$
$$A \;\mapsto\; a_{ij} y^i \partial^j$$
$$x_i \;\mapsto\; \tfrac{1}{2} x_i y_j y^j \partial^i - x_j y_i y^j \partial^i.$$

Consider, for example, the Laplacian $\Delta = \partial_k \partial^k$. In n dimensions:

$$[\lambda, \Delta] \;=\; -2\lambda\Delta$$
$$[A, \Delta] \;=\; -2a_{ij}\partial^i\partial^j = 0$$
$$[x, \Delta] \;=\; (2 - n)x_i \Delta + 2x_i y^i \Delta.$$

This shows that, as an operator from functions to functions, the Laplacian is *never* invariant. However, if $n = 2$, then the only equation preventing invariance is $[\lambda, \Delta] = -2\lambda\Delta$. Unraveling the meaning of '-2' in this equation shows that Δ is invariant provided Δf is interpreted as a density of weight -2. The details are beyond the scope of this review but as a brief indication we remark that the representation of $\mathbf{p}$ given by

$$\begin{pmatrix} \lambda & -x^t & 0 \\ 0 & A & x \\ 0 & 0 & -\lambda \end{pmatrix} \longmapsto w\lambda$$

induces on S^n the bundle of conformal densities of weight w (where induction is defined as in [12], for example). More generally, arbitrary tensor and spinor bundles on S^n are in 1-1 correspondence with the irreducible representations of $\mathbf{p}$. Moreover, the commutation computations as above and more generally may be carried out equivalently in the Lie algebra of $SO(1, n+1)$ rather than as differential operators. This reduces the classification of general $SO(1, n+1)$-invariant operators to an algebraic task in the representation theory of Lie algebras. The appropriate theory is known as the theory of *Verma modules* and its application in classifying invariant differential operators is discussed in [26] for the four dimensional case and in [5] for the general case. The corresponding results for Verma modules are in [11]. We shall end this section by describing these results.

3.1 Even Dimensions

One obtains invariant differential operators according to the following pattern:

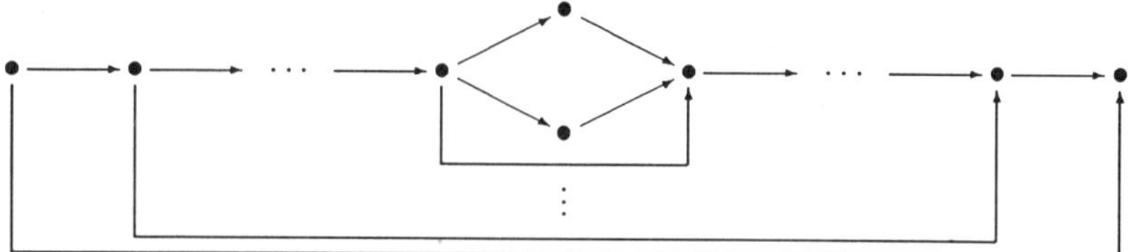

Here, each $\bullet$ represents an irreducible representation of $\mathbf{p}$ and, consequently, a weighted tensor or spinor bundle. Between two such bundles joined by an arrow there is exactly one invariant operator (up to overall scale) and there are no other invariant operators. The simplest example of this pattern occurs for the de Rham sequence where the two stacked $\bullet$'s denote the splitting of middle forms into self-dual and anti-self-dual parts. More generally, without the differential operators denoted by ⌊____⌉ the whole pattern is known as a *Bernstein-Gelfand-Gelfand resolution* and generalizes the de Rham resolution (see [9, 44]). In order to be more precise about what this means it is necessary to describe the irreducible representations of $\mathbf{p}$. This is straightforward but, in general, one is obliged to use the language of representation theory and this requires setting up a certain amount of notation. In [9] such notation is established[7] specifically with a view to describing these representations (amongst others). In terms of the Dynkin diagram notation of [9] the pattern is

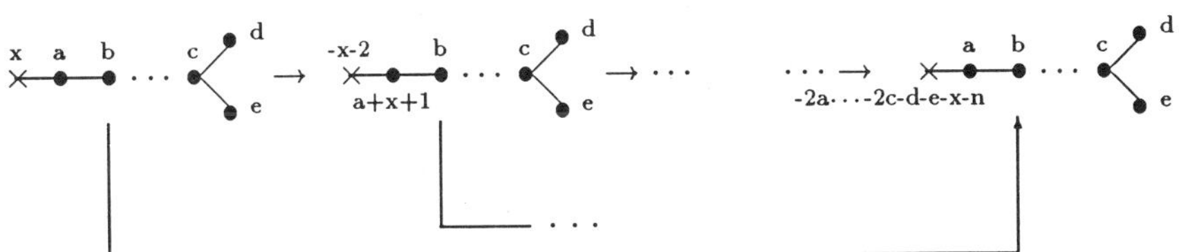

where $x, a, b, \ldots, e \geq -1$ and it is generated by the affine action of the Weyl group of $SO(n+2, \mathbb{C})$. The weight x (simply related to the conformal weight) is required to be integral (see [22]). The computation of this action is explained in [9] (see section 4.3 (especially 4.3.7) and chapter 8). Notice that if $a = -1$, then the first homogeneous bundle in this pattern does not make sense (whereas if $x, b, \ldots, e \geq 0$, then the second one is fine). In this case it is simply omitted from the pattern. More generally, if any of $x, a, b, \ldots, e$ is equal to -1, then there will be omissions (and also repetitions) in the pattern. This case is referred to as *singular*. In the non-singular case (i.e. if $x, a, b, \ldots, e \geq 0$) all bundles makes sense and are distinct[8]. The basic non-singular pattern (when $x = a = \cdots = e = 0$) gives the de Rham sequence [9, p. 34]. All bundles occur in precisely one of these patterns—to determine which one it suffices to act with the Weyl group until all coefficients are ≥ -1. In practise this can be tedious and it probably better to adapt the notation of, for example, [55] for representations of the orthogonal group—the action of the Weyl group should be manifest.

Whilst the language of representation theory is generally unavoidable in higher dimensions, one can rewrite in terms of spinors in dimension four. If $\mathcal{S}(d, e, w)$ denotes the bundle of spinors with d primed symmetric lower spinor indices, e unprimed symmetric lower spinor indices, and of conformal weight w, then the pattern of invariant operators is

---

[7]Notice that [9] is written in the context of *complex* homogeneous spaces. The notation for real homogeneous spaces (such as S^n) is essentially unchanged, the only difference being that, in describing the representations of P, the coefficient over the crossed node need no longer be integral. See also the note by Ed Dunne in this volume concerning notation for the Penrose transform.

[8]These facts follow easily from the geometric action of the Weyl group on the Weyl chambers and, in particular, as to whether weights lie on a wall. A complete explanation can be found in [9].

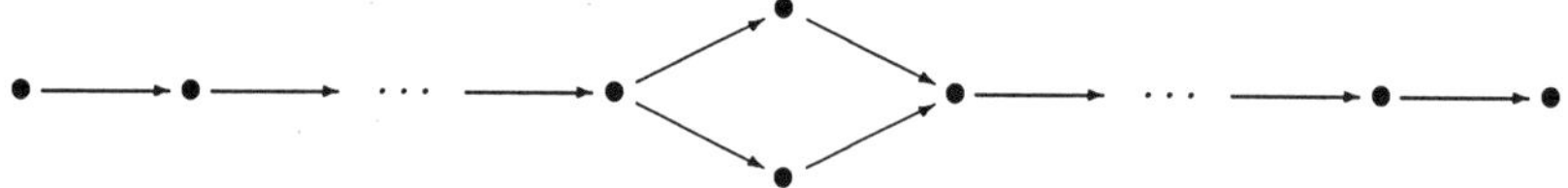

for $x, e, d \geq -1$.

There is another important concept concerned with these patterns.

Definition 3.1 *In each pattern, the operators*

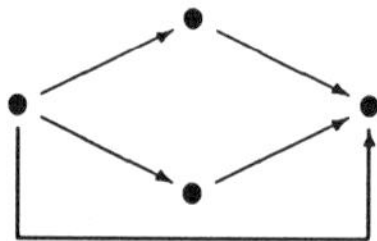

are known as standard *operators. The composition of any two standard operators is also known as standard. All other operators are known as non-standard[9].*

Some remarks concerning this definition are in order. The patterns of invariant operators are not commutative. The central diamond

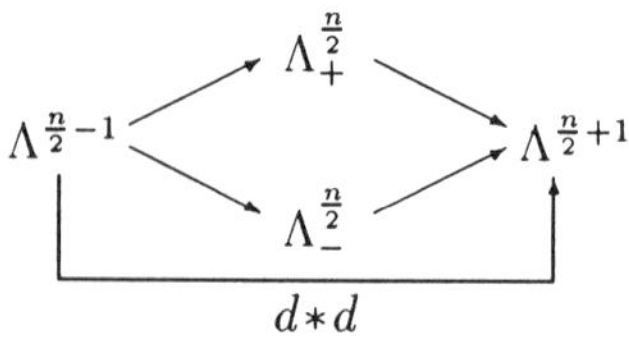

is commutative provided all four positions are non-zero. However, any other composition of standard operators is zero (cf. the BGG resolution in [9]). In the basic non-singular case ($x = a = \cdots = e = 0$), for example, the standard operators are the exterior derivatives with the central diamond

$$
\begin{array}{ccccc}
 & & \Lambda^{\frac{n}{2}}_{+} & & \\
 & \nearrow & & \searrow & \\
\Lambda^{\frac{n}{2}-1} & & & & \Lambda^{\frac{n}{2}+1} \\
 & \searrow & & \nearrow & \\
 & & \Lambda^{\frac{n}{2}}_{-} & & \\
\end{array}
$$
$$d * d$$

---

[9]The terminology is borrowed from the corresponding notions in the theory of Verma modules [44].

where Λ^p denotes the p-forms on M. The non-standard operators in this basic case map between

$$\Lambda^k \to \Lambda^{n-k} \quad \text{for } 0 \le k \le \tfrac{n}{2} - 2$$

and include

$$\Delta^{\frac{n}{2}} : \Lambda^0 \to \Lambda^n \quad \text{(in the flat metric)}$$

where Δ is the Laplacian.

There is another useful piece of terminology concerning the operators which, in the general pattern, map from the extreme left to the extreme right as does $\Delta^{\frac{n}{2}}$ in the basic case.

Definition 3.2 *The differential operators*

for $x, a, b, \ldots, e \ge 0$ are called long *operators*[10].

Notice that long operators are, by definition, confined to the non-singular case. In the singular case, $x = -1$, the first two members of the pattern coincide and so the same operator is repeated in a shorter position. It is, therefore, not genuinely regarded as long.

3.2 Odd Dimensions

In this case, the Bernstein-Gelfand-Gelfand resolutions follow the pattern of the de Rham resolution

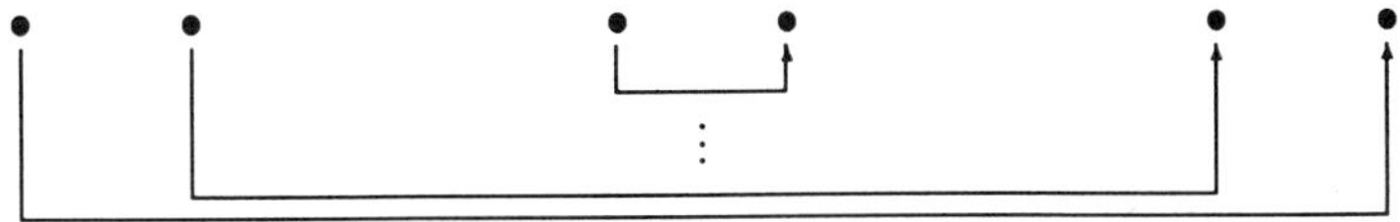

but do not admit additional non-standard operators. However, there are additional operators the patterns for which look like:

---

[10]Again, the terminology is borrowed from representation theory. These two representations are linked by the longest element of the Weyl group.

These operators always act between bundles with half-integral weight while the BGG operators act between bundles with integral weight. In the second pattern, the shortest operator is standard but all others are non-standard. There are no non-trivial operators between singular bundles. Again, the simplest of the non-standard operators are powers of the Laplacian.

Whilst the general case needs special notation, the three dimensional case can be explained in terms of spinors. We shall write $\mathcal{S}(e, w)$ to denote the bundle of spinors with e symmetric lower spinor indices and of conformal weight w. The first pattern of operators is then

$$\mathcal{S}(e, e+x) \longrightarrow \mathcal{S}(e+2x+2, e+x) \longrightarrow \mathcal{S}(e+2x+2, x-1) \longrightarrow \mathcal{S}(e, -x-3)$$

for $x, e \geq 0$ and integral. The second pattern is

$$\mathcal{S}(e, e+x) \qquad \mathcal{S}(e+2x+2, e+x) \qquad \mathcal{S}(e+2x+2, x-1) \qquad \mathcal{S}(e, -x-3)$$

for $x \geq -\frac{1}{2}$ and half-integral and $e \geq 0$ and integral. Typical examples from this second pattern are (in case $x = -\frac{1}{2}$ and $e = 0$)

$$\mathcal{S}(0, -\tfrac{1}{2}) \longrightarrow \mathcal{S}(0, -\tfrac{5}{2}) \quad \text{given by} \quad \phi \mapsto \Delta\phi$$
$$\mathcal{S}(1, -\tfrac{1}{2}) \longrightarrow \mathcal{S}(1, -\tfrac{3}{2}) \quad \text{given by} \quad \phi_A \mapsto \nabla^B_A \phi_B.$$

3.3   Detailed classification

The classification of homomorphisms of Verma modules in [11] is in the *non-singular* case, both integral and non-integral. The *singular* case can be deduced from this by translation. Precisely, careful analysis shows that translation from non-singular to singular *covers* all homomorphisms of Verma modules. Some may translate to zero, but all nonzero operators can be obtained by translation. Actually, the non-integral cases are very simple. It follows from Jantzen's criterion [40] that in the even dimensional case all non-integral Verma modules are *irreducible* and so there are *no* non-trivial invariant operators between bundles with fractional conformal weight. In the odd dimensional case, all Verma modules which are not (half-)integral are likewise irreducible. Those which have half-integral conformal weight give rise to precisely one invariant operator [11, §1.2]—these include the invariant Laplacians.

4 The Curved Case

If one checks various cases by hand, at first it appears that it is always possible to begin with a flat invariant operator and obtain a *curved analogue* by adding suitable *curvature correction terms* as lower order operators. We have already seen this in example 2.2 where the variation in $\nabla^a \nabla_a$ under conformal rescaling on suitably weighted densities corresponds exactly to the variation of the scalar curvature. It is evident from the calculation we made that the operator must act on a density of precisely the right weight to be invariant. It is rather remarkable that (for $n \geq 4$ and even) by increasing the order we can find a scalar invariant operator on *functions*. For instance, in four dimensions the square of the Laplacian is invariant (on the sphere) and it has a curved generalization given by [27, 57]

$$f \to \nabla_b [\nabla^b \nabla^a - 4P^{ab} - 2Pg^{ab}] \nabla_a f$$

with values in the four-forms, identified with densities of weight -4. This is easily verified by direct calculation where one finds that

$$
\begin{aligned}
\hat{\nabla}_b [\hat{\nabla}^b \hat{\nabla}^a - 4\hat{P}^{ab} - 2\hat{P}\hat{g}^{ab}] \hat{\nabla}_a \hat{f} &= \Omega^{-4} \nabla_b [\nabla^b \nabla^a - 4P^{ab} - 2Pg^{ab}] \nabla_a f \\
&\quad + 2\Omega^{-4} \Upsilon^b \nabla^a [\nabla_a \nabla_b - \nabla_b \nabla_a] f \\
&= \Omega^{-4} \nabla_b [\nabla^b \nabla^a - 4P^{ab} - 2Pg^{ab}] \nabla_a f
\end{aligned}
\tag{10}
$$

as required. As we shall see, this operator seems to be on the boundary of those flat operators which admit curved analogues. The existence of a curved analogue to $\Delta^{\frac{n}{2}}$ is quite subtle. Though it is amenable to a more delicate version of the 'curved translation principle' to be discussed below, undoubtedly the best argument is due to Graham, Jenne, Mason, and Sparling [34] who use the Fefferman-Graham ambient metric construction [30]. It would take us too far afield to discuss the details of this construction here (though we come back to it at the end of section 5). Suffice it to say that there is an obstruction to the existence of the ambient metric construction beyond a certain order (it is given in dimension four by the Bach tensor). The argument of [34] uses the ambient metric construction right up to this order!

In the flat case, the operator

$$\Delta^{\frac{n}{2}+1} : L \longrightarrow L^{-n-1}$$

is also invariant (as are all higher powers of the Laplacian when acting on densities of an appropriate weight). However, the argument of [34] breaks down for this operator since the ambient metric is required to a higher order than it actually exists. This is what led Graham and Mason to doubt the existence of a curved analogue of Δ^3 in dimension four.

Theorem 4.1 (Graham [35]) *The cube of the Laplacian in dimension four has no curved analogue.*

This is probably the simplest non-existence result but its proof is quite formidable—twenty nine pages of careful calculation eliminate all possible curvature correction terms! Such computations approach the boundary of human capabilities. In general a better proof is needed.

4.1 The Curved Translation Principle

Fortunately, the proof of the existence of curved analogues, though constructive, is largely free from computation. We should remark that the method of construction (the curved translation principle) is certainly not an effective method of generating formulae for these curved analogues. The computations rapidly get out of hand. In many cases this is remedied by Gover's method in 4.2 below. We shall start with some examples before describing the general construction[11]

The main tool is the local twistor bundle $\mathcal{T}$ and its associated connection which we introduced earlier. Recall that a section of this bundle is a collection

$$\mathbf{f} = \begin{pmatrix} \sigma \\ \mu_a \\ \rho \end{pmatrix}$$

varying according to a certain rule under conformal rescaling. The projection to the first or *primary* component is evidently invariant. There is a splitting to this which is invariantly defined, namely

$$\sigma \mapsto \begin{pmatrix} \sigma \\ \nabla_a \sigma \\ \frac{1}{n}(\nabla^a \nabla_a + P)\sigma \end{pmatrix} \tag{11}$$

which comes from solving the equations (7)–(9) as best we can. In any case it is easy to check invariance. We can compose this with the local twistor connection to reconstruct the operator on L itself which led to $\mathcal{T}$ in the first place:

$$\begin{pmatrix} \sigma \\ \nabla_a \sigma \\ \frac{1}{n}(\nabla^a \nabla_a + p)\sigma \end{pmatrix} \overset{\nabla_b}{\longmapsto} \begin{pmatrix} \nabla_b \sigma - \nabla_b \sigma \\ \nabla_b \nabla_a \sigma + P_{ab}\sigma - \frac{1}{n}g_{ab}(\nabla^c \nabla_c + P)\sigma \\ \frac{1}{n}\nabla_b(\nabla^c \nabla_c + P)\sigma + P_{ab}\nabla^a \sigma \end{pmatrix}$$

$$= \begin{pmatrix} 0 \\ \nabla_{(b}\nabla_{a)}\sigma + P_{ab}\sigma - \frac{1}{n}g_{ab}(\nabla^c \nabla_c \sigma + P)\sigma \\ \frac{1}{n}\nabla_b(\nabla^c \nabla_c + P)\sigma + P_{ab}\nabla^a \sigma \end{pmatrix}$$

---

[11] Our description will unfortunately be somewhat less than complete. Though we hope to make the recipe clear, the fine details require some calculation in the realm of representation theory. These computations are straightforward (using, for example, the machinery of [9]) but, as we are trying to avoid representation theory in this article, cannot be given here. See [6] for an algebraic formalization of this proceedure.

and an invariant operator appears as the secondary part of the result—this is manifestly invariant, since the primary part vanishes and so makes no contribution when we rescale the metric. Notice that a calculation was required here to see that the primary part of $\nabla_a \mathbf{f}$ vanished. In fact, this is unnecessary as it is easy to verify that $\mathcal{T}_a \to \mathcal{O}_{ab}[1]$ defined by

$$\begin{pmatrix} \sigma_a \\ \mu_{ab} \\ \rho_a \end{pmatrix} \longmapsto \mu_{ab} + \nabla_a \sigma_b - \frac{1}{n-1} g_{ab} \nabla^c \sigma_c$$

is invariant[12]. Now we can write the new invariant operator as a composition

$$\mathcal{O}[1] \to \mathcal{T} \xrightarrow{\nabla_a} \mathcal{T}_a \to \mathcal{O}_{ab}[1]$$

followed by taking the trace-free symmetric part. Each part of this composition is conformally invariant.

This shows us that we can get our second order equation by translating the ordinary exterior differential

$$\nabla_a \quad : \quad functions \ \to \ 1\text{-}forms$$

into local twistor terms, i.e. replacing the functions by the local twistors, the 1-forms by the 1-forms with values in the local twistors and using the local twistor connection instead of the exterior differential.

The procedure can be continued by yet again coupling in local twistors and their connection. Each operator in our composition remains invariant under coupling—the computations to check invariance do not see the local twistor coefficients. Thus we obtain an invariant operator:

$$\mathbf{f} \to \nabla_{(a}\nabla_{b)}\mathbf{f} + P_{ab}\mathbf{f} - \tfrac{1}{n}g_{ab}(\nabla^c\nabla_c\mathbf{f} + P\mathbf{f})$$

where now $\mathbf{f}$ is a local section of $\mathcal{T} \otimes L$ and ∇ indicates the local twistor connection. To manufacture an invariant operator on tensors and densities we again need to see that there is an invariant splitting

$$L^2 \quad \to \quad \mathcal{T} \otimes L$$

lifting the natural projection. This takes the same form as (11) except that we must adjust the scalars a little [3]:

$$f \longmapsto \begin{pmatrix} f \\ \tfrac{1}{2}\nabla_a f \\ \frac{1}{2(n+2)}(\nabla^a\nabla_a + 2P)f \end{pmatrix} = \mathbf{f}$$

---

[12]Here, we are using another convenient notation (as in [3, 4, 25]) for simple conformally weighted tensor bundles. The indices on $\mathcal{O}$ indicate the indices on the tensor and the number in the square brackets is the conformal weight.

Applying the local twistor covariant derivative ∇_b yields

$$\begin{pmatrix} \frac{1}{2}\nabla_b f \\ \frac{1}{2}\nabla_b\nabla_a f + P_{ab}f - \frac{1}{2(n+2)}g_{ab}\Box f \\ \frac{1}{2(n+2)}\nabla_b\Box f + \frac{1}{2}P_{ab}\nabla^a f \end{pmatrix}.$$

where we write $\Box f$ for $(\nabla^k\nabla_k + 2P)f$. Apply ∇_c, again. The resulting primary part is

$$-P_{bc}f + \frac{1}{2(n+2)}g_{bc}\Box f$$

and the secondary part is

$$\tfrac{1}{2}\nabla_c\nabla_b\nabla_a f + \nabla_c(P_{ab}f) + \tfrac{1}{2}P_{ac}\nabla_b f - \frac{1}{2(n+2)}\left(2g_{a(b}\nabla_{c)}\Box f + \tfrac{1}{2}g_{ac}P_{db}\nabla^d f\right).$$

To these we must add $P_{bc}\mathbf{f}$, symmetrize over bc and take the trace free part over bc. When we do this, the primary part goes to zero and we are left with the totally trace free part of

$$\tfrac{1}{2}\left(\nabla_{(a}\nabla_b\nabla_{c)}f + 4P_{(ab}\nabla_{c)}f + 2(\nabla_{(a}P_{bc)})f\right). \tag{12}$$

This is an invariant operator. Again, one need not compute the primary part—there is a differential operator which splits it off.

It is not too difficult to check that similar splitting formulae exist more generally and, by iterating the calculations just done, we obtain curved analogues of all the flat space operators

$$f \longmapsto \text{ totally trace free part of } \underbrace{\nabla_{(a}\nabla_b \ldots \nabla_{c)}f}_{s+1 \text{ terms}}$$

which are conformally invariant on densities f of weight s. They are all deduced from the ordinary exterior derivative on functions. In fact, even more operators arise this way—instead of tensoring $\mathcal{T}$ by powers of L we can consider tensor powers of $\mathcal{T}$ itself tensored in with powers of L. The local twistor connection extends in the usual way to such powers of $\mathcal{T}$. From these we obtain all curved variants of the flat space operators

$$f_{d\ldots e} \longmapsto \text{ totally trace free part of } \underbrace{\nabla_{(a}\nabla_b \ldots \nabla_c f_{d\ldots e)}}_{s+1 \text{ terms}}$$

where $f_{d\ldots e} \in \mathcal{O}^{\circ}_{(d\ldots e)}[s+2r]$ has rank r, the superscript 'o' meaning that $f_{d\ldots e}$ is also totally tracefree.

So much for examples. What of the general theory? How do we know, for example, that these compositions of increasingly complicated though invariant differential operators give non-zero results? The answer is that this is exactly how the operators are created in the flat case! Therefore, the symbols are exactly as in the flat case (and, in particular, the operators are non-zero).

The translation principle in the flat case is known as the *Jantzen-Zuckerman* translation principle [56, p. 464] and it is responsible for the results stated in §3. Its operation in four dimensions is explained in [26]. A major ingredient is the notion of *central character* (also explained in [26] for the four-dimensional case). This notion easily produces the splittings we've been using (initially in the flat case but hence in the curved case by using the general form of these splittings[13]). A simpler, though sufficient, series of splitting is given in [26] in dimension four and an algebraically equivalent version of these splittings is in [6] for the general case; see also the discussion of footnote 14.

Thus, if we start with the de Rham sequence, then the curved translation principle produces curved analogues of all the BGG operators described in §3.1 and §3.2[14].

There is a problem, however, in the general application of the translation principle in the curved case. The error in [26] is the assertion (made on page 223) that any invariant operator on tensors necessarily remains invariant if coupled with local twistors. With a simple operator, such as an exterior derivative, this is clear. More generally, operators constructed from the curved translation principle, being compositions of invariant operators each of which may be invariantly coupled, may themselves be invariantly coupled. The problem is that, for example,

$$\nabla_b[\nabla^b\nabla^a - 4P^{ab} - 2Pg^{ab}]\nabla_a : \Lambda^0 \to \Lambda^4 \quad \text{in dimension 4}$$

does not remain invariant when operating on local twistors. Indeed, in the computation (10), the error term

$$[\nabla_a\nabla_b - \nabla_b\nabla_a]\mathbf{f}$$

is precisely the curvature of $\mathcal{T}$ (essentially the Weyl curvature). More gener-

---

[13]Specifically, in the computation of conformal change, only the first derivatives of Υ_a occur and these are compensated by change in P_{ab} exactly as in the flat case—the formulae are algebraically identical in curved versus flat. See also the following footnote.

[14]Strictly speaking, this is not true unless one is speaking only of invariant operators between tensors (in even dimensions this means $d + e$ is even). In order to translate to spinor bundles one is obliged additionally to use a local twistor bundle based on a basic spin representation of $SO(1, n+1)$ (as in [7]) whereas the one we have been using so far is based on the standard representation. This is precisely what is done in [26] which uses Penrose's local twistor for the four dimensional case. Indeed, in four dimensions it is sufficient to use only the Penrose local twistors (cf. footnote 5). Generally, it suffices to use twistors based on the two spin representations. In fact (as hinted in footnote 6) there are advantages to be gained from using these species of local twistors. As in [26], the splitting formulae are first order (rather than the second order splitting operators we've been using in the main text) and, in particular, involve no curvature (and so trivially work in the curved case since they do in the flat). In the four dimensional case they are written out quite explicitly in [26, p. 223]. Therefore, in the curved analogues which this version of the translation principle produces, the curvature correction terms are produced entirely from the curvature P_{ab} occurring in the formula for local twistor transport.

ally, it seems that, in even dimensions, although

$$\Delta^{\frac{n}{2}} : \Lambda^0 \to \Lambda^n$$

has a curved analogue, it is not invariant when acting on $\mathcal{T}$ (certainly the proof in [34] breaks down). This is consistent with the fact that it is impossible to obtain this operator directly by translation from anything simpler—as proved above, any operator obtained from translation is also invariant when coupled with local twistors. In any case we cannot translate from the curved analogue of $\Delta^{\frac{n}{2}}$. However, translation from $\Delta^{\frac{n}{2}}$ is precisely what is needed in the flat case to obtain all the other long operators (see definition 3.2). Bearing in mind the discussion preceding theorem 4.1 concerning the Fefferman-Graham ambient metric construction, we are therefore led to

Conjecture 4.2 *In even dimensions,*

$$\Delta^{\frac{n}{2}} : \Lambda^0 \to \Lambda^n$$

is the only long operator which admits a curved analogue.

We should emphasize that the only case of this conjecture which is actually proved (by Graham) is for Δ^3 in dimension 4. This is almost the simplest operator falling within the conjecture. If one allows spinor fields, then

$$\Delta^2 \nabla_A^{A'} : \mathcal{O}_{A'}[1] \to \mathcal{O}_A[-4]$$

is a slightly simpler operator which according to this conjecture should have no curved analogue.

Another reason for our conjecture is that all other operators are known to have curved analogues. We have indicated how this occurs for the BGG operators. For the remaining operators we can proceed as follows.

4.1.1   Even Dimensions

In this case it suffices to translate from the conformally invariant Laplacian of example 2.2. The computation given in example 2.2 goes through without change even if f is a section of a local twistor bundle and so it is legitimate to translate from this operator. It turns out that all the non-standard operators save for the long operators can be thus obtained:

Theorem 4.3 *On an even dimensional conformal manifold all the flat invariant operators of §3.1 except the long operators have curved analogues. (Also $\Delta^{\frac{n}{2}}$ admits a curved analogue.)*

Whilst the detailed computations are beyond the scope of this review, we shall give an illustrative and quite non-trivial example. This example arises by translating twice using $\mathcal{T}$ from the Laplacian. In the interests of economy, we shall instead just translate once using $\wedge^2 \mathcal{T}$. An element of $\wedge^2 \mathcal{T} \otimes L^{-1}$ may be written as a quadruple

$$\begin{pmatrix} & \kappa_a & \\ \lambda & , & \theta_{bc} \\ & \nu_d & \end{pmatrix} \quad \text{s.t.} \quad \begin{aligned} \hat{\kappa}_a &= \Omega\kappa_a \\ \hat{\lambda} &= \Omega^{-1}[\lambda + \Upsilon^a\kappa_a] \ , \quad \hat{\theta}_{bc} = \Omega[\theta_{bc} + 2\Upsilon_{[b}\kappa_{c]}] \\ \hat{\nu}_d &= \Omega^{-1}[\nu_d + \Upsilon^a\theta_{ad} - \Upsilon_d\lambda + \tfrac{1}{2}\Upsilon^a\Upsilon_a\kappa_d - \Upsilon_d\Upsilon^a\kappa_a] \end{aligned}$$

One may readily check that

$$\kappa \longmapsto \begin{pmatrix} & \kappa & \\ & \frac{1}{n-1}\nabla^a\kappa_a \ , \quad 2\nabla_{[b}\kappa_{c]} & \\ \frac{1}{n-2}\Delta\kappa_a - \frac{n}{(n-1)(n-2)}\nabla_d\nabla^a\kappa_a - \frac{n}{n-2}P_d{}^a\kappa_a & \end{pmatrix}$$

is an invariant splitting $\mathcal{O}_a[1] \to (\wedge^2)\mathcal{T})[-1]$ and that

$$\begin{pmatrix} & \kappa_a & \\ \lambda & , & \theta_{bc} \\ & \nu_d & \end{pmatrix} \longmapsto \begin{aligned} &\nu_d - \tfrac{1}{3}\nabla_d\lambda + \tfrac{1}{n-5}\nabla^a\theta_{ad} + \tfrac{1}{(n-5)(n-6)}\Delta\kappa_d \\ &+ \tfrac{n-8}{3(n-5)(n-6)}\nabla_d\nabla^a\kappa_a - \tfrac{n-4}{(n-5)(n-6)}P\kappa_d + \tfrac{n-8}{(n-5)(n-6)}P_d{}^a\kappa_a \end{aligned}$$

is an invariant splitting $(\wedge^2 \mathcal{T})[-3] \to \mathcal{O}_d[-3]$ (provided that $n \neq 5, 6$). Thus, if $n = 4$ we obtain a conformally invariant differential operator $\mathcal{O}_a[1] \to \mathcal{O}_d[-3]$ as the composition

$$\mathcal{O}_a \to (\wedge^2\mathcal{T})[-1] \xrightarrow{\square} (\wedge^2\mathcal{T})[-3] \to \mathcal{O}_d[-3]$$

where $\square$ is the conformally invariant Laplacian coupled using the local twistor connection on $\wedge^2 \mathcal{T}$. This provides a curved analogue of the flat operator

$$\phi_a \longmapsto 3\Delta^2\phi_d - 4\nabla_d\nabla^a\Delta\phi_a.$$

Using this method actually to compute the curved analogue is difficult. The naïve method of adding curvature correction terms by trial and error is easier, though in this case, is nevertheless an arduous task. This naïve method has been successfully carried through by Robin Graham (who has also found a crucial error in [16], a paper whose purpose is to show the non-existence of a curved analogue to this particular operator). This example really shows the power of the translation method. Another iteration gives an invariant operator as the composition

$$\mathcal{O}^{\circ}_{(ab)}[3] \to (\wedge^2\mathcal{T})_a[1] \to (\wedge^2\mathcal{T})_d[-3] \to \mathcal{O}^{\circ}_{(cd)}[-3].$$

This is surely beyond naïve calculation.

Another interesting aspect of the operator $\mathcal{O}_a[1] \to \mathcal{O}_d[-3]$ is that, as observed in [16], the curved analogue is not unique. There is the option of adding any multiple of $\phi_a \mapsto B_d{}^a\phi_a$, where B_{ac} is the Bach tensor. In general, the question of uniqueness is unknown. As we shall see in Gover's method below, there is a canonical curved analogue for all the BGG operators and we suspect that for these operators the curved analogue is unique. More generally, this is probably true for all standard operators.

4.1.2   Odd Dimensions

We shall finish this section with a few words about the odd dimensional case. This case is rather easier and there are good reasons why this should be anticipated. The first is the simpler results of the flat space classification—there is never more than one invariant operator acting on any particular irreducible bundle. This contrasts with the situation in four dimensions, for example, where there is the exterior derivative and Δ^2 both acting on unweighted functions. A second reason is the much better status of the ambient metric construction of Fefferman and Graham [30]—there are no obstructions to this construction. This is used in [34] to show the existence of curved analogues to every conformally invariant power of the Laplacian:

$$\Delta^k : \mathcal{O}[k - \tfrac{n}{2}] \to \mathcal{O}[-k - \tfrac{n}{2}].$$

(See also [8, §7].)

The appropriate calculations (of central character to determine splittings) show that

Theorem 4.4 *On an odd dimensional conformal manifold all the flat invariant operators of §3.2 have curved analogues.*

For the BGG operators one can translate from the exterior derivative as indicated above. All the rest can be obtained by translating the conformally invariant Laplacian.

4.2   Gover's Method

A second method of proof (at least for the BGG operators) which is computationally far more effective is due to Rod Gover [32, 33]. It is in the spirit of a standard trick in representation theory where one reduces calculations to the case of $\mathrm{sl}(2, \mathbb{C})$. For conformal manifolds this amounts to observing that null geodesics inherit a natural projective structure.

A *projective structure* on a one dimensional manifold γ may be defined as follows[15]. Firstly choose a square root L of the tangent bundle (i.e. a spin structure). Sections of L^w will be called *projective densities* of *weight w*. In particular, a projective density of weight -2 is the same as a 1-form. Suppose χ is a nowhere vanishing local section of L. Such a χ is known as a *projective scale* and determines a local connection ∇ on L by insisting that χ be covariant constant. This induces connections on all L^w. Bearing in mind that 1-forms are projective densities of weight -2, we obtain local differential operators:

$$\nabla : \Gamma(L^w) \to \Gamma(L^{w-2}) \quad \text{for all } w.$$

---

[15]The general definition of projective structure [3] naïvely specializes to be vacuous in dimension one. The definition we adopt here (or, rather, its complex version) is equivalent to that in [36]. Our notation, however, follows [3].

A one dimensional projective manifold is a choice of L together with an assignment of $\Phi \in \Gamma(L^{-4})$ for each projective scale $\chi \in \Gamma(L)$ such that if $\hat{\chi} = \Omega^{-1}\chi$, then

$$\hat{\Phi} = \Phi - \nabla\Upsilon + \Upsilon^2 \tag{13}$$

where $\Upsilon \in \Gamma(L^{-2})$ is defined to be $\Omega^{-1}\nabla\Omega$. The transformation law for ∇ is easily determined:

$$\hat{\nabla} f = \nabla f + \Upsilon f \quad \text{for } f \in \Gamma(L)$$

for then

$$\hat{\nabla}\hat{\chi} = (\Omega^{-1}\hat{\chi}) + \Omega^{-2}(\nabla\Omega)\chi = \nabla\chi = 0.$$

It follows that

$$\hat{\nabla} f = \nabla f + w\Upsilon f \quad \text{for } f \in \Gamma(L^w). \tag{14}$$

A projective structure determines a distinguished family of local coördinates on γ as follows. Given a projective scale, the equation $\nabla(dt) = 0$ determines a parameter $t : \gamma \to \mathbf{R}$ up to affine transformations (the usual affine parameter along a geodesic). Suppose we restrict the projective scale so that the corresponding Φ vanishes. This is always possible by solving the *Ricatti equation*

$$\nabla\Upsilon - \Upsilon^2 = \Phi.$$

Having done this, the remaining freedom in ∇ when acting on 1-forms is

$$\nabla \longmapsto \hat{\nabla} = \nabla - 2\Upsilon$$

where $\nabla\Upsilon - \Upsilon^2 = 0$. Suppose t is a local parameter such that $\nabla(dt) = 0$ and $s = g(t)$ is such that $\hat{\nabla}(ds) = 0$. Then $ds = g'(t)dt$ so

$$0 = \hat{\nabla}(ds) = \nabla(g'dt) - 2\Upsilon g'dt = (g'' - 2\Upsilon g')dt.$$

Substituting back into the Ricatti equation $\nabla\Upsilon - \Upsilon^2 = 0$, gives

$$\nabla\left(\frac{g''}{2g'}\right) - \left(\frac{g''}{2g'}\right)^2 = \frac{g'''}{2g'} - \frac{3}{4}\left(\frac{g''}{g'}\right)^2 = 0.$$

This is the *Schwarzian derivative* of g [36]. It can be written

$$\sqrt{g'}\left(\frac{1}{\sqrt{g'}}\right)'' = 0 \tag{15}$$

so

$$g' = \frac{1}{(ct+d)^2}$$

and s is a Möbius transformation of t:

$$s = \frac{at+b}{ct+d}.$$

These transformations form the group $\mathrm{PSL}(2,\mathbf{R})$ and so we have a distinguished set of local coördinates, up to the action of this group (cf. the definition of projective in [36]).

We employ these facts as follows. Let γ be a null geodesic in M, a conformal manifold. Up to an overall irrelevant constant, we can identify the projective densities on γ with the restrictions of conformal densities on M. To do this, choose a 1-form l_a defined along γ such that l^a is everywhere tangent and such that

$$l^a \nabla_a l_b = 0. \tag{16}$$

By equation (1) this is conformally invariant and so l_a is fixed up to an overall constant. l^a scales under conformal change of metric

$$l^a \to \hat{l}^a = \Omega^{-2} l^a.$$

Thus, if χ is a conformal density of weight 1, then $\chi^2 l^a$ is a genuine tangent vector along γ. In other words, the assignment

$$\chi \longmapsto \chi^2 l^a$$

identifies the bundle L of conformal densities as a square root of the tangent bundle to γ, as required. χ also determines a metric $\chi^{-2} g_{ab}$ in the conformal class. We can define Φ in terms of this metric by

$$\Phi \equiv l^a l^b P_{ab}.$$

Then by equation (4), Φ varies according to (13). So we have

Lemma 4.5 *A conformal structure distinguishes (up to an overall constant) a projective structure on null geodesics*[16].

Otherwise put, we can always choose a metric in the conformal class and an affine coördinate along a null geodesic so that the tangential component of P_{ab} is zero and this coördinate is unique up to Möbius transformations. The differential operator ∇ is just $l^a \nabla_a$.

This reduces the problem of finding conformally invariant differential operators along a null geodesic to understanding the operators on projective densities (on the circle) invariant under $\mathrm{PSL}(2,\mathbf{R})$. This is easily accomplished using the methods of the previous section. Let t be a parameter on which $\mathrm{PSL}(2,\mathbf{R})$ acts by Möbius transformations. If f is a density of weight $s \geq 0$ (using the volume form dt to trivialize projective densities) then only

$$f \to (\frac{d}{dt})^{s+1} f \tag{17}$$

---

[16]It is interesting to note that an arbitrary *nowhere-null* curve in a conformal manifold also inherits a projective structure [1].

is invariant. (This is the 'conformally invariant power of edth' [28]—see also
[36, p203].) There are *no* invariant operators for $s < 0$. A simple proof of
these facts may be found in [33], using (15).

Now suppose, in terms of our definition of projective structure, we change
to an arbitrary χ and associated ∇. The operator (17) becomes

$$\mathcal{D}_s \equiv (\nabla - s\Upsilon)(\nabla + (2 - s)\Upsilon)\ldots(\nabla + s\Upsilon)f. \tag{18}$$

by virtue of equation (14). Here, Υ satisfies the Ricatti equation:

$$\nabla\Upsilon - \Upsilon^2 = \Phi.$$

This means that the right hand side of (18) can be written as a polynomial
$P(\nabla, \Upsilon, \Phi)$ acting on f. The claim is that this does not involve Υ. The reason
is that Ricatti equation does not completely determine Υ and we are free to
choose its value at a point quite arbitrarily whilst holding Φ itself fixed. On
the other hand, as we do this $P(\nabla, \Upsilon, \Phi)f$ remains unchanged. Since this is
true for any f, $P(\nabla, \Upsilon, \Phi)$ is unchanged and therefore cannot depend on Υ.
A typical example of such an operator is

$$\mathcal{D}_2 = \nabla^2 + \Phi$$

acting on projective densities of weight 1. The corresponding equation

$$\nabla^2\phi + \Phi\phi = 0$$

is known as *Hill's* equation[17].

From this we can immediately deduce certain conformally invariant oper-
ators. For, given any tensor $\kappa_{ab\ldots c}$ of rank r and conformal weight q we can
obtain on γ a (projective) density of weight $s = q - 2r$:

$$\kappa = l^a l^b \ldots l^c \kappa_{ab\ldots c}.$$

(Notice that $\kappa_{ab\ldots c}$ may as well be totally trace free and symmetric; any other
irreducible component does not contribute to κ.) Because of (16) and because
it is independent of Υ, $\mathcal{D}_s\kappa$ is homogeneous of degree $q - r + 1$ in l^a. So it
must be of the form

$$l^a l^b \ldots l^d \lambda_{ab\ldots d}$$

for some totally symmetric trace free tensor $\lambda_{ab\ldots d}$ of rank $q - r + 1$ and
conformal weight q. The required differential operator is then

$$\kappa_{ab\ldots c} \longmapsto \lambda_{ab\ldots d}.$$

---

[17]Any homogeneous linear second order equation $y'' + ay' + by = 0$ can be put into the
form of Hill's equation [19, pp. 18,19] and so defines a projective structure; the solutions
of the equation are the covariant constant local twistors.

Example 4.6 The third order invariant operator on projective densities of weight 3 is given by

$$\mathcal{D}_3\kappa = \nabla^3\kappa + 4\Phi\nabla\kappa + 2(\nabla\Phi)\kappa$$

which shows that if $\kappa^{ab...c}$ has conformal weight 3 then

$$\kappa^{ab...c} \longmapsto \text{ttf}\left(\nabla_{(a}\nabla_b\nabla_c\kappa_{de...f)} + 4\Phi_{(ab}\nabla_c\kappa_{de...f)} + 2(\nabla_{(a}\Phi_{bc)}\kappa_{de...f)}\right)$$

is conformally invariant(cf. the operator of (12)). (ttf means 'totally trace-free part of'.)

In this way we get all possible operators of the kind we earlier generated from the first exterior differential. A modification of the method produces all the BGG operators, though not, unfortunately, the others. Especially mysterious in this context are the standard operators on tensors with half integral weight in the odd dimensional case (but see [8]).

5   Other Methods

Since the appearance of the two methods described in the previous section there has been some progress in understanding how to generate invariant operators using either the local twistor connection in a more direct fashion or embeddings into higher dimensional ambient spaces (in the spirit of Fefferman and Graham [30]). This section is given over to a brief description of these methods and the original Fefferman-Graham construction.

Via Lie algebra cohomology

The first of these comes from some general considerations involving Lie algebra cohomology. This arises from methods used to construct the Cartan (or local twistor) connection and is detailed in [7, 8]. It is applicable to all so-called *almost Hermitian symmetric structures* which include conformal, projective and paraconformal structures. In the one dimensional projective case the results are very simple to state and in fact combine with Gover's construction immediately to give simple formulae for invariant operators. In general, the method gives explicit means of calculating invariant operators. It is similar to the Fefferman-Graham method in that the construction of non-standard operators can be obstructed.

Begin by defining the local twistor bundle for a one dimensional projective structure. Sections of this bundle are pairs

$$\begin{pmatrix} r_0 \\ r_1 \end{pmatrix}$$

which transform to

$$\begin{pmatrix} r_0 \\ r_1 + \Upsilon r_0 \end{pmatrix} \tag{19}$$

under a change of scale. Here we are regarding r_0 as a density of weight 1 (*not* the function representing it) and r_1 likewise as a density of weight -1. Secretly, the pair belongs to the self representation of $SL(2,\mathbf{R})$. We obtain a connection on this bundle by employing the equation

$$\nabla^2 r_0 + \Phi r_0 = 0$$

which gives

$$\nabla \begin{pmatrix} r_0 \\ r_1 \end{pmatrix} = \begin{pmatrix} \nabla r_0 - r_1 \\ \nabla r_1 + \Phi r_0 \end{pmatrix}.$$

We want to generalize this to symmetric powers of the original bundle. A section of the kth power is a collection

$$\mathbf{r} = \begin{pmatrix} r_0 \\ r_1 \\ \vdots \\ r_i \\ \vdots \\ r_k \end{pmatrix}$$

where[18] r_i is a density of weight $k - 2i$. The transformation rule for this is best given infinitesimally[19]. For convenience let us write $\mathbf{r}$ horizontally and define the *infinitesimal* projective transform by

$$\Upsilon.\mathbf{r} = \Upsilon(0, r_0, 2r_1, \ldots, (i+1)r_i, \ldots, kr_{k-1}). \tag{20}$$

Then it is easy to check that under (19) $\mathbf{r}$ should transform by

$$\mathbf{r} \longmapsto (\exp \Upsilon).\mathbf{r}$$

which is defined in the obvious way. (The easiest way to do this calculation is to use a little of the structure theory of representations of sl(2).) We shall also define two operators δ and Φ in $\mathbf{r}$ by similar expressions:

$$\delta \mathbf{r} = (kr_1, (k-1)r_2, \ldots, (k-i+1)r_i, \ldots, r_k, 0)$$

---

[18]In terms of two-component spinors, one may write $\mathbf{r}$ as a totally symmetric spinor $r_{A...BC...D}$ and $r_i = o^A \ldots o^B \iota^C \ldots \iota^D r_{A...BC...D}$ (i occurrences of o). Here o^A and ι^A are basis vectors for the local twistor bundle.

[19]Generally, in checking conformal invariance, it suffices to neglect second order terms in Υ_a. This amounts to writing $\Omega^2 = e^{2\phi}$ and rescaling by $e^{2t\phi}$ as t varies from 0 to 1. Differentiating with respect to t effectively gives such an infinitesimal conformal transformation. See, for example, [14, 37, 52, 57].

and

$$\boldsymbol{\Phi}.\mathbf{r} = \Phi(0, r_0, 2r_1, \ldots, (i+1)r_i, \ldots, kr_{k-1}).$$

Then the local twistor connection is just given by

$$\boldsymbol{\nabla} = \nabla - \delta + \boldsymbol{\Phi}.$$

From this, the ith entry of $\boldsymbol{\nabla}\mathbf{r}$ is

$$\nabla r_i - (k-i)r_{i+1} + i\Phi r_{i-1}. \tag{21}$$

Invariance amounts to $[\boldsymbol{\nabla}, \boldsymbol{\Upsilon}] = 0$. The operator δ which appears here is in fact the differential of a certain complex which calculates relative Lie algebra cohomology.

Now here is the crunch. Beginning with r_0 we can seek to complete $\mathbf{r}$ so that as much of $\boldsymbol{\nabla}\mathbf{r}$ as possible is zero. It is evident from (21) that we can make all entries but the last zero and this last is completely determined. Recursively,

$$r_1 = \tfrac{1}{k}\nabla r_0$$
$$r_i = \tfrac{1}{k-i+1}(\nabla r_{i-1} + (i-1)\Phi r_{i-2}) \qquad 2 \le i \le k$$

and the last entry in $\boldsymbol{\nabla}\mathbf{r}$ is

$$\tfrac{1}{(k-1)!}\mathcal{D}_k r_0 = \nabla r_k + k\Phi r_{k-1}.$$

$\mathcal{D}_k$ is evidently an invariant operator on densities of weight k with values in the densities of weight $-k-2$. It is precisely the $\mathcal{D}_k$ of the previous section.

Example 5.1 We obtain $\mathcal{D}_3$ as follows:

$$r_1 = \tfrac{1}{2}\nabla r_0,$$
$$r_2 = \nabla r_1 + \Phi r_0$$
$$ = \tfrac{1}{2}\nabla^2 r_0 + \Phi r_0,$$
$$\mathcal{D}_3 r_0 = 2(\nabla r_2 + 2\Phi r_1)$$
$$\phantom{\mathcal{D}_3 r_0} = \nabla^3 r_0 + 2(\nabla\Phi)r_0 + 4\Phi\nabla r_0.$$

Just to show how easy this is we compute $\mathcal{D}_4$:

$$r_1 = \tfrac{1}{3}\nabla r_0,$$
$$r_2 = \tfrac{1}{2}(\nabla r_1 + \Phi r_0)$$
$$ = \tfrac{1}{2}(\tfrac{1}{3}\nabla^2 r_0 + \Phi r_0),$$
$$r_3 = \nabla r_2 + 2\Phi r_1$$
$$ = \tfrac{1}{6}(\nabla^3 r_0 + 7\Phi\nabla r_0 + 3(\nabla\Phi)r_0),$$
$$\mathcal{D}_4 r_0 = 6(\nabla r_3 + 3\Phi r_2)$$
$$\phantom{\mathcal{D}_4 r_0} = (\nabla^4 + 10(\nabla\Phi)\nabla + 10\Phi\nabla^2 + 3(\nabla^2\Phi) + 9\Phi^2)r_0.$$

The whole formalism of ∇, δ, Φ extends to arbitrary almost Hermitian symmetric manifolds and generates explicit formulae for invariant operators on such spaces [8].

Via a conformal calculus

This method will be fully described in a forthcoming paper of Bailey, Gover, Graham, and one of the authors (MGE). We give here only a brief indication.

Recall that fundamental to Riemannian geometry is the Levi Civita connection:

$$\omega_{j\ldots k} \longmapsto \nabla_i \omega_{j\ldots k}.$$

In conformal geometry a good replacement is provided by the Cartan connection on the bundle $\mathcal{T}$ of local twistors:

$$\phi_{J\ldots K} \longmapsto \nabla_i \phi_{J\ldots K}$$

where we are using capital indices to denote sections of $\mathcal{T}$. However, this is perhaps not the best replacement. If we think of the local twistor bundle as a replacement for the tangent bundle in Riemannian geometry then we should seek invariantly defined operators

$$\phi_{J\ldots K} \longmapsto D_I \phi_{J\ldots K}$$

to give a 'conformal calculus'. Indeed there are naturally defined such operators acting on fields with arbitrarily many local twistor indices and of arbitrary conformal weight(first constructed by Thomas [54]). For f a density of weight w

$$D_I f = \begin{pmatrix} w(n + 2w - 2)f \\ (n + 2w - 2)\nabla_i f \\ (\nabla^i \nabla_i + wP)f \end{pmatrix}$$

and a similar formula applies to $\phi_{J\ldots K}$ provided ∇_i is interpreted as the Cartan connection. When $w = 1$, this is a simple rescaling of equation (11). Now suppose $w = 0$. The formula becomes

$$D_I f = \begin{pmatrix} 0 \\ (n - 2)\nabla_i f \\ (\nabla^i \nabla_i)f \end{pmatrix}$$

and we deduce that $f \mapsto \nabla_i f$ is conformally invariant. If $w = \frac{2-n}{2}$, then

$$D_I f = \begin{pmatrix} 0 \\ 0 \\ (\nabla^i \nabla_i + wP)f \end{pmatrix}$$

and $(\nabla^i \nabla_i + wP)f$, the conformal Laplacian, is invariant. The utility of D_I comes from the fact that it can be iterated. In projective geometry a similar calculus can be constructed and $D_I \cdots D_K$ gives curved analogues for all flat projective invariant operators. The conformal case is more difficult and has yet to be worked out.

Via the Fefferman-Graham construction

Recall (as in §3) that the flat model of conformal geometry is provided by the sphere S^n realized as the generators of the cone in $\mathbf{R}^{1,n+1}$ or, equivalently, as a non-singular quadric in projective space $\mathbf{RP}_{n+1}$. The conformal geometry of S^n is derived from the Lorentzian geometry of $\mathbf{R}^{1,n+1}$ (or, equivalently, the projective geometry of $\mathbf{RP}_{n+1}$). This suggests that, in this flat case, conformally invariant operators can be derived from the well-understood Lorentzian invariant operators on the ambient $\mathbf{R}^{1,n+1}$. This was accomplished for the Laplacian and some others by Hughston and Hurd [38].

Fefferman and Graham [30] sought a replacement for this geometry in the curved case. They found that, at least formally, the bundle of scales over an arbitrary conformal manifold M gave rise to a canonical ambient Lorentzian manifold at least to some order. The requirement which fixes this ambient space is that it be Einstein (and homogeneous with respect to scaling). In the odd dimensional case the construction works to all orders. In the even dimensional case there is an obstruction to extending to order $n/2$. Now one can take any Lorentzian invariant tensor or operator and try to restrict to an invariant of M. For example, if one starts with a density of weight w on M extends off, applies the wave operator, and restricts back to M then, precisely for $w = \frac{2-n}{2}$, the result is independent of choice of extension. This gives the conformal Laplacian. More examples were found by Jenne [41].

Another possibility is to take a density on M and attempt to extend it order by order into the ambient space so as to satisfy the wave equation. It turns out that for critical weights there is an obstruction at a corresponding order. This obstruction must be a conformal invariant. The conformally invariant modifications of powers of the Laplacian arise in this way [34].

There are evident links between this scheme and the local twistor methods. The bundle of local twistors may be regarded as the first order part of the Fefferman-Graham extension. If one takes a density on M, extends it off so as to satisfy the wave equation, applies the ambient Levi Civita connection, and restricts back to M, the result can therefore be regarded as a local twistor. Up to scale, the answer is the operator D_I of the previous section.

6   Conclusion

We have seen that there is a natural family of invariant operators on any conformal manifold. This family arises by first classifying the operators on the flat model invariant under the finite-dimensional group of conformal motions. It then turns out that these operators are invariant under the infinite-dimensional group of conformal rescalings. They can be written in terms of the Levi Civita connection and curvature with respect to any metric in the flat conformal class. It is then remarkable that most of these operators persist in the curved case i.e. they admit *curved analogues* as in theorems 4.3 and 4.4. There is some scope for further investigation. In particular, a *characterization* of these operators would certainly be useful. Also, these operators are far from a complete list of conformal invariants. Rather trivially, composition gives further examples—whilst composition in the flat case is often zero, the composition in the curved case gives Weyl curvature in the symbol. However, there are many more invariant operators besides—they involve Weyl curvature in such a way as to vanish in the flat case. This is a much more difficult question even for scalar invariants [30].

One can ask as to whether this scheme of curved analogues extends to other geometries. Somewhat simpler than the conformal case is the investigation for projective geometries [3, 7]. Also there are the paraconformal geometries of [2] where this scheme works. These are all special cases of almost Hermitian symmetric manifolds [7] and curved analogues of flat invariant operators are discussed in [8]. We suspect that these investigations can also be carried out for some higher order structures, perhaps the most interesting of which are CR-structures. The flat case is straightforward but the curved case requires much more work.

Acknowledgement

This article was written with support from the Australian Research Council and the University of Adelaide. RJB would like to thank the University of Adelaide for hospitality during this time.

References

[1] T.N. Bailey and M.G. Eastwood. Conformal circles and parametrizations of curves in conformal manifolds. Proc. A.M.S. *to appear.*

[2] T.N. Bailey and M.G. Eastwood. Complex paraconformal manifolds— their differential geometry and twistor theory. Preprint (1988).

[3] T.N. Bailey, M.G. Eastwood, and R. Gover. Local twistors for conformal, projective and related structures. Preprint (1989).

[4] T.N. Bailey and M.A. Singer. The Penrose transform in four dimensions. This volume.

[5] R.J. Baston. *The Algebraic Construction of Invariant Differential Operators.* D. Phil. thesis. Oxford University 1985.

[6] R.J. Baston. Verma modules and differential invariants of conformal structure. J. Differential Geometry (1990), *to appear.*

[7] R.J. Baston. Almost Hermitian symmetric manifolds I: local twistor theory. Preprint (1990).

[8] R.J. Baston. Almost Hermitian symmetric manifolds II: Differential invariants. Preprint (1990).

[9] R.J. Baston and M.G. Eastwood, *The Penrose Transform: its Interaction with Representation Theory.* Oxford University Press 1989.

[10] H. Bateman. The transformation of the electrodynamical equations. Proc. L.M.S. **8** (1910), 223–264.

[11] B.D. Boe and D.H. Collingwood. A Comparison Theory for the Structure of Induced Representations I. J. Alg. **94** (1985), 511–545.

[12] R. Bott. Homogeneous vector bundles. Ann. Math. **60** (1957), 203–248.

[13] T. Branson. Conformally covariant equations equations on forms. Comm. P.D.E. **7** (1982), 393–431.

[14] T. Branson. Differential operators canonically associated to a conformal structure. Math. Scand. **57** (1985), 293–345.

[15] T. Branson. Second order conformal invariants I,II. Preprint (1989).

[16] T. Branson. Conformal transformation, conformal change, and conformal covariants. Supp. Rend. Circ. Mat. Pal. **21** (1989), 115–134.

[17] T. Branson and B. Ørsted. Conformal indices of Riemannian manifolds. Composito Math. **60** (1986), 261–293.

[18] T. Branson and B. Ørsted. Conformal deformation and the heat operator. Indiana Univ. Math. J. **37** (1988), 83–110.

[19] J.C. Burkill. *The Theory of Ordinary Differential Equations.* Oliver and Boyd, 1962.

[20] É. Cartan, Les espaces à connexion conforme, *Oeuvres Complètes III.1,* 747–797, Gauthiers-Villars 1955.

[21] M.G. Eastwood. The Penrose transform for curved ambitwistor space. Quart. Jour. Math. **39** (1988), 427–441.

[22] M.G. Eastwood. On the weights of conformally invariant operators. Twistor Newsletter **24** (1987) 20–23.

[23] M.G. Eastwood. A Penrose transform for D_4 and homomorphisms of generalized Verma modules. Preprint (1989).

[24] M.G. Eastwood. The Penrose transform. This volume.

[25] M.G. Eastwood, R. Penrose, and R.O. Wells, Jr.. Cohomology and massless fields. Commun. Math. Phys. **78** (1981), 305–351.

[26] M.G. Eastwood and J.W. Ricc. Conformally invariant differential operators on Minkowski space and their curved analogues. Commun. Math. Phys. **109** (1987), 207–228.

[27] M.G. Eastwood and M.A. Singer. A conformally invariant Maxwell gauge. Phys. Lett. **107A** (1985), 73–74.

[28] M.G. Eastwood and K.P. Tod. Edth—a differential operator on the sphere. Math. Proc. Camb. Phil. Soc. **92** (1982), 317–330.

[29] C. Fefferman. Parabolic invariant theory in complex analysis. Adv. Math. **103** (1979), 131–262.

[30] C. Fefferman and C.R. Graham. Conformal invariants. In: *Élie Cartan et les Mathématiques d'Aujourdui.* Astérisque (1985), 95–116.

[31] H. Friedrich. Twistor connection and normal conformal Cartan connection. Gen. Rel. Grav. **8** (1977), 303–312.

[32] R. Gover. *A Geometrical Construction of Conformally Invariant Differential Operators.* D. Phil. thesis. Oxford University 1989.

[33] R. Gover. Conformally invariant operators of standard type. Quart. J. Math. **40** (1989), 197–207.

[34] C.R. Graham, R. Jenne, L.J. Mason, and G.A.J. Sparling. Conformally invariant powers of the Laplacian. Preprint (1989).

[35] C.R. Graham. Non-existence of curved conformally invariant operators. Preprint (1990).

[36] R.C. Gunning. *Lectures on Riemann Surfaces.* Princeton University Press 1968.

[37] P. Günther and V. Wünsch. On some polynomial conformal tensors. Math. Nach. **124** (1985), 217–238.

[38] L.P. Hughston and T.R. Hurd. A $\mathbf{CP}^5$ calculus for space-time fields. Phys. Rep. **100** (1983), 273–326.

[39] H.P. Jakobsen. Conformal covariants. Publ. RIMS, Kyoto Univ. **22** (1986) 345-364.

[40] J.C.Jantzen. Moduln mit einem höchsten gewicht. Lecture Notes in Mathematics **750**. Springer-Verlag.

[41] R. Jenne. *A Construction of Conformally Invariant Differential Operators*. Ph. D. thesis. University of Washington 1988.

[42] C.R. LeBrun. Spaces of complex null geodesics in complex-Riemannian geometry. Trans. A.M.S. **278** (1983), 209–231.

[43] J.M. Lee and T.H. Parker. The Yamabe problem. Bull. A.M.S. **17** (1987), 37–91.

[44] J. Lepowsky. A generalization of the Bernstein-Gelfand-Gelfand resolution. J. Alg. **49** (1977), 496–511.

[45] B. Ørsted. Conformally invariant differential equations and projective geometry. J. Funct. Anal. **44** (1981), 1–23.

[46] R. Penrose. Non-linear gravitons and curved twistor theory. Gen. Rel. Grav. **7** (1976), 31–52.

[47] R. Penrose and W. Rindler. *Spinors and Space-time*, vol. I: Two-spinor calculus and relativistic fields. Cambridge University Press 1984.

[48] R. Penrose and W. Rindler. *Spinors and Space-time*, vol. II: Spinor and twistor methods in space-time geometry. Cambridge University Press 1986.

[49] R. Penrose and R.S. Ward. Twistors for flat and curved space-time. In: *General Relativity and Gravitation*, vol II (ed. A. Held. Plenum 1980), pp. 283–328.

[50] R. Schimming. Konforminvarianten vom Gewicht −1 eines Zusammenhanges oder Eichfeldes. Zeit. für Anal. u. Anw. **3** (1984), 401–412.

[51] J.A. Schouten. *Ricci-Calculus*. Springer Verlag 1954.

[52] P. Szekeres. Conformal tensors. Proc. Roy. Soc. **A304** (1968), 113-122.

[53] T.Y. Thomas. On conformal geometry. Proc. Nat. Acad. Sci. **11** (1925), 352–359.

[54] T.Y. Thomas. Conformal tensors I. Proc. N.A.S. **18** (1932) 103–112.

[55] V. S. Varadarajan. *Lie Groups, Lie Algebras, and their Representations*. Prentice-Hall 1974.

[56] D.A. Vogan. *Representations of Real Reductive Lie Groups.* Prog. Math. vol. 15. Birkhäuser 1981.

[57] V. Wünsch. On conformally invariant differential operators. Math. Nachr. **129** (1986), 269–281.

[58] G. Zuckerman. Tensor products of finite and infinite dimensional representations of semisimple Lie groups. Ann. Math. **106** (1977) 295-308.

Penrose's Quasi-local Mass

K.P. Tod

1 Introduction

Penrose's definition of a kinematic twistor associated to an arbitrary topologically spherical 2-surface in an arbitrary space-time [14] was given seven years ago. Since then the definition has been a focus of constant attention. The aim of this review is to collect together what is now known and to say what the successes and failures have been.

The definition was originally made in the hope of identifying within the kinematic twistor, $A_{\alpha\beta}$, a total momentum and angular momentum for the space-time threading through the 2-surface S. Most subsequent investigations have concentrated on the less ambitious target of deriving just a momentum or just a scalar invariant, the total mass-energy. One notable exception to this trend has been the study of angular momentum for 2-surfaces S at future-null-infinity or space-like infinity in asymptotically-flat space-times [60, 21].

The dominating fact about the subsequent investigations has been the necessity to distinguish 2-surfaces S in which the data, by which I shall mean the first and second fundamental forms, reveal the presence at S of conformal or Weyl curvature from those in which it does not. Calling the former contorted (and the latter non-contorted) following the suggestion of Penrose [15] we shall see that for non-contorted S there do exist scalar invariants. Furthermore the definitions of total mass-energy which then arise have appealing properties in a large variety of particular cases.

However in the contorted case there is evidence that the original construction requires modification and it is unclear exactly how to proceed. For this reason, if for no others, one cannot at present contemplate finding completely general theorems about, for example, positivity of mass-energy.

The plan of the review will be as follows. In Section 2, I shall motivate and define the original construction and review some simple properties of it. In this section there is a discussion of the contorted/non-contorted dichotomy and the problems which it brings. In Section 3, I gather together the main classes of examples which have been calculated. These are almost entirely for non-contorted surfaces. However, there is a great variety of them and the variety of different manifestations of mass-energy which are successfully captured by the Penrose definition is very wide. This section represents the main successes of the construction. Finally in Section 4, I review what is known in the case of contorted 2-surfaces. This includes the evidence that the construction requires modification and some account of the possible modifications which have been suggested together with particular cases which support these suggestions.

The references are divided into two classes; an initial section which forms what I hope is a fairly complete bibliography of the subject, and the remaining 'background' references.

While most of the material presented here is a review of earlier work, there are a few things which have not appeared in print before. These are the work of R.M. Kelly in 3(ii) which has appeared in Twistor Newsletter, the work on cylindrical waves of 3(vi) which is to appear elsewhere, the calculations and graph for the Kerr solution, figure 4 and the expression (29) involving torsion. I am grateful to J.R. Coupe for assistance with the preparation of figures 2 and 3 and to B.P. Jeffryes for assistance with the preparation of figure 4.

2   The Kinematic Twistor

The general references for the twistor theory in this section are Penrose and Rindler [60, 61] or Huggett & Tod [54].

To motivate Penrose's definition of the kinematic twistor associated to a 2-surface in a general curved space [14] we recall the description of total momentum and angular momentum for a system in special relativity and in linearised general relativity. This is based on what we may conveniently call the fundamental identity:

$$A_{\alpha\beta}Z_1^\alpha Z_2^\beta = Q[k] = \int_\Sigma T_{ab}k^a d\sigma^b = \frac{1}{4\pi G}\int_S R_{abcd}f^{ab}d\sigma^{cd} \qquad (1)$$

To understand this formula, begin with the middle equality. Given a distribution of matter in special relativity determined by a stress-energy- momentum tensor T_{ab} we may define currents $T_{ab}k^a$, one for each Killing vector k^a of the back-ground Minkowski space. If T_{ab} satisfies the conservation equation

$$\nabla^a T_{ab} = 0 \qquad (2)$$

then these currents are conserved and can be integrated over a space-like hypersurface Σ to give a charge $Q[k]$. The value of the charge is independent of the choice of Σ (assuming say a spatially compact source). This is the middle equality. If k^a is a translation then $Q[k]$ is a component of total momentum and if k^a is a rotation then $Q[k]$ is a component of total angular momentum. If k^a is an anti-self-dual or asd rotation, which means to say that the derivative of k^a is an asd bivector, then $Q[k]$ is a component of the asd part of the angular momentum bivector. This has all the information of the angular momentum bivector so that in the middle equality we can obtain all the information of the total momentum and angular momentum by considering only those Killing vectors k^a which have asd derivatives (since of course translations have zero derivative). Now these are precisely the Killing vectors which arise from valence-2 twistors. The valence-2 twistor equations on a symmetric spinor ω^{AB} can be written

$$\nabla^A_{A'}\omega^{BC} = -i\epsilon^{A(B}k^{C)}_{A'}. \tag{3}$$

This equation has a ten (complex) dimensional space of solutions, the valence-2 twistors, and the vector k^a on the right-hand-side is automatically a Killing vector with asd derivative. The solution space of (3) is spanned by fields ω^{AB} of the form $\omega_1^{(A}\omega_2^{B)}$ where each ω_i^A is a valence-1 twistor i.e. ω_i^A satisfies

$$\nabla^A_{A'}\omega^B = -i\epsilon^{AB}\pi_{A'} \tag{4}$$

The solutions of (4) define a 4-complex-dimensional vector space which is twistor space T. Elements of T are written Z^α and so the solutions of (3) lie in the symmetric tensor product $\mathsf{T} \odot \mathsf{T}$ and can be written $B^{\alpha\beta} = B^{(\alpha\beta)}$. This space is spanned by the products $Z_1^{(\alpha}Z_2^{\beta)}$, which enables us to explain the left-most equality in (1): if the Killing vector k^a is chosen to arise from an element of the form $Z_1^{(\alpha}Z_2^{\beta)}$ in T then the corresponding charges define an element $A_{\alpha\beta}$ of the dual $(\mathsf{T} \odot \mathsf{T})^*$. $A_{\alpha\beta}$ is the *kinematic twistor* of the given source. Since the relevant vector-space is 10-complex dimensional, $A_{\alpha\beta}$ apparently has too much information in it. It should only contain the 10 real components of momentum and angular momentum. This is a point which I shall come back to.

For the right-most equality in (1) we need to bring in linearised general relativity: T_{ab} satisfying (2) is thought of as a source for a linearised gravitational field. In this case, one anticipates being able to measure the total momentum and angular momentum from the field or in other words form the linearised Riemann tensor. This is what the right-most equality does where

$$f_{ab} = \omega_{AB}\epsilon_{A'B'}$$

and ω_{AB} is related to k^a by (3). The right-most equality now follows from the divergence theorem when S is a 2-surface on the 3-surface Σ surrounding the source.

According to Penrose [14], the problem of defining momentum and angular momentum in a curved space is the problem of carrying over some part of the fundamental identity.

Before we consider that problem we note some other properties of the kinematic twistor $A_{\alpha\beta}$ which follow from (1). For these we need some of the standard machinery of twistor theory in Minkowski space. We introduce coordinates on T as usual:

$$Z^\alpha = (\Omega^A, P_{A'}) \tag{5}$$

and define the pseudo-Hermitian norm

$$\Sigma(Z,\overline{Z}) = \Sigma_{\alpha\alpha'}Z^{\alpha}\overline{Z}^{\alpha'} = \Omega^{A}\overline{P}_{A} + \overline{\Omega}^{A'}P_{A'}. \qquad (6)$$

The norm defines the map from the complex conjugate $\overline{\mathsf{T}}$ to the dual T^{*}:

$$\Sigma : (\Omega^{A}, P_{A'}) \longrightarrow (\overline{\Omega^{A'}}, \overline{P}_{A}) \longrightarrow (\overline{P}_{A}, \overline{\Omega}^{A'})$$
$$Z^{\alpha} \longmapsto \overline{Z}^{\alpha'} \longmapsto \overline{Z}_{\alpha} = \Sigma_{\alpha\alpha'}\overline{Z}^{\alpha'} \qquad (7)$$

which is the way one usually thinks of complex conjugation in twistor theory.

Finally there is the infinity twistor for Minkowski space $I^{\alpha\beta}$ which we think of as a map of T^{*} to T:

$$I : (\lambda_{A}, \mu^{A'}) \longmapsto (\lambda^{A}, 0)$$
$$W_{\alpha} \longmapsto Z^{\alpha} = I^{\alpha\beta}W_{\beta}.$$

The norm and infinity twistor fit together to define a bar-hook operation from T to T (those familiar with Penrose's diagrammatic notation will see the reason for the name):

$$(\Omega^{A}, P_{A'}) \longmapsto (\overline{P}^{A}, 0)$$
$$Z^{\alpha} \longmapsto I^{\alpha\beta}\Sigma_{\beta\beta'}\overline{Z}^{\beta'} \qquad (8)$$

Equation (3) has a 3-complex-parameter family of solutions with zero k^{a}. These are the constant spinors ω^{AB} and clearly they give zero in (1). The corresponding (valence-2) twistors turn out to be of the form

$$Z_{1}^{(\alpha}Z_{2}^{\beta)} = I^{(\alpha|\gamma|}W_{1\gamma}I^{\beta)\delta}W_{2\delta}$$

for arbitrary $W_{1\alpha}, W_{2\beta}$ so that $A_{\alpha\beta}$ must satisfy

$$A_{\alpha\beta}I^{\alpha\gamma}I^{\beta\delta} = 0.$$

Next, if k^{a} is a real translation then $Q[k]$ is real. These Killing vectors come from solutions of (3) spanned by (valence-2) twistors of the form

$$Z^{(\alpha}I^{\beta)\gamma}\overline{Z}_{\gamma} \qquad (9)$$

The charge $Q[k]$ of (1) in this case gives the component

$$A_{\alpha\beta}Z^{\alpha}I^{\beta\gamma}\overline{Z}_{\gamma} \qquad (10)$$

of A. If this is to be real then the combination $A_{\alpha\beta}I^{\beta\gamma}$ must be Hermitian in the sense that

$$A_{\alpha\beta}I^{\beta\gamma} = \overline{A}^{\gamma\beta}I_{\beta\alpha} \qquad (11)$$

Note that, as is usual in twistor theory, in (11) primed twistor indices are eliminated by using the norm $\Sigma_{\alpha\alpha'}$. We may write (11) with explicit appearances of the norm as

$$A_{\alpha\beta}I^{\beta\gamma}\Sigma_{\gamma\alpha'} = \overline{A}_{\alpha'\beta'}I^{\beta'\gamma'}\Sigma_{\gamma'\alpha} \tag{12}$$

In this form we see that this Hermiticity condition uses only the barhook combination. If we break $A_{\alpha\beta}$ down into components compatibly with the coordinatisation (5) then we easily find that there is a 10-real-parameter family of A satisfying (11). In other words it is the Hermiticity condition that restricts $A_{\alpha\beta}$ to the right number of components.

In the particular case of a twistor of the form (9), the Killing vector k^a is a null and future-pointing translation. If the stress-energy-momentum tensor T_{ab} of the source satisfies the Dominant Energy Condition [50, 60], then the integrand in the third term of (1) is positive and therefore so is the quantity (10). In summary, there is a *Hermiticity condition* on $A_{\alpha\beta}$ phrased in terms of the bar-hook operation which reduces the space of $A_{\alpha\beta}$ to 10-real-dimensions and a *positivity condition* on $A_{\alpha\beta}$ involving the same expression when the source satisfies a suitable positivity condition. This positivity condition corresponds to the total momentum being time-like and future-pointing: with the coördinates (5) for Z^α, (10) is $2P^{AA'}P_{A'}\overline{P}_A$ where $P^{AA'}$ is the total momentum.

Although the fundamental identity (1) was originally thought of as a description of conserved quantities for a source in flat space and linearised general relativity, we can just as well use it to describe systems in other spaces of constant curvature, and general relativity linearised about other spaces of constant curvature. The difference in the twistor theory is that in these spaces, namely de Sitter space and anti-de Sitter space, the appropriate infinity twistor is not simple. Replacing (8) we have

$$\begin{aligned} (\Omega^A, P_{A'}) &\longmapsto (\overline{P}^A, \Lambda\overline{\Omega}_{A'}) \\ Z^\alpha &\longmapsto I^{\alpha\beta}\Sigma_{\beta\beta'}\overline{Z}^{\beta'} \end{aligned} \tag{13}$$

where Λ is negative for de Sitter space and positive for anti-de Sitter space.

The physical difference is that we now have the de Sitter or anti-de Sitter group instead of the Poincare group and so we cannot classify the 10 real charges as linear momentum or angular momentum. We still have Hermiticity for $A_{\alpha\beta}$ according to (11) or (12) and this still reduces the space of possible $A_{\alpha\beta}$ to 10 real dimensions, but the question of positivity depends on the sign of Λ. For anti-de Sitter space the dominant energy condition does imply positivity of (10) but for de Sitter space it does not. Really this is to be expected since de Sitter space has compact space-sections and the 3-surface integral in (1) will give zero if taken over the whole compact 3-surface. Since the 2-surface integral doesn't distinguish between inside and outside it now cannot satisfy a positivity condition. (Anything positive on the 'inside' would be negative on the 'outside').

We now return to the problem of carrying (10) over to curved space. The basic idea is to take the outer two terms:

$$A_{\alpha\beta} Z_1^\alpha Z_2^\beta = \frac{1}{4\pi G} \int_S R_{abcd} f^{ab} d\sigma^{cd} \tag{14}$$

and use the full Riemann tensor in the 2-surface integral on the right.

To define the bivector f^{ab} we need to be able to define an analogue on S of the solutions ω^{AB} of the valence-2 twistor equation. Penrose's proposal [14] is to take for these the product $\omega_1^{(A}\omega_2^{B)}$ of spinor fields which satisfy (4) projected tangent to S. In other words, one takes the parts of (4) which involve only differentiations in directions lying in S.

Four equations result from this projection. To write these equations out, we choose a normalised spinor dyad $(o^A, \iota^A; o_A \iota^A = 1)$ defined by S in that $\ell^a = o^A \bar{o}^{A'}$, $n^{AA'} = \iota^A \bar{\iota}^{A'}$ are the 2 null normals to S and $m^{AA'} = o^A \bar{\iota}^{A'}$ is a complex null tangent to S. We want the parts of (4) in the m^a and $\bar{m}^a$ directions and these, in the GHP formalism [46, 60] are [14]

$$\eth\omega^1 - \sigma\omega^0 = 0; \quad \eth'\omega^0 - \sigma'\omega^1 = 0 \tag{15}$$

$$\eth'\omega^1 - \rho\omega^0 = -i\pi_{0'}; \quad \eth\omega^0 - \rho'\omega^1 = -i\pi_{1'}, \tag{16}$$

The first pair (15) define an elliptic system on the components of ω^A. For the generic topologically-spherical 2-surface S in the generic space-time this sytem has a 4-complex-dimensional family of solutions (i.e. the index is 4) and this family defines the 2-surface twistor space $\mathsf{T}(S)$. (On a surface of arbitrary genus g the index is $4(1-g)$, (see [3]), so that there are generically no solutions if $g > 0$.) On a spherical surface where the index is 4 it is possible to choose the data so that there exist more than 4 linearly independent solutions to (15) (see [7]).

Symmetrized products of these 2-surface twistors are taken to define the bivector f^{ab} and then (14) defines a kinematic twistor $A_{\alpha\beta}$ as an element of the space $(\mathsf{T}(S) \odot \mathsf{T}(S))^*$ i.e. a valence-2, symmetric dual twistor.

The second pair of equations (16) define $\pi_{A'}$ from ω^A. Note that by commuting derivatives on ω^A we may deduce that

$$\begin{aligned}
\eth\pi_{0'} + \rho\pi_{1'} &= i(\psi_2 - \phi_{11} - \Lambda)\omega^1 + i(\psi_1 - \phi_{01})\omega^0 \\
\eth'\pi_{1'} + \rho'\pi_{0'} &= i(\psi_3 - \phi_{21})\omega^1 + i(\psi_2 - \phi_{11} - \Lambda)\omega^0
\end{aligned} \tag{17}$$

Unfortunately there is no similar expression for the other 2 components $\eth'\pi_{0'}$, $\eth\pi_{1'}$, of the derivative of $\pi_{A'}$.

With these preliminaries, the definition (14) can be written out as

$$A_{\alpha\beta} Z_1^\alpha Z_2^\beta = \frac{1}{4\pi G} \int R_{ABcd} \omega_1^A \omega_2^B d\sigma^{cd} \tag{18}$$

$$= \frac{1}{4\pi G}\int \{(\phi_{01}-\psi_1)\omega_1^0\omega_2^0 + (\phi_{11}+\Lambda-\psi_2)(\omega_1^0\omega_2^1+\omega_1^1\omega_2^0)$$
$$+ (\phi_{21}-\psi_3)\omega_1^1\omega_2^1\}dS \qquad (19)$$
$$= \frac{-i}{4\pi G}\int (\pi_{0'}^1\pi_{1'}^2 + \pi_{1'}^1\pi_{0'}^2)dS. \qquad (20)$$

The second formula (19) is (18) written out in the GHP formalism and then (20) follows from (19) by using (17).

This then is the construction. What can we say about its properties and the extent to which it resembles the kinematic twistor for a system in Minkowski space or de Sitter or anti-de Sitter space?

First of all we note that $A_{\alpha\beta}$ depends only on the data of S by which we mean the first and second fundamental forms of S, [32]. The rest of the space-time doesn't enter into the construction. In particular this makes the comparison of the $A_{\alpha\beta}$ at two different surfaces difficult since in general we don't know how to relate the different 2-surface twistor spaces.

We may ask whether the 2-surface twistor space has the structures found in the twistor space of spaces of constant curvature. In particular given the spinor fields $(\omega^A, \pi_{A'})$ we may form the quantity

$$\Sigma = \omega^A\overline{\pi}_{A'} + \overline{\omega}^{A'}\pi_{A'}. \qquad (21)$$

In a space of constant curvature, this is constant even though constructed from spinor fields which vary from point to point, and this constant is the norm (6) (at least in the conventional coordinatisation). However, for a general 2-surface in a general space-time there is no reason to expect (21) to be constant. One can show that it is constant for every 2-surface twistor *if and only if the 2-surface with its data can be embedded in a conformally flat space-time* (modulo certain conditions of genericity) [32, 6]. Following the suggestion of Penrose [15], we shall call a 2-surface with this property *non-contorted* and say that on the contrary a 2-surface is *contorted* if its data can detect the presence of conformal curvature. Then for a non-contorted 2-surface we have a norm (21) and the corresponding map (7).

We might next seek an infinity twistor for $\mathsf{T}(S)$. However, from what we already have we can see that this would be optimistic. The question is whether the 2-surface with its data can determine an $I^{\alpha\beta}$ and in particular can make a decision between the 3 different types which we have met so far. In fact for a general 2-surface no clear candidate presents itself. We might still hope for the combined bar-hook operation and there is one class of 2-surfaces where this is what we have in a natural way (see Section 3(i)). In general however the way is blocked. We always have a definition of kinematic twistor, (18), but other structures on $\mathsf{T}(S)$, which we need to have if we want $A_{\alpha\beta}$ to have any properties, are hard to find and seem to rely on special properties of S.

Most of the rest of this review is concerned with results found for non-contorted 2-surfaces. In this case we have the norm and so in particular we can calculate a scalar invariant from $A_{\alpha\beta}$ namely

$$M_N^2 = -\tfrac{1}{2} A_{\alpha\beta}\overline{A}^{\alpha\beta} = -\tfrac{1}{2} A_{\alpha\beta}\overline{A}_{\alpha'\beta'}\Sigma^{\alpha\alpha'}\Sigma^{\beta\beta'}. \tag{22}$$

For an $A_{\alpha\beta}$ which arises from a system in Minkowski space this is the length-squared of the total-momentum vector and so is positive given the positivity condition.

In the non-contorted case we also have a preferred 4-form on $\mathbf{T}(S)$. This is defined as follows: given 4 linearly independent 2- surface twistors $(\omega_i^A, \pi_{iA'})$ labelled by Z_i^α, consider the 4×4 matrix whose rows are $(\omega_i^0, \omega_i^1, \pi_{i0}, \pi_{i1})$. The determinant of this matrix

$$\nu = \left(\omega_i^0, \omega_i^1, \pi_{i0}, \pi_{i1}\right) \tag{23}$$

turns out to be constant for non-contorted 2-surfaces and therefore defines a 4-form:

$$\nu = \epsilon_{\alpha\beta\gamma\delta} Z_1^\alpha Z_2^\beta Z_3^\gamma Z_4^\delta. \tag{24}$$

We could make another scalar invariant from the determinant of $A_{\alpha\beta}$ using this 4-form:

$$M_D^4 = \tfrac{1}{4} \det A_{\alpha\beta}. \tag{25}$$

For an $A_{\alpha\beta}$ which arises from a system in Minkowski space, M_N and M_D are the same. However, for an $A_{\alpha\beta}$ which arises from a system in de Sitter or anti-de Sitter space, they are two different scalar invariants.

The invariants (22) and (25) are what we shall call the quasi- local mass. When there is no confusion we shall write it M_P.

To end this section I remark that the notion of non-contortedness can usefully be extended to spaces of dimension other than 2. We may look for 3-surfaces Σ which admit solutions of the 3-surface twistor equation (see [30]), by which I mean (4) projected into Σ. We find that the 3-surface twistor equation has a 4-complex-dimensional family of solutions (a 3-surface twistor space $\mathbf{T}(\Sigma)$) iff Σ with its first and second fundamental forms can be embedded in a conformally flat space-time. In this case, the 2-surface twistor space $\mathbf{T}(S)$ for any S lying on Σ is obtained by restricting the 3-surface twistors, and any such S is non- contorted. In particular this gives back the middle term in the fundamental identity (1) and expresses the kinematic twistor in terms of 3-surface integrals involving the stress-energy-momentum tensor T_{ab}. (The vector k^a is however not a Killing vector in this case). We may call such Σ a non-contorted 3-surface.

Finally, in a conformally-flat space-time (such as a Friedman-Robertson-Walker cosmology) there are solutions of (4) throughout the space-time. Any 2-surface twistor arises by restricting a '4-space twistor' and the kinematic twistor is once again written in terms of flux integrals. We may think of this as a 'non-contorted 4-space'.

3   Examples

i) Asymptotically flat-space time

In a space-time which is asymptotically flat at spatial infinity, one has a definition of total momentum, namely the ADM momentum (see e.g. [37]). Subject to extra conditions on the space-time one also has the total angular momentum. These definitions follow from the quasi-local kinematic twistor calculated on large spheres tending suitably towards infinity, as was shown by Shaw [27]. Also one has positivity by the Witten argument [69].

For space-times asymptotically flat at null infinity $\mathcal{I}^+$ one has the Bondi momentum (see e.g. [61]) but the appropriate definition of angular momentum is less clear (see e.g. Winicour [68]). Two-surfaces at $\mathcal{I}^+$ are contorted but one has a natural bar-hook operation. Essentially this is because σ vanishes so that equations (15) decouple. The anti-linear map $(\omega^0, \omega^1) \rightarrow (\overline{\omega}^1, 0)$ preserves solutions of (15) and is the desired bar-hook. With respect to this operation $A_{\alpha\beta}$ is Hermitian as in (12) and the momentum defined by (10) is the Bondi momentum. Furthermore positivity follows by the modifications of the Witten argument [57, 53, 62]. Here one uses 2-surface twistors in the image of bar-hook as the boundary values for the Witten spinors (since these are the 'asymptotically constant' ones).

One can extract a definition of total angular momentum from $A_{\alpha\beta}$ which differs from earlier definitions and has a greater appearance of naturalness (see e.g. Penrose and Rindler [61]).

ii) Asymptotically de Sitter and anti-de Sitter spaces

This case has been investigated by R.M. Kelly and the results are in his thesis but have not been previously published (though see Kelly [11]). A definition of asymptotically anti-de Sitter spaces has been given by Ashtekar and Magnon [38]. It turns out that their conditions give a time-like $\mathcal{I}$ on which 3-surface twistors exist (see also [49, 30]). The kinematic twistor can therefore be written in terms of 3-surface integrals of fluxes. In this form the components of $A_{\alpha\beta}$ are equal to the 10 real quantities constructed from the anti-de Sitter group at $\mathcal{I}$ by Ashtekar and Magnon. There is a natural infinity twistor like (13) and $A_{\alpha\beta}$ is Hermitian with respect to it. Now if one can find a complete space-like hypersurface Σ inside the space-time and bounded by a cut S of $\mathcal{I}$ then one can repeat the modified Witten-argument of Gibbons *et al* [48]: solve the modified Witten equation on Σ with boundary values equal to a 3-surface-twistor at S, then the dominant energy condion inside the spacetime implies that

$$A_{\alpha\beta}Z^\alpha I^{\beta\gamma}\overline{Z}_\gamma \geq 0$$

with equality only for anti-de Sitter space. This gives a more natural setting of the result of Gibbons *et al*, as the positive-definiteness of the Hermitian matrix $A_{\alpha\beta}I^{\beta\gamma}$. This positivity implies positivity of various scalar invariants, for example

$$M_N \geq M_D \geq 0$$

in the language of (22, 25).

One can define asymptotically de Sitter spaces and treat them in the same sort of way. Now future infinity $\mathcal{I}^+$, is a space-like surface with 3-surface twistors. The kinematic twistor $A_{\alpha\beta}$ is again Hermitian with respect to a naturally arising infinity twistor. However, one does not this time get positivity which is after all to be expected from the discussion in Section 2.

iii) Spherically Symmetric 2-surfaces

A spherically symmetric 2-surface together with its data is characterised by just 2 numbers. One of these is the area and the other turns out to be the quasi-local mass (Tod [29]). A definition of mass for spherically symmetric systems in general relativity has been given by Cahill and McVittie [40] based on different ideas, and also by Hernandez and Misner [52]. Fortunately these definitions are all in agreement. As particular examples of the formula we find:

a) for all spheres of symmetry in the Schwarzschild solution the quasi-local mass M_P is the Schwarzschild mass M regardless of the radius of the sphere.

b) for a sphere of symmetry in the Reissner-Nordstrom solution

$$M_P = M - \frac{e^2}{2r}$$

where e is the charge and r the area-radius of the sphere.

c) for a sphere of symmetry S in Friedman-Robertson-Walker cosmologies the mass is

$$M_P = \rho V$$

where ρ is the density and V is the volume of a sphere in flat-space with the same area as S. Note how this distinguishes between the different spatial metrics: for $k = 1$, M_P rises with the radius to a maximum value and then drops to zero; for $k = 0$, M_P is proportional to the volume within S; and for $k = -1$, M_P grows with radius much faster than the volume.

d) One class of Kantowski-Sachs cosmologies have space-sections homeomorphic to $S^1 \times S^2$. As pointed out by Tipler [28] one can assign a quasi-local mass to one of the S^2's, but since it can't be shrunk to a point, one cannot think of this as the mass of the matter 'inside'. Tipler argued that in this case one should only think of the mass between two such spheres, and defined by the difference between the quasi-local quantities, as significant. We shall see in subsection (iv) on non-contorted 3-surface that, on the contrary, some very appealing results are associated with 2-spheres which can't be shrunk if we stick to the quasi-local mass definition without this modification. Indeed even in the case at hand one finds the 'conservation law'

$$M'_P = -pV'.$$

Here M_P is the quasi-local mass of a sphere of symmetry in the Kantowski-Sachs metric, p is the pressure, V is the volume of a sphere in flat-space with the same area and $'$ is differentiation with respect to the canonical time coordinate (Tod [33]).

iv) Non-contorted 3-surfaces

Any spherically symmetric 3-surface Σ in the Schwarzschild metric is non-contorted (Tod [32]). From the discussion in Section 2, this means that the $A_{\alpha\beta}$ for a 2-surface S on Σ is unchanged as S is moved about on Σ and that any such S is non-contorted. In particular if S can be shrunk to a point on Σ then the associated $A_{\alpha\beta}$ is zero, while if S can be moved to be a 2-sphere of symmetry then the associated M_P is M (by (iii)(a) above).

There are explicit many-black-hole solutions of the constraints for the time-symmetric initial value problem (see e.g. Wheeler [66]). This initial value surface is non-contorted and so again $A_{\alpha\beta}$ for a 2-surface only involves the question of how many holes are linked. For the example of figure 1 there is a mass for each hole and for each pair of holes and so on.

By looking at these expressions at large separations of the holes we find expressions resembling potential energy. For example, for the surfaces as shown in figure 1 the quasi-local masses are related by

$$M_{12} = M_1 + M_2 - \frac{M_1 M_2}{d} + O(\frac{1}{d^2})$$

where d is a measure of the separation (Tod [29]).

In the LRS Bianchi cosmologies and in the Kantowski-Sachs cosmologies the surfaces of constant time are non-contorted (Tod [33]). This makes it straightforward to calculate the quasi-local mass for cylindrical 2-surfaces in these metrics. Typically one obtains

$$M_P = \rho \times \text{ volume } \times f(L) \times \chi.$$

Here ρ is the density, f gives the dependence on length L over and above what is in the volume term and χ contains two sorts of term: first, curvature terms like those in the FRW cases 3(iii)(c) and, second, kinetic energy terms

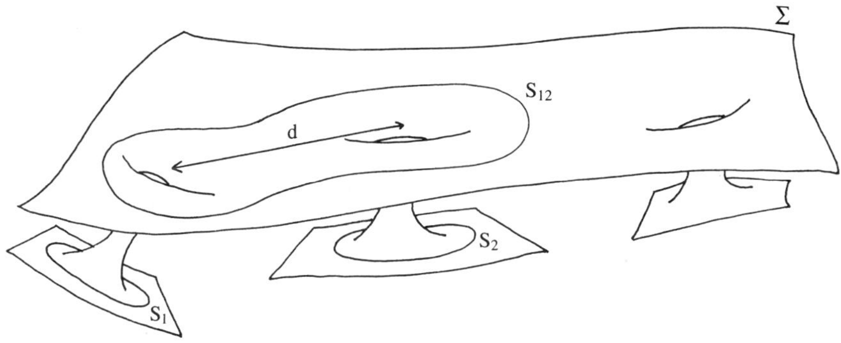

Figure 1: Time-symmetric vacuum initial data

when the fluid flow is tilted with respect to the surfaces of homogeneity. This can happen in LRS Bianchi type V and in this case the matter is flowing along the cylinder.

The dependence on length $f(L)$ takes one of three forms which are universal for cylinders of symmetry (Tod [35]). These are

$$f(L) = 1, \quad \frac{\sin \omega L}{\omega L}, \quad \frac{\sinh \omega L}{\omega L}$$

where ω is an invariant of the data of the cylinder. We may then say that cylinders fall into three 'types': parabolic, elliptic and hyperbolic respectively. The interpretation of type is in terms of potential energy: if a cylinder has hyperbolic type then doubling the length more than doubles the mass, so the cylinder has positive potential energy; an elliptic cylinder has negative potential energy and a parabolic cylinder has zero potential energy. We shall meet this phenomenon again in 3(vi).

v) The Post-Newtonian Limit

There are various formalisms in which to discuss the Newtonian limit of General Relativity [67, 45, 43]. They all involve considering a family of perfect-fluid space-times depending on a parameter λ and such that the family tends to a Newtonian space time (Cartan [41, 42]; Trautman [64]) as λ tends to zero.

Jeffryes [8] has calculated the quasi-local kinematic twistor for a 2-surface S in such a family when the 2-surface lies in a 3-surface of constant Newtonian time in the Newtonian limit. It turns out that the 3-surface is non-contorted to a certain power of λ and therefore that the twistor-space norm is well-defined to a certain power of λ. This allows Jeffryes to calculate a mass from the norm of the kinematic twistor. In the Newtonian limit this mass is just the conserved mass defined by the conservation equation on the perfect-fluid source:

$$\nabla_a(N(\rho, P)u^a) = 0 \quad \Rightarrow \quad M_C = \int N u_a d\Sigma^a = \text{const}$$

where N is the number density and u_a is the fluid 4-velocity.

The next two terms in an expansion in λ are also invariantly defined. The first is a correction to M_C but when the next term is included the result takes the form

$$M_P^2 = P^a P_a + O(\lambda^7)$$

where P_a is a total momentum vector defined previously by Will [67] and Damour [43] (among others). They make the definition in the context of post-Newtonian calculations in relativity, since this vector is conserved to an appropriate order and is physically motivated as a total momentum.

In summary, the quasi-local-mass is the norm of the total momentum vector defined by earlier workers on the post-Newtonian formalism. It seems very likely that one could define an infinity twistor at the appropriate order in λ and actually obtain the same momentum vector from the kinematic twistor but this hasn't yet been done. It is very striking that the quasi-local-mass, which is constructed to be correct in the limit of linearised relativity should also be correct in the limit of Newtonian gravity.

vi) Cylindrical gravitational waves

With the aim of detecting gravitational wave energy quasi-locally rather than globally, as with the Bondi mass (3(i)) I considered cylindrically symmetrical space-times [35]. I considered only solutions with 2 commuting, hypersurface-orthogonal Killing vectors since for these a cylinder of symmetry, with 'flat' ends orthogonal to the z-Killing vector and curved sides ruled by both Killing vectors, is non-contorted. As remarked in 3(iii), the quasi-local-mass for such a cylinder has a dependence on coordinate length which is one of three types. The different types are related to the sign of potential energy. For example, a cylindrically-symmetric, static charged-dust solution in which the electro-static repulsion is exactly balanced by gravitational attraction always has parabolic type, which is consistent with the identification of this type with zero potential energy. Static perfect-fluid space-times always have hyperbolic type and therefore positive potential energy.

The dependence of the quasi-local-mass on the radius of the cylinder depends on the details of the solution. For static perfect-fluids, one can show that it is always positive. For gravitational waves one is forced to use an approximation. There is a solution corresponding to an incoming wave-packet which passes through the axis to become an outgoing wave-packet

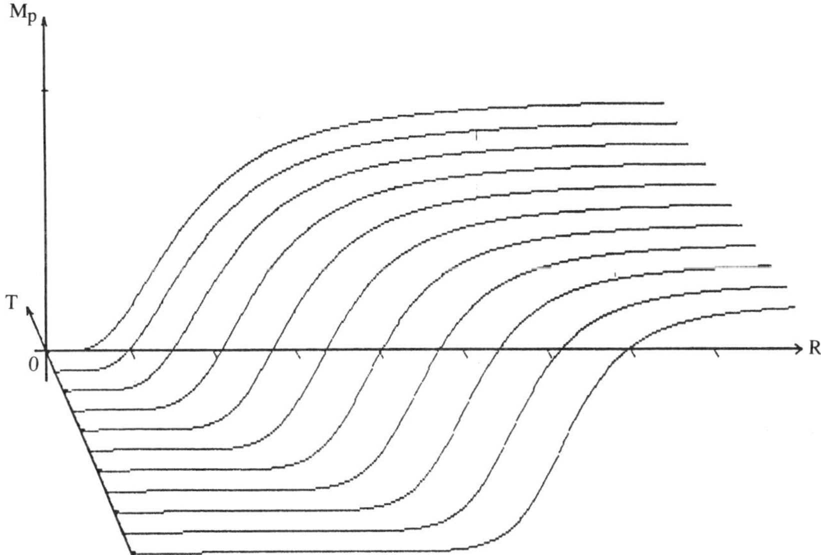

Figure 2: The quasi-local mass for a cylindrical wave-pulse; for negative T there is an incoming wave close to $T + R = 0$; the graph is even in T.

([65, 39]; the solution somewhat resembles an elementary state in the language of twistor theory). For this solution the quasi-local-mass as a function of radius and time is shown in figure 2, with the approximation mentioned above. As anticipated, the mass at a fixed radius goes up as the wave passes on its inward path and drops back to zero as the wave passes on its outward path. The plateau value is related to the conicality or deficit angle round the symmetry axis as seen from large space-like separation. One can show that this behaviour is quite general for waves of this nature. In figure 3 is shown a notional density equal to mass over radius-squared for the same wave. Again one sees an incoming-outgoing wave of energy-density.

vii) Black hole in a box

The idea here is to consider a stationary black hole deformed by external matter (the 'box') and to take the cross-over 2-surface on the Killing horizon as S. The data of S reduce in this case just to its metric and one function in the extrinsic curvature. In general S is contorted but if the one function vanishes, which corresponds to a static black hole, then S is non-contorted. In this case, the quasi-local mass is a functional of the metric alone (Tod [31, 32]).

Considerations based on the Cosmic Censorship Hypothesis [58] lead one

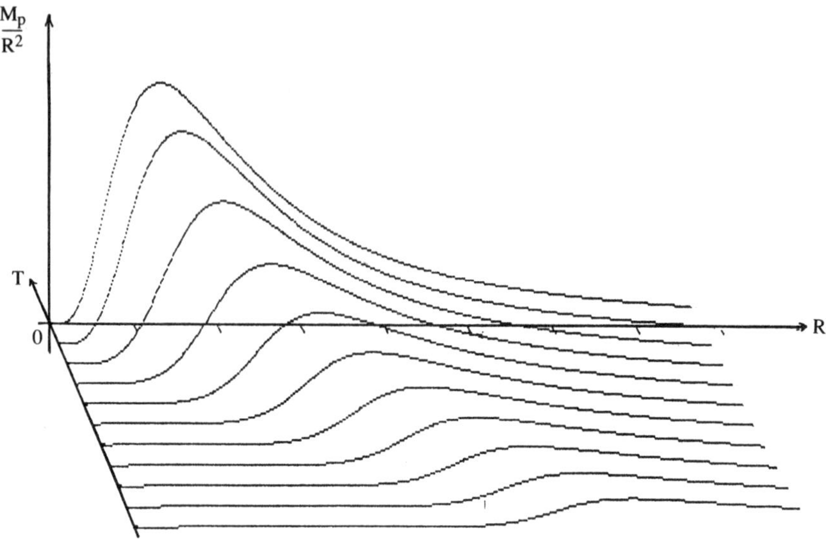

Figure 3: The corresponding wave of 'energy-density'

to expect that a black hole should satisfy the inequality

$$16\pi M^2 \geq A$$

between its area and mass, where the mass can be defined. This is some-times called the *isoperimetric inequality for black holes* (see e.g. [47]). In the case under consideration this inequality reduces to an inequality for positive smooth functions Ω on the unit S^2

$$\left(\int \Omega(1 - \nabla^2 \log \Omega)\right)^2 \geq 4\pi \int \Omega^2 \qquad (26)$$

where the integration is over the unit S^2 and ∇^2 is the unit-S^2 Laplacian.

Inequality (26) has arisen elsewhere as a prediction of Cosmic Censorship (Penrose). It was proved by Tod [31, 32]. Yet another way to prove it is to apply the usual isoperimetric inequality to the solid in $\mathbf{R}^3$ defined by $r = \Omega(\theta, \phi)$.

If the black hole is stationary but not static, then this gives a very simple example of a contorted 2-surface. We shall return to a specific example, namely the cross-over 2-surface in the Kerr solution, in Section 4.

viii) Other Examples

Various other cases which haven't yet been mentioned have been calculated.

In the Robinson-Trautman metrics, one can calculate the quasi-local-mass at 2-surfaces of constant u and r. (Tod [29]). (These coordinates are invariantly defined by the space-time). These surfaces are contorted but they have a natural bar-hook operation (since σ vanishes) and are very similar to 2-surfaces at $\mathcal{I}^+$. One obtains the Bondi momentum as the quasi-local-momentum for every value of r at fixed u. This is consistent with the picture one has of the RT metrics as representing purely outgoing radiation.

A calculation for 'large spheres' has been done by Shaw [27] as an expansion in powers of $1/r$ away from a 2-surface at $\mathcal{I}^+$. There are some ambiguities in the definition of the norm, but the calculations are interesting because of the appearance of the NP conserved quantitites (see e.g. Penrose and Rindler [61]). Numerical calculations have been done by Jeffryes [9] for the mass at various surfaces in the static, axisymmetric Weyl solutions. He obtains a graph of mass against radius for families of surfaces in particular Weyl solutions. It would be interesting to relate these graphs to the question of existence of sources for the Weyl solutions satisfying, say, the Dominant Energy Condition.

We may summarise this section by listing different manifestations of mass-energy which are successfully 'detected' by the quasi-local-mass construction. These are

1. total mass-energy; in the sense of the ADM or Bondi mass.

2. gravitational wave energy at infinity; in the Bondi mass.

3. rest-mass-energy; in the k = 0 FRW cosmology.

4. matter kinetic-energy; in the tilted Bianchi V cosmology.

5. gravitational potential-energy; in the time-symmetric initial-value problem, in the k = 1 FRW cosmologies and in the cylindrically-symmetric examples.

6. electro-static field energy; in the Reissner-Nordstrom solution.

7. gravitational wave energy 'quasi-locally'; in the cylindrically symmetric gravitational waves.

8. the post-Newtonian conserved energy.

There is some arbitrariness in the way one classifies the properties of e.g. the FRW cosmologies but none-the-less this is an impressive list of successes.

4   The Contorted Case

In this section I shall summarise and comment on what is known about the quasi-local-mass construction on contorted 2-surfaces. By contrast with the case of non-contorted 2-surfaces, what is known is rather less.

Recall from Section 2 that, on a contorted 2-surface S one has the 2-surface twistor space $\mathsf{T}(S)$ and the kinematic twistor as usual, but the standard definition of norm (21) does not define a function on $\mathsf{T}(S)$. Without this, or better still a bar-hook operation, one does not know how to extract information from the kinematic twistor.

The response to this problem, in general terms, has been to seek, and seek to justify, some modification to the construction when the 2-surface S is contorted. Such a modification would be required to disappear when S becomes non-contorted. There are of course many possible modifications but here we will concentrate mostly on one centering on the determinant ν defined in (23, 24). This quantity is a constant on S when S is non-contorted. When S is contorted, ν is not obliged to vary and one can define a sub-class of contorted 2-surfaces on which ν is constant and study these. (This condition has been considered by Jeffryes [10]. Note in particular, following the discussion in Section 3(i) that 2-surfaces at $\mathcal{I}^+$ in the presence of radiation, which are contorted but have a bar-hook, also have constant ν). However, if ν does vary then we can be sure that the surface is contorted.

The first modification using ν was suggested by Penrose [15] following a consideration of properties of the Schwarzschild solution. The suggestion is to replace the definition (18) by

$$A_{\alpha\beta}Z_1^{\alpha}Z_2^{\beta} = \frac{1}{4\pi G}\int R_{ABcd}\omega_1^A\omega_2^B\eta d\sigma^{cd} \tag{27}$$

where η is a constant multiple of ν (Penrose and Rindler [61]). The connection with the Schwarzschild solution is as follows: there exists a valence-2 twistor i.e. a solution of (3) throughout the Schwarzschild solution; if this twistor lies in the 2-surface twistor space $\mathsf{T}(S) \odot \mathsf{T}(S)$ and one adopts the modified definition (27) then the quasi-local mass is zero for any 2-surface which does not go round the hole. (In fact, the whole $A_{\alpha\beta}$ would be zero on such a 2-surface). This would extend the result of Section 3(iii) in a particularly nice way.

Taking this modification seriously, we need some other examples of contorted 2-surfaces to try it on. One class was considered in the 'small spheres' calculation and its subsequent 'small ellipsoids' modification [12, 36]. The idea in both these calculations is to calculate $A_{\alpha\beta}$ for a sequence of 2-surfaces $S(r)$ which shrink to a point as the parameter r goes to zero. For the 'small spheres' we choose a point p and a time-like unit vector t^a at p and consider the surfaces of constant affine-parameter r on the null cone N of p. (The significance of t^a is that, to define r we fix the scale of the null generators

l^a of N by $t^a l_a = 1$ at p). For the 'small ellipsoids' one chooses p and t^a as before but considers a sequence of ellipsoids on a space-like surface through p orthogonal to t^a.

In both cases, the leading non-zero term in $A_{\alpha\beta}$ is $O(r^3)$ in the presence of matter and $O(r^5)$ in vacuum. The norm defined by (18) is constant in leading order (up to $O(r)$) but has varying terms at $O(r^2)$. In the presence of matter the leading term in the norm of $A_{\alpha\beta}$ is the Minkowski norm of the vector

$$P_a = \tfrac{4}{3}\pi r^3 T_{ab} t^b$$

This satisfactory result is probably just a reflection of the fact that the construction is correct for linear theory.

In vacuum, on the other hand, the situation is quite unsatisfactory. Even in the Schwarzschild solution, the norm of $A_{\alpha\beta}$ can be negative in leading order. The determinant of $A_{\alpha\beta}$ will be positive in the Schwarzschild solution, but if we try small spheres in the Kerr solution the determinant can be complex.

To attempt to deal with these pathologies we recall that these surfaces are contorted and we try the modification (27). We choose the scale of η so that

$$\eta = 1 + O(r^2).$$

Then for the small spheres, the entire modified $A_{\alpha\beta}$ is zero at $O(r^5)$ and only appears at $O(r^6)$. On the other hand for the small ellipsoids, most of the components of the modified $A_{\alpha\beta}$ are zero at $O(r^5)$ and in particular the norm and determinant vanish.

There is some comfort in this situation: the modification (27) which has some justification from the Schwarzschild solution removes the $O(r^5)$ pathology from the small surfaces calculation. However, there is still the question of what is the leading non-zero term in $A_{\alpha\beta}$? This has defied calculation. A related problem emphasised by Woodhouse [36], is that the remaining $O(r^5)$ term in $A_{\alpha\beta}$ for the small ellipsoids will mean that the $O(r^6)$-calculation, if it is ever done, would need more than the leading term in the norm. Since we have no general prescription for defining this on a contorted 2-surface, more ambiguities will arise at the next order.

Another simple class of contorted 2-surfaces follows from the discussion in Section 3(vii): for a stationary but non-static 'black-hole-in-a-box' the crossover 2-surface is contorted. Such a 2-surface S is characterised by the complex curvature K of S in the sense of Penrose [60]. This takes the form

$$K = \tfrac{1}{2}(k + iq)$$

where k is the Gaussian curvature of S, and q is an extrinsic curvature quantity which integrates to zero on S. It turns out that on such an S, the determinant ν is automatically of the form $e^{i\theta}$ where θ is a potential for q:

$$\nabla^2 \theta = -q$$

We can remove the ambiguity in ν by requiring θ to integrate to zero on S, and then take ν in (27) to be $e^{i\theta}$. (There is still an ad hoc element in this since ν will not always have this simple form).

Given the modified $A_{\alpha\beta}$, we can now calculate a mass from its determinant or we can seek a modified norm. In general, one doesn't have a procedure for modifying the norm but for this class of surfaces the quantity

$$\omega^A \overline{\pi}_A + \overline{\omega}^{A'} \pi_{A'} + \Upsilon_{AA'} \omega^A \overline{\omega}^{A'} \tag{28}$$

is constant where

$$\Upsilon_a = \tfrac{1}{2} h_a^b \nabla_b \theta$$

(i.e. Υ_a is half the gradient of θ projected into the surface S). This is the kind of modification to the norm which would result from a complex conformal rescaling with conformal factor $e^{i\theta/2}$. This suggests intriguing possibilities which we shall return to but it is not known to lead to anything concrete.

Returning to the quasi-local-mass, one now has four possibilities for its definition: one can use the norm or the determinant and the modified or unmodified $A_{\alpha\beta}$. For the particular case of the Kerr solution one can readily calculate all four. Both definitions using the unmodified $A_{\alpha\beta}$ look reasonable, but if we regard the lesson of the 'small surfaces' as being the necessity for modifying $A_{\alpha\beta}$, then we must discard these. Using the norm and the modified $A_{\alpha\beta}$ leads to an M_P^2 which is negative for a range of the Kerr parameter a. Consequently we are reduced to the determinant definition with the modified $A_{\alpha\beta}$. A graph of this is given in figure 4 expressed as the ratio M_P/M of the quasi-local mass to the Kerr mass parameter M, which is also the ADM mass, and plotted against the dimensionless angular momentum parameter a/M. For comparison we include in the graph the irreducible mass M_i defined in terms of the area of A of the surface S by

$$16\pi M_i^2 = A.$$

As we see from the graph according to this definition much more of the ADM mass of the Kerr solution is in the space outside the hole as compared to the irreducible mass definition. The other notable feature is the local minimum in M_P which occurs at about $a/m = 0.965$. This may be a kind of 'preferred minimum-energy' configuration for Kerr black holes. (Notice, however, that it differs from the preferred value $a/m = 0.998$ found by Thorne [63] on the basis of detailed calculation of accretion.)

There have been other proposals for a definition of the mass of a Kerr black hole (e.g. [56]) but none that I know of agrees with this one.

To end this section, and this review, I shall mention some of the other attempts to deal with contortedness by modifying (18).

The first of these picks up the comment around (28): the contorted cross-over has very much the appearance of a non-contorted cross-over subjected

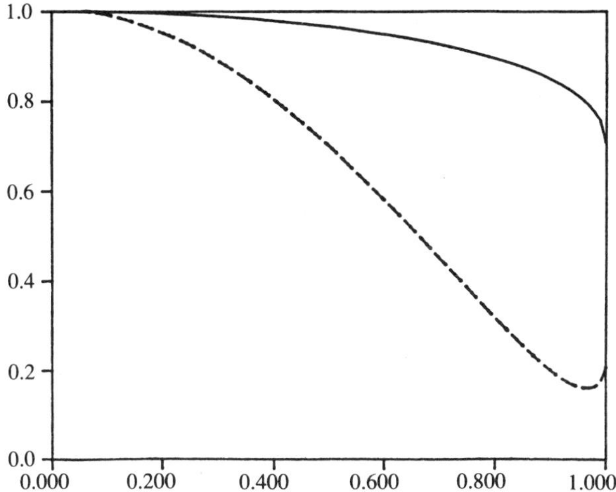

Figure 4: The quasi-local mass at the Kerr cross-over (dashed) and the irreducible mass (solid) as a fraction of the Kerr mass-parameter M plotted against the dimensionless angular momentum parameter.

to a complex conformal rescaling:

$$\epsilon_{AB} \rightarrow \hat{\epsilon}_{AB} = e^{i\theta/2}\epsilon_{AB}.$$

Complex conformal rescalings were considered by Penrose [59] and connected with torsion. If one wants to make some connection between contortedness and torsion then one might begin by modifying the fundamental identity (1) in the presence of torsion.

For the case of the linearised Einstein-Cartan-Sciama-Kibble (ECSK) theory (taking conventions and definitions as given in Penrose and Rindler [60]), one finds that the quantity

$$J^a = E^{ab}k_b + \tfrac{1}{2}S^{bca}\nabla_b k_c \tag{29}$$

is a conserved current for any Killing vector k_a. Here E^{ab} is the energy-momentum tensor (not necessarily symmetric) and S^{abc} is the spin-density of the matter content.

If we take (29) in place of $T^{ab}k_b$ in (1) then we can obtain a modification to the integrand in the 2-surface integral and redefine $A_{\alpha\beta}$. However it is not clear that this is the right thing to do. On the one hand, (29) is only conserved in linearised ECSK theory: there is a more complicated current available in the full theory (Hehl and Hecht [51]). On the other hand, even if we take one of these currents the twistor equation itself is modified in the presence of torsion and we would have to contemplate changing (15) as well.

While it may be possible to make progress in this direction, nothing concrete has yet emerged.

A modification like (27) but with the aim of preserving the form (20) was suggested by Penrose [18]. The new definition is

$$A_{\alpha\beta}Z_1^\alpha Z_2^\beta = \frac{-i}{2\pi G}\int \hat{\pi}_{0'}\hat{\pi}_{1'}dS$$

where

$$
\begin{aligned}
\hat{\pi}_{0'} &= \mu\pi_{0'} + i\omega^1\eth'\mu \\
\hat{\pi}_{1'} &= \mu\pi_{1'} + i\omega^0\eth\mu \\
\mu^2 &= \eta
\end{aligned}
$$

(Note that this is almost the same as (27) after integration by parts).

Applied to the small spheres calculation, this gives a positive M_P^2 proportional to the square of the magnetic part $H_{ab} = {}^*C_{acbd}t^c t^d$ of the Weyl tensor. However the motivation is rather slender.

A number of possible modifications involving η have been suggested by Mason [13] based on an attempt to understand the quasi-local-mass in terms of a Hamiltonian analysis of general relativity. The idea is that the components of the kinematic twistor are the values of the Hamiltonians generating motion of a 3-surface spanning the 2-surface S. For a 2-surface at infinity, this procedure is well defined and well understood but for a finite 2-surface there is still some obscurity. However this is an exciting suggestion, since it would connect the quasi-local-construction to these different notions of mass and also offer the possibility of a proof of positivity.

5   Conclusion

In conclusion the position is as follows: For non-contorted surfaces one has a definition of total mass-energy which has produced many appealing results. There is not a definition of momentum or angular momentum even in this case. Also there are no general proofs of Hermiticity or positivity though these are available in particular cases.

For contorted surfaces there are indications both that modifications are necessary and as to what they might be. However there is no clear way forward.

To end on a positive note, one might aim only to define a quasi- local momentum $A_{\alpha\beta}I^{\beta\gamma}\Sigma_{\gamma\alpha'}$. There is a way to do this which has recently been discovered. It avoids the problem of disentangling momentum and angular momentun and offers a more direct connection with the canonical analysis of general relativity and the Witten proof of positivity of total mass-energy. Whether the new definition shares the virtues of the old without sharing its vices remains to be investigated.

References

[1] **A Bibliography of Quasi-local Mass**

[2] M.A. Awada, G.W. Gibbons & W.T. Shaw (1986) *Conformal supergravity, twistors and the super BMS group*, Ann. Phys. **171**, 52–107.

[3] R.J. Baston (1984) *The index of the 2-twistor equation*, Twistor Newsletter **17**.

[4] T. Dray (1984) *Momentum flux at null infinity*, Class. Quantum Grav. **2**, L7–L10.

[5] T. Dray & M. Streubel (1985) *Angular momentum at null infinity*, Class. Quantum Grav. **1**, 15–26.

[6] B.P. Jeffryes (1984) *Two-surface twistors and conformal embedding*, in [44].

[7] B.P. Jeffryes (1986) *'Extra' solutions to the 2-surface twistor equations*, Class. Quantum Grav. **3**, L9–L12.

[8] B.P. Jeffryes (1986) *The Newtonian limit of Penrose's quasi-local mass*, Class. Quantum Grav. **3**, 841–852.

[9] B.P. Jeffryes (1986) *Quasi-local mass for the Weyl solutions*, privately circulated report.

[10] B.P. Jeffryes (1987) private communication.

[11] R.M. Kelly (1985) *Asymptotically anti-de-Sitter space-times*, Twistor Newsletter **20**.

[12] R.M. Kelly, K.P. Tod & N.M.J. Woodhouse (1986) *Quasi-local mass for small surfaces*, Class. Quantum Grav. **3**, 1151–1167.

[13] L.J. Mason (1989) *Hamiltonian interpretations of Penrose's quasi-local mass*, Class. Quantum Grav. **6**, L7.

[14] R. Penrose (1982) *Quasi-local mass and angular momentum in general relativity*, Proc. R. Soc. London **A381**, 53–63.

[15] R. Penrose (1984) *New improved quasi-local mass and the Schwarzschild solution*, Twistor Newsletter **18**.

[16] R. Penrose (1984) *Integrals for general relativistic sources: a development from Maxwell's electromagnetic theory*, in Maxwell Symposium Volume ed. M.S. Berger (North-Holland, Amsterdam).

[17] R. Penrose (1984) *Mass and angular momentum at the quasi-local level in general relativity*, in [44].

[18] R. Penrose (1985) *A suggested modification to the quasi-local formula*, Twistor Newsletter **20**.

[19] R. Penrose (1988) *Aspects of quasi-local angular momentum*, in [55].

[20] W.T. Shaw (1983) *Spinor fields at space-like infinity*, Gen. Rel. Grav. **15**, 1163–1189.

[21] W.T. Shaw (1983) *Twistor theory and the energy-momentum and angular momentum of the gravitational field at spatial infinity*, Proc. R. Soc. London **A390**, 191–215.

[22] W.T. Shaw (1984) *Twistors, asymptotic symmetries and conservation laws at null and spatial infinity*, [44].

[23] W.T. Shaw (1984) *Symplectic geometry of null infinity and 2-surface twistors*, Class. Quantum Grav. **1**, L33–L37.

[24] W.T. Shaw (1985) *Witten identities for rotations, spinor boundary value problems and new gauge conditions for asymptotic symmetries*, Class. Quantum Grav. **2** 189–217.

[25] W.T. Shaw (1986) *Total angular momentum for asymptotically flat space-times with non-vanishing stress tensor*, Class. Quantum Grav. **3**, L77–L81.

[26] W.T. Shaw (1986) *The asymptopia of quasi-local mass and momentum: 1; general formalism and stationary space-times*, Class. Quantum Grav. **3**, 1069–1104.

[27] W.T. Shaw (1988) *Quasi-local mass for large spheres*, in [55].

[28] F.J. Tipler (1985) *Penrose's quasi-local-mass in the Kantowski-Sachs closed universe*, Class. Quantum Grav. **2**, L99–L104.

[29] K.P. Tod (1983) *Some examples of Penrose's quasi-local mass construction*, Proc. R. Soc. London A 388, 457–477.

[30] K.P. Tod (1984) *Three-surface twistors and conformal embedding*, Gen. Rel. Grav. **16**, 435–443.

[31] K.P. Tod (1985) *Penrose's quasi-local mass and the isoperimentric inequality for static black holes*, Class. Quantum Grav. **2**, L65–L68.

[32] K.P. Tod (1986) *More on Penrose's quasi-local mass*, Class. Quantum Grav. **3**, 1169–1189.

[33] K.P. Tod (1987) *Quasi-local mass and cosmological singularities*, Class. Quantum Grav. **4**, 1457–1468.

[34] K.P. Tod (1988) *The integral constraint vectors of Traschen and 3-surface twistors*, Gen. Rel. Grav. **20**, 1297–1308.

[35] K.P. Tod (1989) *Penrose's quasi-local mass for cylindrically symmetric space-times*, in preparation.

[36] N.M.J. Woodhouse (1987) *Ambiguities in the definition of quasi-local mass*, Class. Quantum Grav. **4**, L121–L123.

Other References

[37] A. Ashtekar (1980) in 'General Relativity and Gravitation: One Hundred Years after the Birth of Albert Einstein Vol. 2' ed. A. Held (New York, Plenum Press).

[38] A. Ashtekar & A. Magnon (1984) Class. Quantum Grav. **1**, L39.

[39] W.B. Bonnor (1957) J. Math. and Mech. **6**, 203.

[40] M.E. Cahill & G.C. McVittie (1970) J. Math. Phys. **11**, 1382.

[41] E. Cartan (1923) Ann. Ec. Norm. Sup. **40**, 325.

[42] E. Cartan (1924) Ann. Ec. Norm. Sup. **41**, 1.

[43] T. Damour (1987) in '300 years of Gravitation', ed. W. Israel & S.W. Hawking, (Cambridge, CUP).

[44] F.J. Flaherty, ed. (1984) 'Asymptotic behaviour of mass and space-time', Springer Lecture Notes in Physics 202.

[45] T. Futamase & B. Schutz (1983) Phys. Rev. **D 28**, 2363.

[46] R. Geroch, A. Held & R. Penrose (1973) J. Math. Phys. **14**, 874.

[47] G.W. Gibbons (1984) in 'Global Riemannian Geometry' ed. T. Willmore & N.J. Hitchin, Ellis Horwood.

[48] G.W. Gibbons, S.W. Hawking, G.T. Horowitz & M.J. Perry (1983) Comm. Math. Phys. **88**, 295.

[49] S.W. Hawking (1983) Phys. Lett. **126B**, 175.

[50] S.W. Hawking & G.F.R. Ellis (1973) 'The Large-scale Structure of Space-time', (CUP, Cambridge).

[51] F.W. Hehl & R. Hecht (1986) private communication.

[52] W.C. Hernandez & C.W. Misner (1966) Ap. J. **143**, 152.

[53] G.T. Horowitz & M.J. Perry (1982) Phys. Rev. Lett. **48**, 371.

[54] S.A. Huggett & K.P. Tod (1985) 'An Introduction to Twistor Theory', (CUP, Cambridge).

[55] J. Isenberg, ed. (1988) 'Mathematics and General Relativity': Contemporary Mathematics 71 (Amer. Math. Soc., Providence RI).

[56] R. Kulkarni, V. Chellathurai & N. Dadhich (1988) Class. Quantum Grav. **5**, 1443.

[57] M. Ludvigsen & J.A.G. Vickers (1982) J. Phys. **A 15**, L67.

[58] R. Penrose (1973) Ann. N.Y., Acad. Sci. **224**, 125.

[59] R. Penrose (1983) Found. of Phys. **13**, 325.

[60] R. Penrose & W. Rindler (1984) 'Spinors and Space-time' Vol. 1, (CUP, Cambridge).

[61] R. Penrose & W. Rindler (1986) 'Spinors and Space-time' Vol. 2, (CUP, Cambridge).

[62] O. Reula & K.P. Tod (1984) J. Math. Phys. **25**, 1004.

[63] K.S. Thorne (1974) Ap. J. **191**, 507.

[64] A. Trautman (1966) in 'Perspectives in Geometry and Relativity', ed. B. Hoffman (Indiana U.P., Bloomington).

[65] J. Weber & J.A. Wheeler (1957) Rev. Mod. Phys. **29**, 509.

[66] J.A. Wheeler (1964) in 'Relativity, Groups and Topology', ed. de Witt & de Witt (Gordon and Breach; New York).

[67] C.M. Will (1981) 'Theory and Experiment in Gravitational Physics' (CUP, Cambridge).

[68] J. Winicour (1980) in 'General Relativity and Gravitation: One Hundred years after the Birth of Albert Einstein' Vol. 2, ed. A. Held (Plenum Press, New York).

[69] E. Witten (1981) Comm. Math. Phys. **80**, 381.

The Sparling 3-form, Ashtekar Variables and Quasi-local Mass

L.J. Mason   J. Frauendiener

In this article we review some basic results connnected with the use of the 'Sparling 3-form', a 3-form defined on the spin bundle of space-time. It is closed if and only if the vacuum equations are satisfied. This structure unifies, at least partially, some of the major spinorial developments in GR.

Ashtekar's spinorial variables for general relativity are extended to covariant variables for space-time and the Einstein-Hilbert action is given in terms of these. The reduction to the canonical formalism yields the Sparling 3-form as the gravitational Hamiltonian derived from the Einstein-Hilbert action. It is the gravitational part of Witten's integrand for the ADM energy. We derive Penrose's quasi-local mass formula from the associated boundary term when the form is restricted to appropriate sections of the spin bundle. The components of the angular momentum twistor for a 2-surface $\mathcal{S}$ arise as Hamiltonians generating motions of a spanning 3-surface that tend to motions generated by quasi-Killing vectors obtained from solutions of the 2-surface twistor equation on $\mathcal{S}$. Connections with super-gravity are discussed.

We present some related ideas. The 3-form extends to one of a collection of 3-forms on the bundle of general frames. When pulled back to space-time using a coordinate section of the frame bundle, this collection gives the pseudo-energy-momentum tensor of the gravitational field.

1   Introduction

Over the last fifteen years there have been many distinct major theoretical advances in general relativity in which the use of spinors played an unexpectedly deep rôle. The highlights are:

(a) The solution of the self-dual vacuum equations by twistor methods [20].

(b) The extension of gravity to super-gravity (see for example [34]).

(c) The Witten proof of positivity of the ADM energy [35].

(d) Penrose's quasi-local mass construction ([22, 33]).

(e) Ashtekar's simplification of canonical gravity using spinor variables [2].

In (c) and (d) the use of spinors is essential in that primary use is made of a spinor with one index. For (a) and (e) it is perhaps possible to state the results without use of spinor indices, but the rôle of spinors is clearly fundamental. For (e) the use of one of the spin connections as coordinates on the gravitational phase space is the principal innovation, and for (a) the equivalence of the self-dual vacuum equations with the condition of flatness of one of the spin connections is what leads to the solubility. Super-gravity goes outside the confines of classical general relativity. It is an extension of general relativity which includes an anticommuting spin $\frac{3}{2}$ field; it has a local 'gauge' symmetry extending the diffeomorphism group which mixes the extended metric field with the spin $\frac{3}{2}$ field.

It seems likely that there are some structures unifying some or all these ideas. The developments (c-e) are certainly closely related. This relation is perhaps most naturally expressed using a 3-form, Γ, on the spin bundle (which can alternatively be represented by a 3-form on space-time depending on a spinor, its complex conjugate and their first derivatives). This was first introduced by Sparling [30] who showed that Γ satisfies the important identity:

$$\Gamma = dW^- - \tfrac{1}{2}l^a G_{ab}\mathbf{X}^b$$

where $\Gamma = id\pi_{A'}\wedge d\bar{\pi}_A \wedge \theta^{AA'}$, $W^- = i\pi_{A'}d\bar{\pi}_A\wedge\theta^{AA'}$ is a two-form, $l^a = \bar{\pi}^A\pi^{A'}$ is a null vector, G_{ab} is the Einstein tensor, $\theta^{AA'}$ is the solder form and

$$\mathbf{X}^b = \tfrac{1}{6}\varepsilon^b_{cde}\theta^c \wedge \theta^d \wedge \theta^e$$

is the 3-volume element pulled back from space-time (see section 3 for definitions of the various quantities; when acting on abstractly indexed quantities, d is the covariant exterior derivative, otherwise it is the regular exterior derivative). The vacuum equations can be simply characterized as $d\Gamma = 0$.

First of all Γ is a Hamiltonian density for canonical general relativity. The derivation of this is particularly natural in the context of Ashtekar's new variables which we give in section 5 using a canonical decomposition of the gravitational Lagrangian (see [19, 12, 29, 10] for similar derivations although the presentation given here has some new features). Using this result, dW^- is shown to be the total Hamiltonian density of general relativity together with the matter fields; it is also the 3-surface integrand in the Witten positive energy proof. The 2-form W^- is therefore the integrand in the boundary 2-surface integrals for the 'conserved quantities', momentum and angular momentum, of the gravitational field at space-like and null infinity. When used 'quasi-locally', W^- gives rise to Penrose's formula for the angular momentum twistor of the quasi-local mass construction [17]. As further evidence that these ideas are required for treatments of conserved quantities in general relativity, we show that the Sparling 3-form can be extended to be one of a collection of 3-forms on the bundle of general linear frames which, when pulled back to space-time, give rise to classical formulae for the pseudo-energy-momentum tensor of the gravitational field [8].

We do not yet have a theory which unites these ideas fully, although the detailed interconnections reviewed in this article present strong evidence that such a theory exists. Developments (c) and (d) have a direct connection with super-gravity; the positivity result was motivated from the corresponding result in super-gravity, and solutions of the twistor equation used in the quasi-local mass construction have the interpretion as global super-symmetry generators in super-gravity. Furthermore, the Lagrangian for super-gravity is chiral and Ashtekar's variables can be extended naturally and elegantly to super-gravity [11] (see also [5] for an approach to super-gravity in the spirit of this article). It is, however, unsatisfactory to have to go to an extension of general relativity in order to explain constructions which have a self contained expression within the confines of classical general relativity.

Ideas from twistor theory are certainly indicated, although no precise connection has appeared. Indeed the Sparling 3-form itself appears to be remarkably resistant to a twistorial formulation. However there are tentative connections between the hypersurface twistor space construction and Ashtekar's variables which may play a prominent rôle in the final theory underlying these ideas. Furthermore Ashtekar, Jacobson & Smolin [3] have given a remarkably simple description of the self-duality equations in terms of the spinorial variables (see also [18]). The fact that chiral descriptions of ambidextrous space-time fields not only exist but also simplify the formalism provides mathematical evidence that the chiral description necessitated by the use of twistors is not unnatural and may even exist! (The work of Capovilla, Dell & Jacobson [4] provides yet more evidence. They give a formulation of full general relativity using only the anti-self-dual spin connection—the metric is eliminated.)

Sparling's original papers, [30, 31, 32] have not yet made it into print although the 3-form does now appear in a book [25] where it is used to clarify the presentation of the Witten argument for positivity of the ADM energy. We give only a very brief discussion of the positivity argument here, and refer the reader to [25] for full details. We shall instead concentrate on other results which have not been covered so fully in the literature.

The 3-form was first discovered while investigating the differential geometry of hypersurface twistor Cauchy-Riemann manifolds. These manifolds are most easily represented as the restriction of the spin bundle to a hypersurface in space-time. An important discovery in these investigations was that the Fefferman metric which is naturally associated to any such CR manifold is the restriction to the hypersurface twistor space of a naturally defined degenerate metric on the spin bundle. The 3-form was then identified as another such canonically defined structure. However it is not Lie derived along the null geodesic spray even in flat space-time, and so does not descend to the space of null geodesics. It therefore does not have a direct twistorial interpretation even in flat space-time (although it may well find one as, say, the primary part of some larger structure).

2   Indexed Forms on the Spin Bundle

Abstract index notation [25] was developed in order to eliminate the ambiguity in expressions such as $\nabla_2 V_3$ which can be interpreted as either the 23 component of the covariant derivative, $\nabla_a V_b$, of V_a or the ordinary derivative of the function V_3 along the 2 direction. An important feature is that it allows the convenience of using an index notation for tensorial computations without the mathematical inelegance of choosing coordinates or an orthonormal frame.

The basic idea is that the index serves as a marker to indicate the vector bundle in which the tensor takes its values. Identities such as $V^a = V^{AA'}$ make sense, where the index a signifies membership of the tangent bundle, $\mathbf{T}^a$, to a 4-dimensional Lorentzian space-time, and AA' signifies membership of the tensor product of the unprimed spin bundle, $\mathbf{S}^A$, and the primed spin bundle, $\mathbf{S}^{A'}$—the two vector bundles are equal by construction of the spin bundles. The indices do not take on numerical values so there is no difficulty with the explicit implementation of the formulae.

If it is desired to write out formulae in terms of standard indices, a local trivialization of the vector bundle must be chosen, say $(\delta_0^a, \delta_1^a, \delta_2^a, \delta_3^a)$ in the case of T^a. This frame is denoted $\delta_{\underline{a}}^a$, with dual frame $\delta_a^{\underline{a}}$, where $\underline{a}$ is a *concrete* index ranging over 0,1,2,3. The components of V^a are then $V^{\underline{a}} = \delta_a^{\underline{a}} V^a$.

In this paper it will be convenient to work with abstractly indexed forms on the spin bundle. When working with forms we shall generally suppress form indices. On space-time M introduce first the indexed 1-form:

$$\theta^a = \theta^{AA'} = \delta_{\underline{a}}^{AA'} dx^{\underline{a}}.$$

where $x^{\underline{a}}$ are coordinates on M and $\delta_{\underline{a}}^{AA'}$ is the corresponding coordinate frame of $\mathbf{T}^{AA'}$. This can be thought of as being just the Kronecker delta, it takes a vector V to the corresponding element of the tangent space T^a, $V^a = V \lrcorner \theta^a$. One can also think of $\theta^{AA'}$ as a solder form.

The indexed form, $\theta^{AA'}$, can be pulled back to the total space of the spin bundle, $\mathbf{S}_{A'}$ where we shall abuse notation by denoting it again by $\theta^{AA'}$. On $\mathbf{S}_{A'}$ we have further canonically defined structures. Firstly we have the tautological spinor field $\pi_{A'}$ on $\mathbf{S}_{A'}$. At the point $(x, \eta_{A'}) \in \mathbf{S}_{A'}$, $\pi(x, \eta_{C'})_{A'} = \eta_{A'}$ so that $\pi_{A'}$ is a canonically defined abstractly indexed 'coordinate up the fibre' of $\mathbf{S}_{A'}$.

The exterior derivative d on $\mathbf{S}_{A'}$ can be extended so as to act on indexed quantities by using the pullback to $\mathbf{S}_{A'}$ of the space-time connection. On quantities with concrete indices or no indices at all, d will act as the ordinary exterior derivative. However, on abstractly indexed quantities d is the covariant derivative so that d^2 will therefore *not* vanish but will instead produce curvature according to the usual formula:

$$d^2\lambda^{A'\cdots}_{A\cdots} = (\nabla_c\nabla_d\lambda^{A'\cdots}_{A\cdots})\theta^c\wedge\theta^d \tag{1}$$

$$= \tfrac{1}{2}R_{B'}{}^{A'}\lambda^{B'\cdots}_{A\cdots} + \cdots - \tfrac{1}{2}R^B_A\lambda^{A'\cdots}_{B\cdots} - \cdots + \tfrac{1}{2}T^c\nabla_c\lambda^{A'\cdots}_{A\cdots} \tag{2}$$

where $R_A{}^B = \theta^c\wedge\theta^d R_{cdA}{}^B$, $T^a = T^a{}_{bc}\theta^b\wedge\theta^c$, $R_{cdA}{}^B$ is the curvature of the unprimed spin connection and $T^a{}_{bc}$ is the torsion, $\nabla_a\nabla_b f = \tfrac{1}{2}T^c{}_{ab}\nabla_c f$. We have:

$$R_{cdA}{}^B = \{\varepsilon_{C'D'}\Psi_{CDA}{}^B - 2\Lambda\varepsilon_{A(C}\varepsilon_{D)}{}^B + \varepsilon_{(C}{}^B\phi_{D)A} + \phi_{(C}{}^B\varepsilon_{D)A}\} + \varepsilon_{CD}\Phi_{C'D'A}{}^B$$

where each of the curvature spinors is symmetric on all indices of the same type. When the connection is torsion free, we have $\Phi_{ABA'B'} = \overline{\Phi}_{A'B'AB}$, $\Lambda = \overline{\Lambda}$ and $\phi_{AB} = 0$.

When the connection has torsion, $\nabla_{[a}\nabla_{b]}f = \tfrac{1}{2}T_{ab}{}^c\nabla_c f$, we also have:

$$d(\theta^a) = \tfrac{1}{2}T^a = \theta^c\wedge\theta^d\tfrac{1}{2}T_{cd}{}^a$$

so that in the absence of torsion $d(\theta^a) = 0$.

We can now introduce the indexed forms $d\pi_{A'}$ and $d\overline{\pi}_A$ which are the covariant exterior derivatives of $\pi_{A'}$ and $\overline{\pi}_A$ respectively. If we choose a primed spin frame, $\varepsilon_{A'}{}^{\underline{A}'} = (\varepsilon_{A'}{}^{0'}, \varepsilon_{A'}{}^{1'})$, then the components of $d\pi_{A'}$ are $\varepsilon_{\underline{A}'}{}^{A'}d\pi_{A'} = d\pi_{\underline{A}'} - \gamma_{\underline{A}'}{}^{\underline{B}'}\pi_{\underline{B}'}$ where $\gamma_{\underline{A}'}{}^{\underline{B}'}$ is the primed spin connection form in the frame $\varepsilon_{A'}{}^{\underline{A}'}$:

$$\gamma_{\underline{A}'}{}^{\underline{B}'} = \varepsilon_{A'}{}^{\underline{B}'}d\varepsilon_{\underline{A}'}{}^{A'} = \theta^{\underline{c}}\varepsilon_{A'}{}^{\underline{B}'}\nabla_{\underline{c}}\varepsilon_{\underline{A}'}{}^{A'}.$$

The forms $(\theta^{AA'}, d\pi_{A'}, d\overline{\pi}_A)$ are an indexed coframe for $\mathbf{S}_{A'}$. We have:

$$d(d\pi_{A'}) = d^2\pi_{A'} = -\tfrac{1}{2}R_{A'}{}^{B'}\pi_{B'}$$

$$d(d\overline{\pi}_A) = d^2\overline{\pi}_A = -\tfrac{1}{2}R^B_A\overline{\pi}_B$$

$$d(\theta^{AA'}) = \tfrac{1}{2}T^{AA'}.$$

It will be convenient also to introduce notation for the indexed forms obtained from wedging $\theta^{AA'}$ with itself. We define:

$$\Sigma^{AB} = \varepsilon_{A'B'}\theta^{AA'}\wedge\theta^{BB'}$$

$$\Sigma^{A'B'} = \varepsilon_{AB}\theta^{AA'}\wedge\theta^{BB'}$$

$$\mathbf{X}^{AA'} = \tfrac{i}{3}\varepsilon_{BC}\varepsilon_{B'C'}\theta^{AB'}\wedge\theta^{BA'}\wedge\theta^{CC'}$$

$$= \tfrac{1}{6}\varepsilon^a_{bcd}\theta^b\wedge\theta^c\wedge\theta^d$$

$$\nu = \tfrac{1}{24}\varepsilon_{abcd}\theta^a\wedge\theta^b\wedge\theta^c\wedge\theta^d.$$

These satisfy the relations

$$\Sigma^{A'B'}\wedge\theta^{CC'} = 2i\mathbf{X}^{C(A'}\varepsilon^{B')C'}$$

and

$$\Sigma^{AB}\wedge\Sigma^{CD} = 4i\varepsilon^{A(C}\varepsilon^{D)B}\nu.$$

3   The Sparling 3-form

In the notation established above, we define the Sparling 3-form to be the 3-form on the total space of the spin bundle $\mathbf{S}_{A'}$:

$$\Gamma = id\pi_{A'} \wedge d\overline{\pi}_A \wedge \theta^{AA'}.$$

We also use the following pair of 2-forms:

$$W^+ = -i\overline{\pi}_A d\pi_{A'} \wedge \theta^{AA'}, \quad W^- = i\pi_{A'}d\overline{\pi}_A \wedge \theta^{AA'};$$

these have become known as the Witten-Nester forms. Note that

$$W^- - W^+ = id(\pi_{A'}\overline{\pi}_A\theta^{AA'})$$

in the absence of torsion. The important identity that the Sparling 3-form satisfies is:

$$\Gamma = dW^\pm - \tfrac{1}{2}G_b^{AA'}\pi_{A'}\overline{\pi}_A\mathbf{X}^b \tag{3}$$

when $T^a = 0$. Here $\mathbf{X}^a = \tfrac{1}{6}\varepsilon^a_{bcd}\theta^b \wedge \theta^c \wedge \theta^d$ and $G_{ab} = -2\Phi_{ab} - 6\Lambda g_{ab}$ is the Einstein tensor. Clearly Γ is closed when $G_{ab} = 0$. Indeed, even when the connection has torsion, we have:

Proposition 3.1 ([30]) *The 3-form Γ is closed if and only if the vacuum equations hold, $G_{ab} = 0$, $T^a_{\ bc} = 0$.*

Proof This follows by taking the exterior derivative of Γ. It is first convenient to compute $d(d\overline{\pi}_A \wedge \theta^{AA'})$. This gives:

$$\begin{aligned}
d(d\overline{\pi}_A \wedge \theta^{AA'}) &= -i\mathbf{X}^b\{\Phi_{BB'}{}^{AA'} + \varepsilon_{B'}{}^{A'}(3\Lambda\varepsilon_B{}^A + \phi_B{}^A)\}\overline{\pi}_A \\
&\quad -d\overline{\pi}_A \wedge T^{AA'}{}_{bc}\theta^b \wedge \theta^c \\
&= -\tfrac{i}{2}\mathbf{X}^b\{G_{BB'}{}^{AA'} + \varepsilon_{B'}{}^{A'}\phi_B{}^A\}\overline{\pi}_A - d\overline{\pi}_A \wedge T^{AA'}{}_{bc}\theta^b \wedge \theta^c
\end{aligned}$$

so that (3) follows when $T^a_{bc} = 0$ and

$$\begin{aligned}
d\Gamma &= \tfrac{1}{2}(\{\overline{G}_b^a + \varepsilon_B{}^A\overline{\phi}_{B'}{}^{A'}\}\pi_{A'}d\overline{\pi}_A + \{G_b^a + \varepsilon_{B'}{}^{A'}\phi_B{}^A\}\overline{\pi}_A d\pi_{A'}) \wedge \mathbf{X}^b \\
&\quad -d\pi_{A'} \wedge d\overline{\pi}_A \wedge T^{AA'}{}_{bc}\theta^b \wedge \theta^c.
\end{aligned}$$

Clearly this vanishes if and only if $T^a_{\ bc} = 0 = G_{ab}$. $\square$

The second major result is the following.

Proposition 3.2 ([30]) *The 3-form Γ generates a closed differential ideal if and only if there exists a conformal rescaling for which the resulting metric is vacuum.*

Proof The condition that Γ generates a closed differential ideal is the condition that $d\Gamma = \Gamma \wedge \alpha$ for some 1-form α. By examining the expression for $d\Gamma$ above, we see that this can only happen if $G_{ab} = 0$, $\phi_{AB} = 0 = \phi_{A'B'}$ and $T^a{}_{bc} = \delta^a_{[b}v_{c]}$ for some v_c. When the torsion is of this form, we have

$$\phi_{AB}\varepsilon_{A'B'} + \bar{\phi}_{A'B'}\varepsilon_{AB} = \nabla_{[a}v_{b]}$$

which vanishes. Thus $v_a = \nabla_a \log \Omega$ for some function Ω.

If we rescale the metric, $g_{ab} \to \Omega^2 g_{ab}$, but keep the same connection it is now torsion free with respect to the new metric and, furthermore, G_{ab} remains zero. The original space-time is therefore conformal to vacuum. $\square$

The Sparling 3-form can be pulled back to the bundle of orthonormal frames, and extended from there to the bundle of general linear frames. This will be discussed in section 7 where these ideas will be used to make contact with pseudo stress-energy tensors.

The Sparling 3-form has analogues in arbitrary dimension for which similar results hold [31, 6]. The results can be expressed in terms of forms on the frame bundle (see section 7) or more elegantly in terms of the Dirac spinors in n-dimensions.

Consider an n-dimensional space-time, M^n. Let γ_a, where $a, b, \dots$ are tangent space indices for M, denote the Dirac matrices which generate the Clifford algebra according to:

$$\gamma_a\gamma_b + \gamma_b\gamma_a = 2g_{ab}I.$$

Define γ by $\gamma = \frac{1}{n!}\varepsilon^{a_1 \cdots a_n}\gamma_{a_1}\gamma_{a_2}\cdots\gamma_{a_n}$. Then the analogue of the Witten-Nestor form is the $(n-2)$-form:

$$W_n = \overline{\psi}\gamma(\theta^a\gamma_a)^{n-3} \wedge d\psi$$

where again d is the covariant exterior derivative on spinors. The analogue of the Sparling form is the $(n-1)$-form:

$$\Gamma_n = d\overline{\psi} \wedge \gamma(\theta^a\gamma_a)^{n-3} \wedge d\psi.$$

We then have identities of the form:

$$\Gamma_n = dW_n - \tfrac{1}{2}G^{ab}(\overline{\psi}\gamma_a\psi)\varepsilon_{bc_1\cdots c_{n-1}}\theta^{c_1} \wedge \theta^{c_2} \wedge \cdots \wedge \theta^{c_{n-1}}$$

generalizing equation (3).

4 The 4-covariant Form of Ashtekar Variables

4.1 The Basic Variables

First we introduce some basic variables in place of the metric and its derivatives. These 4-covariant generalizations of Ashtekar's variables have appeared previously in the literature [26] and [9]. The use of these variables is explored in greater detail in [5]. Geometrically we think of the variables which replace the metric as defining a 3-dimensional sub-bundle $\wedge^{2-}$ of the bundle of 2-forms $\wedge^2 M$ on M together with a volume form for $\wedge^{2-}$ represented by a tensor $\varepsilon_{ijk} = \varepsilon_{[ijk]}$ where the index i denotes membership of $\wedge^{2-}$. This will turn out to be the bundle of anti-self-dual 2-forms of a metric, and ε will give the scale for the inner product between two such 2-forms.

The bundle $\wedge^2 M$ has a natural conformal structure given simply by the wedge product. If α and β are 2-forms, define $h(\alpha, \beta)$ to be the coefficient of some volume form ν in the expression $\alpha \wedge \beta$. This is clearly defined up to multiples of itself. The sub-bundle $\wedge^{2-}$ therefore determines its orthogonal complement $\wedge^{2+}$, the 3-dimensional sub-bundle of $\wedge^2 M$ consisting of 2-forms each of which is orthogonal to each section of $\wedge^{2-}$. The bundle $\wedge^{2+}$ will turn out to be the bundle of self-dual forms of some metric.

This bundle $\wedge^{2-}$ can be represented by a triple of 2-forms, Σ^i, where $i = 1, \ldots, 3$; these are determined up to complex linear combinations with unit determinant (because of the ε_{ijk}). This data can be thought of as the basic variables for (complex) general relativity. It determines and is determined by the metric. One can choose an arbitrary but generic triple of 2-forms, Σ^i, $i = 1, 2, 3$ with the only proviso that they be linearly independent, and that the conformal structure h be nondegenerate on this triple.

As a plausibility argument, this corresponds to a choice of 18 (complex) numbers per space-time point. If we are now allowed to make arbitrary $SL(3, \mathbb{C})$ transformations, $\Sigma^i \rightarrow A^i{}_j \Sigma^j$, this removes 8 degrees of freedom, leaving the ten degrees of freedom of the metric.

In order to reduce to the 10 degrees of freedom of the metric, we can remove the 8 degrees of freedom from the 18 degrees of freedom for Σ^i in two stages. The expression $\Sigma^i \wedge \Sigma^j = h^{ij} \nu$, determines a symmetric tensor h^{ij}. (Recall that in order that this procedure be nonsingular, the forms should be chosen so that h^{ij} is nondegenerate; this will be true for a generic choice of Σ^i.) We can remove 5 of the 8 degrees of gauge freedom by reducing h^{ij} to some standard form. There then remains a 3-dimensional $SO(3, \mathbb{C})$ (whose spin group is $SL(2, \mathbb{C})$) gauge freedom. This gauge freedom is the choice of an unprimed spin frame. We define $\Sigma^{AB} = \Sigma^{(AB)}$ by putting

$$\Sigma^i = (\Sigma^{00}, \Sigma^{01}, \Sigma^{11})$$

with

$$(\Sigma^{00})^2 = (\Sigma^{11})^2 = \Sigma^{00} \wedge \Sigma^{01} = \Sigma^{11} \wedge \Sigma^{01} = 0$$

and

$$\Sigma^{00} \wedge \Sigma^{11} = -(\Sigma^{01})^2 = \nu. \tag{4}$$

The residual $SO(3, \mathbb{C})$ invariance is represented by the action of $SL(2, \mathbb{C})$

$$\Sigma^{AB} \rightarrow l^A{}_C l^B{}_D \Sigma^{CD}, \text{ with } \varepsilon_{AB} l^A{}_C l^B{}_D = \varepsilon_{CD}.$$

4.2 *Reconstruction of the Metric*

Let us assume that we have $\Sigma^{\underline{i}} = (\Sigma^{00}, \Sigma^{01}, \Sigma^{11})$ defined as above. Then the relations $(\Sigma^{00})^2 = (\Sigma^{11})^2 = 0$ imply that the forms Σ^{00} and Σ^{11} are simple, so that they can be expressed as the wedge product of pairs of 1-forms. Put

$$\Sigma^{00} = \theta^{00'} \wedge \theta^{01'}$$

for some pair of 1-forms $(\theta^{00'}, \theta^{01'}) = \theta^{0\underline{A}'}$ defined up to $SL(2,\mathbb{C})$ transformations on the $\underline{A}'$ index. Similarly we obtain

$$\Sigma^{11} = \hat{\theta}^{10'} \wedge \hat{\theta}^{11'}$$

for some pair of 1-forms $(\hat{\theta}^{10'}, \hat{\theta}^{11'}) = \hat{\theta}^{1\underline{A}'}$. The relations $\Sigma^{01} \wedge \Sigma^{00} = 0$ and $\Sigma^{01} \wedge \Sigma^{11} = 0$ imply that

$$\Sigma^{01} = \theta^{00'} \wedge (a\hat{\theta}^{10'} + b\hat{\theta}^{11'}) + \theta^{01'} \wedge (c\hat{\theta}^{10'} + d\hat{\theta}^{11'}),$$

where $ad - bc = 1$ follows from $\Sigma^{00} \wedge \Sigma^{11} = -(\Sigma^{01})^2 = \nu$. Now define

$$\theta^{1\underline{A}'} = (c\hat{\theta}^{10'} + d\hat{\theta}^{11'}, a\hat{\theta}^{10'} + b\hat{\theta}^{11'}).$$

The tetrad of 1-forms $\theta^{A\underline{A}'} = (\theta^{0\underline{A}'}, \theta^{1\underline{A}'})$ is therefore determined by the $\Sigma^{\underline{A}\underline{B}}$ up to $SL(2,\mathbb{C})$ transformations on the primed indices. The $\Sigma^{\underline{A}\underline{B}}$ are in turn determined only up to $SL(2,\mathbb{C})$ rotations on the unprimed spinor indices. Thus the tetrad is determined up to complex Lorentz transformations on the concrete indices so that the true degrees of freedom are just the metric. The $\Sigma^{\underline{A}\underline{B}}$ are, by construction, $\Sigma^{\underline{A}\underline{B}} = \varepsilon_{\underline{A}'\underline{B}'}\theta^{A\underline{A}'} \wedge \theta^{B\underline{B}'}$, and are therefore automatically the anti-self-dual forms of the metric.

Reality Conditions

When a Lorentzian real space-time is required, we must also satisfy the reality condition:

$$\Sigma^i \wedge \overline{\Sigma^j} = 0$$

so that $\wedge^{2+}$, the orthogonal complement of $\wedge^{2-}$ in $\wedge^2$ relative to h, is the complex conjugate of $\wedge^{2-}$. When we consider the field equations, it will be convenient to also impose the condition that $\Sigma^i \wedge \Sigma_i$ be pure imaginary, although for the purposes of reconstructing a real metric it is inessential[1]. For Euclidean reality conditions $\wedge^{2-}$ should be real and h should be positive, and for (2,2) signature $\wedge^{2-}$ should be real and h should have indefinite signature.

---

[1] The phase of the metric can be scaled to unity to yield a real metric. However, when the field equations are discussed below, this rescaling cannot be done with impunity since the connection will preserve the metric with the complex scale, and will therefore have torsion relative to the Levi-Civita connection of the real metric. This alters the field equations.

The First Derivative Variables

Rather than use the full connection as the first derivative variables, we use the unprimed spin connection, $\gamma^A{}_B$. This is determined from σ^{AB} by the structural equation:

$$d\Sigma^{\underline{AB}} - 2\gamma^{(A}{}_{\underline{C}}\Sigma^{\underline{B})\underline{C}} = 0.$$

Note that this is 12 equations for 12 unknowns so that $\gamma_{\underline{A}}{}^{\underline{B}}$ is uniquely determined. This structural equation provides a more convenient starting point for the computation of a connection from a metric than the regular null tetrad formalism.

5   Derivation of Γ as Hamiltonian Density

See [19, 12, 29, 10] for similar derivations of the following material.

5.1   *The Lagrangian and the Variational Equations*

The Einstein-Hilbert action, $S = \int R\nu$, with R the scalar curvature, and ν the metric volume form, can be represented in terms of the variables introduced above as follows. Let $R_A{}^B$ be the curvature 2-form of the unprimed spinor connection. We have:

$$R\nu = R_{AB} \wedge \Sigma^{AB}.$$

Thus the action can be represented simply in terms of the variables discussed above[2].

    The field equations are obtained by varying the connection arbitrarily, and the two-forms Σ^{AB} subject to the constraint

$$\Sigma^{(AB} \wedge \Sigma^{CD)} = 0. \tag{5}$$

This constraint arises from the condition that $\Sigma^{\underline{AB}}$ satisfies the conditions (2) (whose invariant expression is equation (4.1)) that guarantee that $\Sigma^{AB} = \theta^{AA'} \wedge \theta^{BB'}\varepsilon_{A'B'}$. The variation with respect to the connection yields the 'torsion free' equation for the connection

$$d\Sigma^{AB} = 0.$$

The constraint on Σ^{AB} implies that the variation of Σ^{AB} satisfies

$$\delta\Sigma^{(AB} \wedge \Sigma^{CD)} = 0$$

---

[2]Note that the 1st order form of the Lagrangian has the following attractively simple form in these variables:

$$S = \int \gamma_{\underline{AB}} \wedge \gamma^{\underline{B}}{}_{\underline{C}} \wedge \Sigma^{\underline{AC}}.$$

Here $\gamma_{\underline{AB}}$ is thought of as determined from the first derivatives of $\Sigma^{\underline{AB}}$ by solving the structural equation. The action is spin frame independent only up to a divergence.

so that

$$\delta\Sigma^{AB} = \phi^{(A}{}_{C}\Sigma^{B)C} + \eta\Sigma^{AB} + \eta^{AB}_{A'B'}\Sigma^{A'B'}$$

for some arbitrary spinors $\phi\ldots,\eta\ldots$ and function η. The action is invariant under $\delta\Sigma^{AB} = \phi^{(A}{}_{C}\Sigma^{B)C}$. From the vanishing of the variation with $\delta\Sigma^{AB} = \eta\Sigma^{AB}$ we obtain the equation $R = 0$, and with $\delta\Sigma^{AB} = \eta^{AB}_{A'B'}\Sigma^{A'B'}$ we obtain

$$R_{AB} \wedge \eta^{AB}_{A'B'}\Sigma^{A'B'} = \eta^{ABA'B'}\Phi_{ABA'B'}(\Sigma^{C'D'} \wedge \Sigma_{C'D'}) = 0$$

so that we obtain the full vacuum equations $R_{ab} = 0$.

As a curiousity, one can also write the Lagrangian in terms of the Σ^{i}:

$$S = \int \Sigma^{i} \wedge R^{j}{}_{k}\varepsilon_{ijl}h^{kl}$$

where h^{ij} is determined by $\Sigma^{i} \wedge \Sigma^{j} = h^{ij}\nu$ for some volume form ν such that h^{ij} has unit determinant with respect to ε_{ijk}. The connection, of which $R^{i}{}_{j}$ is the curvature must preserve h_{ij} so that $R^{i}{}_{j}h_{ik} = -R^{i}{}_{k}h_{ij}$.

Following Jacobson we can enforce the constraint (5) by including a Lagrange multiplier $\Psi_{ABCD} = \Psi_{(ABCD)}$ in the action:

$$S = \int R_{AB} \wedge \Sigma^{AB} - \tfrac{1}{2}\Psi_{ABCD}\Sigma^{AB} \wedge \Sigma^{CD}$$

where now the variation with respect to Ψ_{ABCD} forces the constraint on Σ^{AB}, the variation with respect to the connection yields the torsion free equation, and the variation with respect to Σ^{AB} now yields:

$$R_{AB} = \Psi_{ABCD}\Sigma^{CD}. \tag{6}$$

This implies that all the Ricci tensor terms in R_{AB} vanish.[3]

---

[3] [4] goes further, eliminating Σ^{AB} entirely by solving (6) for Σ^{AB} from R^{AB} when Ψ_{ABCD} is invertible as a 3×3 matrix. This works in a much more striking fashion than one might have originally expected. Their action is:

$$S = \int \mu(R^{AB} \wedge R^{CD})(R_{AC} \wedge R_{BD})$$

where the basic variables are a spin connection, of which R_{AB} is the curvature and an inverse density, μ, so that the integrand in the action is a single density. This action yields the field equations:

$$(R^{AB} \wedge R^{CD})(R_{AC} \wedge R_{BD}) = 0 \text{ and } D\{\mu(R^{AB} \wedge R^{CD})R_{BC}\} = 0$$

and the forms Σ^{AB} are recovered from the equation $\Sigma^{AD} = \mu(R^{AB} \wedge R^{CD})R_{BC}$. The torsion free equation, $d\Sigma^{AB} = 0$ follows from the second field equation, and the equation $R_{AB} = \psi_{ABCD}\Sigma^{CD}$ follows by definition of Σ^{AB} where $\psi_{(AB)(CD)}$ is the inverse of the matrix $\mu(R^{(C}{}_{(A} \wedge R_{B)}{}^{D)})$ thought of as 3×3 matrices with indices clumped as bracketed. The matrix ψ_{ABCD} turns out to be totally symmetric on the spinor indices as a consequence of the μ field equation. The constraint $\Sigma^{AB} \wedge \Sigma^{CD} = 0$ follows from some remarkable algebraic identities which can be derived from the fact that these 3×3 matrices satisfy their characteristic polynomial.

5.2   Canonical Decomposition of the Lagrangian

The decomposition given here will be given without the choice of spin frame, that is using abstract indices throughout. This is only meant as a curiosity, as it is sometimes asserted that this cannot be done. It is a straightforward matter to reintroduce a choice of spin frame.

Choose a foliation of M by spacelike hypersurfaces $\mathcal{H}_t$ of constant time t, and a vector field $L = l\partial/\partial t$ transverse to this foliation (where l is the lapse function, $l = L(t)$), which is chosen so as to be null:

$$L\lrcorner\theta^{AA'} = L^{AA'} = \lambda^A\lambda^{A'}.$$

Here and in the following, $\lrcorner$ denotes the contraction of a vector with a form.

We can now decompose the action into the standard form

$$S = \int dt\,(p\dot{q} - H)$$

for the canonical formalism:

$$
\begin{aligned}
S &= \int_M R_{AB} \wedge \Sigma^{AB} = \int \frac{dt}{l} \int_{\mathcal{H}_t} L\lrcorner(R_{AB} \wedge \Sigma^{AB}) \\
&= \int \frac{dt}{l} \int_{\mathcal{H}_t} (L\lrcorner R_{AB}) \wedge \Sigma^{AB} + R_{AB} \wedge \lambda^A\lambda_{B'}\theta^{BB'}.
\end{aligned}
$$

At this stage we wish to write $L\lrcorner R_{AB}$ in terms of the Lie derivative of the connection. We therefore need to introduce a derivation of spinors along L so as to act on spinor indexed quantities. This can be done in an invariant fashion as follows. Since L is not a Killing vector its action on spinor indexed quantities is not canonically defined, and so we write down the general such derivation by defining $\mathcal{L}_L\alpha^A = L\lrcorner d\alpha^A + \gamma(L)^A{}_B\alpha^B$ where $\gamma(L)_{AB} = \gamma(L)_{(AB)}$ can be specified freely. (A reasonable choice, for example, is that $\gamma(L)_{AB}$ should be the anti-self-dual part of the curl of L, $\gamma(L)_{AB} = \nabla^{A'}_{(A}L_{B)A'}$ so that $\mathcal{L}_L$ agrees with Lie derivation when L is a Killing vector). The Lie derivative of the connection, $(\mathcal{L}_L\gamma^A{}_B)$, is then invariantly defined by

$$(\mathcal{L}_L\gamma^A{}_B)\alpha^B = \mathcal{L}_L(d\alpha^A) - d(\mathcal{L}_L\alpha^A)$$

for some spinor α^A. Thus, using the formula $\mathcal{L}_L\beta = L\lrcorner d\beta + d(L\lrcorner\beta)$ where β is a general form, we obtain:

$$
\begin{aligned}
(\mathcal{L}_L\gamma^A{}_B)\alpha^B &= L\lrcorner d^2\alpha^A + \gamma(L)^A{}_B d\alpha^B + d(L\lrcorner d\alpha^A) \\
&\quad -d(L\lrcorner d\alpha^A + \gamma(L)^A{}_B\alpha^B) \\
&= L\lrcorner R^A{}_B\alpha^B - d(\gamma(L)^A{}_B)\alpha^B.
\end{aligned}
$$

If we substitute this relation[4] into the decomposition of the action (using also $R_{AB}\lambda^B = d^2\lambda_A$) we obtain:

$$
\begin{aligned}
S &= \int \frac{dt}{l} \int_{\mathcal{H}_t} (L\lrcorner R_{AB}) \wedge \Sigma^{AB} + d^2\lambda_A \wedge \lambda_{A'}\theta^{AA'} \\
&= \int \frac{dt}{l} \int_{\mathcal{H}_t} \mathcal{L}_L(\gamma_{AB}) \wedge \Sigma^{AB} + d\gamma(L)_{AB} \wedge \Sigma^{AB} + d^2\lambda_A \wedge \lambda_{A'}\theta^{AA'}.
\end{aligned}
$$

We integrate by parts on the last term, and cast off the resulting boundary integral to obtain:

$$
S = \int \frac{dt}{l} \int_{\mathcal{H}_t} \mathcal{L}_L(\gamma_{AB}) \wedge \Sigma^{AB} + d(\gamma(L)_{AB}) \wedge \Sigma^{AB} + d\lambda_A \wedge d(\lambda_{A'}\theta^{AA'}).
$$

The first term corresponds to the $p\dot{q}$ term in the Lagrangian and the second two terms give the Hamiltonian. The first term, $\mathcal{L}_L\gamma_{AB} \wedge (\Sigma^{AB})$, shows that the restriction of the anti-self-dual spin connection (say represented by the spin connection 1-forms in a given spin frame, γ_{AB}) to $\mathcal{H}_t$ is canonically conjugate to the restriction to $\mathcal{H}_t$ of the 2-forms Σ^{AB}.

The first Hamiltonian term, $d(\gamma(L)_{AB}) \wedge \Sigma^{AB}$, generates spin frame rotations, and the second, $d\lambda_A \wedge d(\lambda_{A'}\theta^{AA'})$, generates translations along the vector field $L^{AA'} = \lambda^A\lambda^{A'}$. This second term is the Sparling 3-form.

Remarks

1. In the first analysis we obtain $(\mathcal{L}_L\gamma_{AB}) \wedge \Sigma^{AB}$ as the $p\dot{q}$ term suggesting perhaps that the connection γ_{AB} should be regarded as the position coordinate and Σ^{AB} as the canonically conjugate momentum coordinate on the gravitational phase space. Indeed this analogy is taken up by Ashtekar who embeds the gravitational constraint surface into the Yang-Mills phase space by regarding Σ^{AB} as the electric field (time derivative) of the connection. However, for general relativity the connection should be thought of as the momentum, and Σ^{AB} should be thought of as the position variables and the $\dot{p}q$ term arises rather than a $p\dot{q}$ term because the 4-dimensional covariant form of the Lagrangian involves time derivatives of the p variables, i.e. the form in footnote 2 which would have yielded $p\dot{q}$ has had the total derivative $(pq)\dot{}$ added to it in order to make it manifestly covariant. The evolution equations for the gravitational variables certainly do not have a direct relation with those for Yang-Mills fields and the identification ignores the status of

---

[4] This relation could also be obtained from the concrete indexed formulae (where γ is the spin connection in this frame)

$$
L\lrcorner(d\gamma_{\underline{AB}} - \gamma_{\underline{AC}} \wedge \gamma^{\underline{C}}{}_{\underline{B}}) = \mathcal{L}_L(\gamma_{\underline{AB}}) - d(\gamma(L)_{\underline{AB}})
$$

with $\gamma(L)_{\underline{AB}} = L\lrcorner\gamma_{\underline{AB}}$.

the connection as a structure derived from the first derivatives of Σ^{AB}. Nevertheless, the analogy turns out to be fruitful.

However, see footnote 3, or Capovilla, Dell & Jacobson [4] for a remarkable formulation of general relativity in which the metric has been eliminated and the spin connection really does play the primary rôle. (This formulation of G.R. is however degenerate when space-time is self-dual or when the anti-self-dual part of the Weyl curvature as an indexed 2-form does not span a 3-dimensional space of 2-forms.)

2. Note that at no stage do we need to use the primed spin connection— the last term, $d\lambda_A \wedge d(\lambda_{A'}\theta^{AA'})$, has been bracketed in such a way as to show that d only acts on unprimed indices.

3. In the above derivation we threw off the boundary terms in order to obtain the answer we required. When $\mathcal{H}_t$ is asymptotically flat, some care is required to select the correct boundary term for the Hamiltonian in order that the formalism give the correct equations of motion at the boundary. This is crucial for obtaining the correct value of the Hamiltonian, since, when the constraints are satisfied, the boundary term is the only contribution to the numerical value of the Hamiltonian. The Sparling 3-form is the correct Hamiltonian density in the asymptotic regime and shall henceforth be assumed to be the correct Hamiltonian density for gravity locally also.

See [27] and [19] for a discussion of these issues. In particular Nester shows that the Sparling 3-form is correct asymptotically. The criteria used in those papers does not fix the precise local structure of the Hamiltonian density; one can add any divergence that vanishes fast enough at infinity without disrupting the equations of motion. One would need a quasi-local canonical formalism in order to justify rigorously the precise local structure of the Hamiltonian density.

4. A peculiar feature is that the Hamiltonian depends on more than just the lapse-shift vector $L^a = \lambda^A \lambda^{A'}$; it depends on the phase of its constituent spinors $(\lambda_A, \lambda_{A'})$ also. In general, the value of the integral will change if the phase of the spinors is altered even though L^a is unchanged. This feature is even present asymptotically: it is an essential feature of the Witten positivity argument that the phase of the constituent spinors of the asymptotic translation must be asymptotically constant, otherwise the integral may differ from the ADM energy.

5.3   *The Constraints and Ashtekar's Variables*

The restriction of Σ^{AB} to $\mathcal{H}_t$ is equivalent to Ashtekar's densitized triads. This can be seen by observing that Ashtekar's densitized triad has the index structure $\sigma^{ABi}_{[klm]}$ where i, k, l, m are tangent space indices for $\mathcal{H}_t$. The forms Σ^{AB} are then just

$$\Sigma^{AB}_{[ij]} = 3\sigma^{ABk}_{[ijk]} \quad \text{and} \quad \sigma^{ABi}_{[klm]} = \Sigma^{AB}_{[kl}\delta^i_{m]}.$$

The restriction of the unprimed spin connection to $\mathcal{H}_t$, represented relative to a choice of spin frame by γ_{AB}, is Ashtekar's 'Sen-Witten' connection, which he denotes A_{AB}.

As usual, the quantities, $\gamma_{AB}(L)$ and $L^{AA'}$ are Lagrange multipliers whose variations give rise to the constraints:

$$d(\Sigma^{AB}) = 0 \quad \text{and} \quad R_{AB} \wedge \theta^{BA'} = G_A{}^{A'}{}_b\mathbf{X}^b = 0.$$

The first is the condition that the connection is compatible with Σ^{AB} (which Ashtekar refers to as the Gauss law constraint) and the second contains both the usual Hamiltonian and momentum constraint: $G_{00} = 0$ and $G_{0i} = 0$.

Polynomial Form of the Constraints

An very important feature of Ashtekar's variables is that the constraints can be made to be polynomial in the spinorial variables. For the first this is transparent. However, one must do a little further work to see that the second constraint is polynomial in the basic variables, as $\Sigma^{AB}(L) = \lambda^{(A}\lambda_{A'}\theta^{B)A'}$ requires components of Σ^{AB} in directions transverse to $\mathcal{H}_t$. In order to see that the second set of constraints is polynomial in these variables we claim that the following form is equivalent:

$$\Sigma_{AB[ij}R^{BC}_{k][l}\Sigma^{DA}_{mn]} = 0.$$

This is quartic in the basic variables, and can be shown to be equal to $(T^D_{B'}R^C{}_B\wedge\theta^{BB'})\otimes T_d\mathbf{X}^d$ where T^a is the normal to $\mathcal{H}_t$ with length 2, $T_aT^a = 2$. The factor of proportionality between the form above and the polynomial form is, roughly speaking, $\sqrt{\det(\Sigma^{AB})}$ as can be seen from the fact that the expression above is a double density, whereas $T^D_{B'}R^C{}_B \wedge \theta^{BB'}$ is a single density.

In order to see how this compares to Ashtekar's version, note first that Ashtekar has:

$$Tr(\tilde{\sigma}^i F_{ij}) = 0 \text{ and } Tr(\tilde{\sigma}^i\tilde{\sigma}^j F_{ij}) = 0$$

where the trace is taken over the spinor indices which are all suppressed. His F_{ij} is our $R^A{}_{Bij}$, and his $\tilde{\sigma}^i$ is our $\Sigma^{AB}_{[jk}\delta^i_{l]}$ where the skew triple of indices, $[jkl]$, indicates that $\Sigma^{AB}_{[jk}\delta^i_{l]}$ is now to be thought of as a densitized vector. In order to see that these are obtained from $\Sigma_{AB[ij}R^{BC}_{k][l}\Sigma^{DA}_{mn]} = 0$, for the first constraint, symmetrize over CD. After a certain amount of work, we find that

$$\Sigma_{AB[ij}R^{B(C}_{k][l}\Sigma^{D)A}_{mn]} = \left(\Sigma_{AB[ij}R^{AB}_{k][l}\right)\Sigma^{CD}_{mn]}.$$

The vanishing of this latter quantity can easily be seen to be equivalent to the vanishing of $\Sigma_{AB[ij}\delta^l_{k]}R^{AB}_{lm}$ alone. This is Ashtekar's version of the momentum constraint. If we multiply $\Sigma_{AB[ij}R^{BC}_{k][l}\Sigma^{DA}_{mn]}$ by ε_{CD} we easily see that we obtain Ashtekar's version of the Hamiltonian constraint.

This formulation has the substantial advantage that it does not explicitly require the 3+1 decomposition of the lapse shift vector, and indeed the expression

$$\Sigma_{AB[ij}R^{BC}_{k][l}\Sigma^{DA}_{mn]}T^{C'}_D$$

is proportional to $R^C{}_B \wedge \theta^{C'B}$, and so is Lorentz covariant on the CC' indices when restricted to $\mathcal{H}_t$. It thus unites the momentum and Hamiltonian constraints in the appropriate fashion.

However this form of the constraints has the defect that it makes the momentum constraint appear as difficult as the Hamiltonian constraint, whereas in Ashtekar's formulation it is clearly much simpler, being linear in the $\tilde{\sigma}$'s. Furthermore, Ashtekar has pointed out that the structure coefficients for the commutators of the constraints in this form are no longer polynomial in the basic variables.

6   Integrals for 'Conserved' Quantities and Quasi-local Mass

The term 'conserved quantity' will be used in the following to refer to components of the momentum or angular momentum of both the gravitational and matter fields contained within a 2-surface. These are not in general conserved in the strong sense of being independent of 2-surface in some homotopy class in the source free region. In full nonlinear general relativity, the gravitational field carries away energy, momentum and angular momentum. These quantities are however conserved in the weak sense that they are independent of spanning 3-surface [22]. See Tod's article in this volume for an account of the quasi-local mass construction, and a review of its successes and failures.

The aim of the following material is to provide a derivation of the quasi-local mass construction from the point of view of the canonical formalism of general relativity. This viewpoint has served to clarify substantially the nature of conserved quantities at space-like and null infinity (see [1] for the case of null infinity). A major problem with the quasi-local mass construction is that, since it was motivated from linearized theory, there are several modifications which all yield the same answer in the linearized limit. If one had a derivation in the context of full general relativity one might be able to see what was wrong with the original formulation and which, if any, modification was correct. This programme has not, however, been realized.

The paradigm for the derivation is the moment map. Consider a Hamiltonian system (Γ, ω, E) where Γ is a $2n$-dimensional phase space, ω is a symplectic form, and E is a Hamiltonian which generates the evolution. If this system has a symmetry group, say for instance the Poincaré group $\mathcal{P}$, then each element p of the Lie algebra of $\mathcal{P}$ gives rise to a vector field X_p on Γ which preserves the symplectic form and E. Since each vector field X_p preserves ω, X_p is a Hamiltonian vector field with Hamiltonian, say H_p. The condition that X_p preserves the Hamiltonian E is also the condition that H_p is preserved by the evolution. Therefore the H_p are conserved quantities. When p is a translation, H_p is the corresponding component of the momentum. When p is a rotation, H_p is the corresponding component of the angular momentum. For a given trajectory, $x(t)$ in Γ of the evolution, the conserved quantities form an element of the dual of the Lie algebra of $\mathcal{P}$, since the map $p \to H_p(x(t))$ is constant and linear in p.

First we derive the total Hamiltonian density for a given lapse shift vector $L^{AA'} = \lambda^A \lambda^{A'}$ for which the decomposition into spinors is given—as noted in remark 4 of section 5 the value of the Hamiltonian depends on the phase of λ^A, and not just $L^{AA'}$ alone. However, the value depends only on the boundary contribution, so we need only construct our candidate symmetry generators on the bounding 2-surface $\mathcal{S}$ of the region $\mathcal{H}$ whose energy momentum we wish to measure. We give a definition of candidate Poincaré group generators on $\mathcal{S}$ which are tangent vectors to M at $\mathcal{S}$ with a given decomposition into spinors obtained from solutions of the twistor equation. The corresponding Hamiltonians are the components of the angular momentum twistor.

6.1 The Total Hamiltonian

The gravitational Hamiltonian is the Sparling 3-form, and the matter Hamiltonian is the integral of certain components of the energy-momentum tensor which, when the field equations are satisfied, equals the Einstein tensor. The total Hamiltonian which generates motion along $L^{AA'}$ is:

$$H_{tot}(L) \;=\; \int_{\mathcal{H}} \tfrac{1}{8\pi G}\Gamma + \tfrac{1}{2}L^a T_{ab}\mathbf{X}^b$$

which reduces when the field equations are satisfied ('on shell') to

$$H_{tot}(L) \;=\; \frac{1}{8\pi G}\int_{\mathcal{H}}\Gamma + \tfrac{1}{2}L^a G_{ab}\mathbf{X}^b = \frac{1}{8\pi G}\int_{\mathcal{H}} dW^-$$
$$=\; \frac{1}{8\pi G}\oint_{\partial\mathcal{H}} W^- .$$

In order to evaluate the Hamiltonian for the evolution along $L^{AA'} = \lambda^A \lambda^{A'}$, we need only compute the boundary integral on $\partial\mathcal{H}$. This implies the 'weak conservation' referred to above.

6.2 Quasi-local Mass

In order to obtain Penrose's original quasi-local mass formula [22] from this Hamiltonian viewpoint, we consider a finite piece of hypersurface $\mathcal{H}$ with boundary $\mathcal{S}$. We assume that dW^- is the correct Hamiltonian density even in this quasi-local context (cf. remark 3 in section 5). We then define a family of lapse shift vectors together with their decompositions into spinors (cf. remark 4 in section 5) on $\mathcal{S}$ which will be thought of as quasi-local symmetry generators (i.e. *quasi-Killing vectors*, or *quasi-conformal Killing vectors*). The conserved quantity associated to each of these quasi-symmetries is then obtained as the value of the Hamiltonian which generates the symmetry. This is given by inserting the constituent spinors for the quasi-symmetry into the boundary integral for the Hamiltonian. We shall describe two choices for the quasi-symmetries. These are both built out of solutions to the 2-surface twistor equation on $\mathcal{S}$.

The twistor equation is:

$$\nabla^{A'(A}\omega^{B)} = 0. \tag{7}$$

This equation is conformally invariant. In Minkowski space there are four independent solutions. In curved space there are no solutions in general as it is overdetermined.

The 2-surface twistor equation is obtained as the pair of equations obtained by looking at only the components of (7) that are tangent to $\mathcal{S}$. These are two equations on two unknowns and so are precisely determined. If o^A and ι^A are a spin frame for which $o^A o^{A'}$ and $\iota^A \iota^{A'}$ are orthogonal to $\mathcal{S}$, $o^{A'}\iota^A$ and $\iota^{A'} o^A$ are complex null vectors tangent to $\mathcal{S}$, so the two equations are $\iota^{A'} o^A o^B \nabla_{AA'}\omega_B = 0$ and $o^{A'}\iota^A \iota^B \nabla_{AA'}\omega_B = 0$. When $\mathcal{S}$ is topologically a 2-sphere, the index theorem implies that there are generically precisely four solutions. In Minkowski space therefore, the four solutions to the 2-surface twistor equation on a generic 2-surface with 2-sphere topology must be the restriction of the four global solutions. See [22, 25] or Tod's article in this volume for further details.

Solutions of the twistor equation relate to symmetries of Minkowski space, **M**, in two distinct ways. These can be found in chapter 6 of Penrose & Rindler, vol. 2 where they are related by means of the *moment sequence*. First, introduce the four solutions to the twistor equation, ω_α^A where $\alpha = 0, \ldots, 3$ can be thought of as a concrete index throughout the following.

Associated to each ω_α^A is a primed spinor $\pi_{\alpha A'}$ defined by:

$$d\omega_\alpha^A = -i\pi_{\alpha A'}\theta^{AA'}; \tag{8}$$

indeed this equation serves both as the twistor equation and the definition of $\pi_{\alpha A'}$; the 2-surface twistor equation, and the definition of $\pi_{\alpha A'}$ for its solutions can be obtained by restricting (pulling back) this equation to the two surface $\mathcal{S}$. The definition of $\pi_{\alpha A'}$ is not conformally invariant.

At this stage we introduce the canonical structures on the solution space T^α to the twistor equation. The first two structures are conformally invariant;

they are a Hermitian bilinear form of signature (2,2) and an alternating tensor

$$h_{\alpha\bar{\alpha}} = \omega_\alpha^A \bar{\pi}_{\bar{\alpha}A} + \pi_{\alpha A'}\bar{\omega}_{\bar{\alpha}}^{A'} \quad \text{and} \quad \varepsilon_{\alpha\beta\gamma\delta} = \varepsilon_{AB}\varepsilon_{A'B'}\omega_{[\alpha}^A \omega_\beta^B \pi_\gamma^{A'} \pi_{\delta]}^{B'}. \tag{9}$$

The third is only Poincaré invariant; it is a skew degenerate twistor, the *infinity twistor*

$$I_{\alpha\beta} = \pi_\alpha^{A'} \pi_{\beta A'}.$$

These are all constant on Minkowski space. The Hermitian form $h_{\alpha\bar{\alpha}}$ can be used to eliminate all barred indices. However, on a 2-surface in curved space these structures are in general not constant, and therefore do not endow the solution space to the twistor equation T^α with any structure in a direct way.

Quasi-conformal Killing Vectors

On **M** the general conformal Killing vector can be represented as

$$C^{AA'} = C^{\alpha\bar{\alpha}}\omega_\alpha^A \bar{\omega}_{\bar{\alpha}}^{A'}$$

where the components of $C^{\alpha\bar{\alpha}}$ are constant. The space of conformal Killing vectors is fifteen dimensional, and $C^{\alpha\bar{\alpha}}$ has 16 components, one of which $h^{\alpha\bar{\alpha}}$ (the inverse of $h_{\alpha\bar{\alpha}}$ as defined above) automatically yields zero.

Quasi-Killing Vectors

The general Killing vector on **M** can be represented by

$$K^{AA'} = Re\{K^{\alpha\beta}\omega_\alpha^A \pi_\beta^{A'}\} = \tfrac{1}{2}\{K^{\alpha\beta}\omega_\alpha^A \pi_\beta^{A'} + \overline{K}^{\bar{\alpha}\bar{\beta}}\bar{\omega}_{\bar{\alpha}}^{A'}\bar{\pi}_{\bar{\beta}}^A\}$$

with $K^{\alpha\beta} = K^{(\alpha\beta)}$. There is also a redundancy in this description; $K^{\alpha\beta}$ has 10 complex components and the space of Killing vectors is 10 real dimensional.

These two definitions can be related by means of the infinity twistor with one of its indices converted into a barred index $I_\alpha{}^{\bar{\alpha}} = h^{\beta\bar{\alpha}}I_{\alpha\beta}$ since this enjoys the property $\pi_\alpha^{A'} = I_\alpha{}^{\bar{\alpha}}\bar{\omega}_{\bar{\alpha}}^{A'}$ (see Tod's article in this volume for further discussion of $I_\alpha{}^{\bar{\alpha}}$, the so-called bar-hook operation). The Killing vector $K^{AA'} = 2Re\{-iK^{\alpha\beta}\omega_\alpha^A \pi_\beta^{A'}\}$ is necessarily also a conformal Killing vector and can therefore be represented in the form $K^{AA'} = K^{\alpha\bar{\alpha}}\omega_\alpha^A \bar{\omega}_{\bar{\alpha}}^{A'}$ where $K^{\alpha\bar{\alpha}} = K^{\alpha\beta}I_\beta{}^{\bar{\alpha}} + K^{\bar{\alpha}\bar{\beta}}\bar{I}_{\bar{\beta}}{}^\alpha$. This relation determines the 10 degrees of freedom of $K^{\alpha\beta}$ that are redundant (i.e. give rise to the zero Killing vector) as those whose corresponding $K^{\alpha\bar{\alpha}}$ vanish. (Such $K^{\alpha\beta}$ are necessarily of the form $G_\gamma^{(\alpha}I^{\beta)\gamma}$ for some $G_\beta^\alpha = \overline{G}_\beta^\alpha$.) The infinity twistor also distinguishes the (conformal) Killing vectors that are translations by $T^{\alpha\bar{\alpha}} = I_\beta{}^{\bar{\alpha}}\bar{I}_{\bar{\beta}}{}^\alpha C^{\beta\bar{\beta}}$ for some $C^{\beta\bar{\beta}}$. See Penrose & Rindler chapter 6 for full details.

These definitions can be made at a 2-surface in terms of 2-surface twistors in a curved space-time. We can proceed to define a 'conserved quantity' for each quasi-Killing vector and quasi-conformal Killing vector since the above definition provides us not only with the vectors, but also a canonical factorization into sums of products of spinors with constant coefficients. We can therefore insert these expressions into those for the total Hamiltonian.

Since, for each Killing vector, the corresponding conserved quantity is a number, the collection of conserved quantities can be represented by an element of the dual space to the space of quasi-(conformal)-Killing vectors. Thus we obtain for the quasi-conformal Killing vectors the conserved quantities $E_{\alpha\bar{\alpha}}$ given by:

$$E_{\alpha\bar{\alpha}}C^{\alpha\bar{\alpha}} = \frac{i}{8\pi G} \oint C^{\alpha\bar{\alpha}}\omega_\alpha^A d\bar{\omega}_{\bar{\alpha}}^{A'} \wedge \theta_{AA'}$$

and for the quasi-Killing vectors we obtain the 'angular momentum twistor' $A_{\alpha\beta}$ given by

$$A_{\alpha\beta}K^{\alpha\beta} = \frac{i}{8\pi G} \oint K^{\alpha\beta}\omega_\alpha^A d\pi_\beta^{A'} \wedge \theta_{AA'}$$

so we can put

$$E_{u\bar{u}} - \frac{i}{8\pi G} \oint \omega_\alpha^A d\bar{\omega}_{\bar{\alpha}}^{A'} \wedge \theta_{AA'} \tag{10}$$

and

$$A_{\alpha\beta} = \frac{i}{8\pi G} \oint \omega_\alpha^A d\pi_\beta^{A'} \wedge \theta_{AA'}. \tag{11}$$

K.P. Tod has proposed also a 'momentum twistor',

$$P_{\alpha\bar{\alpha}} = \frac{i}{8\pi G} \oint \bar{\pi}_{\bar{\alpha}}^A d\pi_\alpha^{A'} \wedge \theta_{AA'},$$

consisting of those conserved quantities dual to the quasi-translations. In linearized theory we have $P_{\alpha\bar{\alpha}} = I_\alpha^{\bar{\beta}}\bar{I}_{\bar{\alpha}}^{\beta} E_{\beta\bar{\beta}}$.

In order to see that (11) agrees with Penrose's original definition (see section 9.9 of Penrose & Rindler) we must insert equation (8) into (11) to get

$$A_{\alpha\beta} = \frac{1}{8\pi G} \oint \omega_\alpha^A d^2\omega_{\beta A} = \frac{1}{16\pi G} \oint R_{AB}\omega_\alpha^A \omega_\beta^B.$$

6.3 Remarks

1. *The Modifications.* Like Penrose's original formulation, these ideas are still too strongly motivated from linearized theory, in particular with respect to the choice of the quasi-Killing vectors and their decomposition into products of spinors. This leads to freedom in the generalization to curved space. This has meant that when the quasi-local mass definition has run into difficulties, it has been possible to introduce modifications which improve the situation. However, once one modification has been introduced it takes little imagination to introduce many more, and, without some new principle, it is not clear which formulation

is correct. Penrose's first modification was to introduce an additional factor of $\eta = \varepsilon_{0123}$ into the integrand. This quantity is constant and can be chosen to be 1 when the 2-surface can be conformally embedded into $\mathbf{M}$ (that is $\mathcal{S}$ is non-contorted), but in general is a nontrivial function. Its effect was to remove problems with space-like momenta that arose from calculations of the angular momentum twistor for small spheres in the Schwarzschild solution in [13]. This modification can be incorporated by replacing the spinor ω^A_α by $\eta\omega^A_\alpha$ in the expression for the quasi-Killing vector and in the first expression for $A_{\alpha\beta}$. Penrose's subsequent modifications [23, 24] can also be incorporated by modifying the definition of the quasi-Killing vectors or their factorization into spinors. All of these different modifications coincide when the 2-surface together with its intrinsic and extrinsic data can be conformally embedded in $\mathbf{M}$. See Tod's article in this volume for a detailed discussion of the various modifications.

2. *Ashtekar Variables.*   The quantities $B_{\alpha\bar\alpha}$ are effectively ambidextrous in that no choice of chirality is made and both spin connections are required for its definition. However, the definition of $A_{\alpha\beta}$ is chiral, and indeed fits in very neatly with the spirit of Ashtekar's variables. The ω^A_α are the solutions of an equation that only involves the unprimed spin connection, and $A_{\alpha\beta}$ is an integral of the curvature of the unprimed spin connection contracted with ω's. All expressions involving derivatives of primed spinors can be bracketed in such a way as to exhibit the connection acting only on unprimed indices.

3. *The Ten Vanishing Integrals.*   When the infinity twistor $I_\alpha{}^{\bar\alpha}$ is constant on $\mathcal{S}$ (such as at space-like or null infinity and in linearized theory) ten of the twenty integrals vanish. We have

$$
\begin{aligned}
A_{\alpha\beta} &= \frac{i}{8\pi G} \oint \omega^A_{(\alpha} d\pi^{A'}_{\beta)} \wedge \theta_{AA'} \\[2mm]
&= \frac{i}{8\pi G} \oint \omega^A_{(\alpha} d(I_{\beta)}{}^{\bar\alpha} \overline{\omega}^{A'}_{\bar\alpha}) \wedge \theta_{AA'} \\[2mm]
&= \frac{i}{8\pi G} I_{(\alpha}{}^{\bar\alpha} \oint \omega^A_{\beta)} d\overline{\omega}^{A'}_{\bar\alpha}) \wedge \theta_{AA'} \\[2mm]
&= I_{(\alpha}{}^{\bar\alpha} E_{\beta)\bar\alpha}
\end{aligned}
$$

where $E_{\alpha\bar\alpha}$ is given by equation (10). Since $E_{\alpha\bar\alpha}$ is automatically real, this implies that $A_{\alpha\beta}$ satisfies reality conditions: $A_{\alpha\beta}\bar{I}_{\bar\alpha}{}^\beta = \overline{A}_{\bar\alpha\bar\beta}I_\alpha{}^{\bar\beta}$.

4. *Positivity.*   It has not thus far been possible to find a positivity proof in general. When $\bar{I}_{\bar\alpha}{}^\beta$ is constant on $\mathcal{S}$, $A_{\alpha\beta}\bar{I}_{\bar\alpha}{}^\beta$ is Hermitian, the

desired positivity condition can be expressed as

$$A_{\alpha\beta}\overline{I}_{\overline{\alpha}}{}^{\beta}Z^{\alpha}\overline{Z}^{\overline{\alpha}} \geq 0$$

for all constant Z^{α}: the Hermitian matrix $A_{\alpha\beta}\overline{I}_{\overline{\alpha}}^{\beta}$ should be positive semi-definite. If we write $\omega^{A} = Z^{\alpha}\omega_{\alpha}^{A}$ and $\pi_{A'} = Z^{\alpha}\pi_{\alpha A'}$, this can be expressed as follows

$$\begin{aligned}
A_{\alpha\beta}\overline{I}_{\overline{\alpha}}{}^{\beta}Z^{\alpha}\overline{Z}^{\overline{\alpha}} &= \frac{i}{8\pi G}\oint_{\mathcal{S}}\overline{\pi}^{A}d\pi^{A'}\wedge\theta_{AA'} \\
&= \frac{i}{8\pi G}\int_{\mathcal{H}}d(\overline{\pi}_{A}d\pi_{A'}\wedge\theta^{AA'}) \\
&= \frac{i}{8\pi G}\int_{\mathcal{H}}d\overline{\pi}_{A}\wedge d\pi_{A'}\wedge\theta^{AA'} - \tfrac{1}{2}G_{ab}\overline{\pi}^{A}\pi^{A'}\mathbf{X}^{b}
\end{aligned}$$

where $\mathcal{H}$ is some space-like spanning surface for $\mathcal{S}$ and $\pi_{A'}$ is any extension of $\pi_{A'}$ from $\mathcal{S}$ to $\mathcal{H}$. The second term is positive definite as a consequence of the dominant energy condition and the first term is positive definite when the spinor $\pi_{A'}$ has been extended over $\mathcal{H}$ so as to satisfy the Sen-Witten equation $D_{A'}^{B'}\pi_{B'} = 0$ with boundary value the given $\pi_{A'}$ on $\mathcal{S}$. Here $D_{A'B'} = T_{(A'}^{A}\nabla_{B')A}$ and $T^{AA'}$ is a normal to $\mathcal{H}$. The $\pi_{A'}$ spinor can only in special cases be extended so as to satisfy the Witten equation, in particular when $\mathcal{S}$ is taken to be at spacelike infinity or when $\mathcal{S}$ is a cut of null infinity (see [25, chapter 6] and [28, 35]).

5. *Further Problems.* It should be clear to the reader that the above ideas are still tentative and should be thought of as work in progress. In curved space the definition of quasi-symmetries suffer from certain defects. For the conformal Killing vectors, it is no longer possible to show that the 16th one with $C^{\alpha\overline{\alpha}} = h^{\alpha\overline{\alpha}}$ vanishes since $h^{\alpha\overline{\alpha}}$ itself is no longer constant and no longer defines an element of $\mathsf{T}^{\alpha\overline{\alpha}}$. For the quasi-Killing vectors in the curved case, as $K^{\alpha\beta}$ varies, the span of $Re\{K^{AA'}\}$ can in general be 20-dimensional—it can no longer be shown that ten of the twenty quasi-Killing vectors vanish. The argument above breaks down since $I_{\alpha}{}^{\overline{\alpha}}$ is no longer constant. Similar problems arise for K.P.Tod's momentum twistor $P_{\alpha\overline{\alpha}}$ which in general cannot be shown to have only 4 rather than 16 components. A further irritation is that in flat space $E_{\alpha\overline{\alpha}}$ does not vanish as defined so that its interpretation as a 'conserved quantity' is unclear.

6.4 *Towards a Derivation of Quasi-local Mass*

The above presentation goes some way to providing a derivation of the quasi-local mass construction in the context of full general relativity. However, it is a long way from being rigorous and complete in the same way as, for instance, the modern treatment of conserved quantities at space-like infinity.

What is required is a version of the canonical formalism in which the dynamics of the finite 2-surface $\mathcal{S}$ plays a significant rôle. For instance one might introduce a canonical formalism in which one considers data on a 3-surface spanning $\mathcal{S}$ without any particular boundary conditions on the data so that the boundary would correspond to a finite 2-surface. One may then be able to derive the quasi-Killing vectors as genuine symmetries of the appropriate gravitational phase space or perhaps from some weaker requirement. In this context, a more rigorous derivation of the correct boundary term for the Hamiltonian is required that would fix the quasi-local ambiguity.

6.5 *Connections with Super-gravity*

The twistor equation is an ingredient which would seem difficult to derive from considerations connected with the canonical formalism for general relativity alone. It may be possible to see that the quasi-Killing vectors as defined above are indeed symmetries of some generalization of the gravitational phase space, but it seems less likely that one could derive the twistor equation itself from that kind of structure.

However, solutions of the twistor equation do indeed play a fundamental rôle in conformal *super-symmetry* as the spinor fields which give rise to conformal super-symmetry generators. The anticommutators of these give rise to conformal Killing vectors of space-time. This can be represented on superspace by the anticommutation relations

$$\{\omega^A D_A, \overline{\omega}^{A'} D_{A'}\} = \omega^A \overline{\omega}^{A'} \nabla_{AA'}$$

where the derivations $D_A = \frac{\partial}{\partial \theta^A} + \theta^{A'} \nabla_{AA'}$ generate super-symmetry transformations in superspace on which $(\theta^A, \theta^{A'})$ are anticommuting coordinates. This provides a derivation (of sorts) for the form of the quasi-conformal Killing vectors *and* their spinorial factorization.

It is less clear how to derive the particular form of the quasi-Killing vectors in this context. However, some insight may be obtained from the version of the canonical formalism for super-gravity using Ashtekar type variables [11, 5]. This is obtained by adding the Lagrangian for an anticommuting spin $\frac{3}{2}$ Rarita-Schwinger field to the version of the Lagrangian given in 4. The spin $\frac{3}{2}$ field is represented by an indexed 1-form, ψ_A and its conjugate field, a 2-form:

$$\chi^A = \theta^{AA'} \wedge \overline{\psi}_{A'}$$

which satisfies the algebraic constraint $\Sigma^{(AB} \wedge \chi^{C)} = 0$. Any indexed 2-form χ^A satisfying this algebraic constraint is of the form $\theta^{AA'} \wedge \tilde{\psi}_{A'}$ for some indexed 1-form $\tilde{\psi}_{A'}$. The super-gravity action in terms of these fields is:

$$S = \int iR_{AB} \wedge \Sigma^{AB} - i\chi^A \wedge d\psi_A.$$

Its variation with respect to the basic fields subject to the various constraints yields the regular super-gravity field equations and its reduction to the canonical formalism yields a version of the super-gravity canonical formalism in terms of Ashtekar type variables.

The supersymmetries are given by:

1. *Right-handed:* $\delta\psi_A = d\epsilon_A$, $\delta\Sigma^{AB} = \epsilon^{(A}\chi^{B)}$, no variation in the connection or χ^A.

2. *Left-handed:* $\delta\chi^A = d(\epsilon_{A'}\theta^{AA'})$, $\delta\Sigma^{AB} = \epsilon_{A'}\theta^{A'(A} \wedge \psi^{B)}$, no variation in the connection or ψ_A.

In order to obtain quasi-Killing vectors, one can first start at the classical section with $\psi_A = 0 = \chi^A$ and perform a right handed super-symmetry using a general solution of the 2-surface twistor equation ω^A. If we then perform a left handed super-symmetry with $\epsilon_{A'}\pi_{A'}$, then the restriction of Σ^{AB} to $\mathcal{S}$ is left invariant. Let $\epsilon^A = \omega^A$, then the subsequent left handed super-symmetry will yield

$$\delta\Sigma^{AB} = \pi_{A'}\theta^{A'A} \wedge d\omega^B = -i\pi_{A'}\theta^{A'(A} \wedge \pi_{B'}\theta^{B)B'} = -i\tfrac{1}{2}\pi_{A'}\pi^{A'}\Sigma^{AB} = 0.$$

This is not a characterization of the form of the quasi-Killing vectors but it does indicate that some connection does exist. It should in particular be noted that this transformation is not just a motion of space-time, but is a combination of a space-time motion and a super-symmetry; only the anticommutator of two super-symmetry transformations is a pure translation.

These ideas are not fully developed, but they do provide scope for a substantial review of the ideas underlying the quasi-local mass construction. It is perhaps also worth noting in this context that Sparling [32] has discovered an analogue of the original Sparling 3-form in super-space descriptions of supergravity, at least in the classical section. Again he finds that closure of this form implies and is implied by the field equations. Its properties with respect to the canonical formalism for supergravity have yet to be analyzed.

7   The Sparling Form and Energy Momentum Pseudotensors

As we have seen in section 3 the Sparling form has a very simple form on the spin bundle of space-time. Moreover, the spin bundle seems the 'smallest' manifold where we can write down an identity between 3-forms that involves the Einstein tensor and is of the form

'exact 2-form = Sparling form + Einstein form'.

Therefore it is natural to regard the spin bundle as the 'home' of the Sparling form. However, if we enlarge this 8-dimensional manifold to the 20-dimensional manifold L of the bundle of linear frames on space-time we find another interesting property of the Sparling form [8].

In order to obtain an expression for the Sparling form on L we proceed as follows. We first extend it to a form on the bundle of orthonormal frames, B. The bundle B can be thought of as the bundle of primed and unprimed spin frames $\varepsilon_{\underline{B}}{}^{A}$ and $\varepsilon_{\underline{B}'}{}^{A'}$. Using the constituent spinors of these spin frames, we can write down the collection of Sparling 3-forms:

$$\Gamma_{\underline{BB'}} = id\varepsilon_{\underline{B}A} \wedge d\varepsilon_{\underline{B}'A'} \wedge \theta^{AA'}.$$

These can alternatively be written as

$$\Gamma_{\underline{BB'}} = i\gamma_{\underline{AB}} \wedge \overline{\gamma}_{\underline{A'B'}} \wedge \theta^{\underline{AA'}}$$

where $\gamma_{\underline{AB}}$ are the unprimed spin connection forms on the spin bundle:

$$\gamma_{\underline{A'}}{}^{\underline{B'}} = \varepsilon_{A'}{}^{\underline{B'}} d\varepsilon_{\underline{A}'}{}^{A'}.$$

Similarly W^- extends to the forms $W_{\underline{a}}^- = W_{\underline{AA'}}^- = i\gamma_{\underline{AB}} \wedge \theta^{\underline{B}}_{\underline{A'}}$, and $W_{\underline{AA'}}^+ = -i\overline{\gamma}_{\underline{A'B'}} \wedge \theta^{\underline{B'}}_{\underline{A}}$. Note that $d(ImW_{\underline{a}}^+) = 0$ follows from the first structural equation on the frame bundle, hence defining $W_{\underline{a}} = (W_{\underline{a}}^- + W_{\underline{a}}^+)$ we obtain

$$dW_{\underline{a}} = 2\Gamma_{\underline{a}} + G_{\underline{a}}^b \mathbf{X}_b,$$

where $\mathbf{X}_b = \frac{1}{6}\varepsilon_{bcde}\theta^c \wedge \theta^d \wedge \theta^e$.

We will now drop the convention of underlining for concrete indices, and assume that all subsequent indices in this section will be concrete—since we are working on the bundle of orthonormal frames, we have a natural parallelization. On bundles of frames we have the globally constant frame rotations under which the forms W_a transform in the following way:

$$W_a \to A^b{}_a W_b$$

where $A^b{}_a$ is in $SO(3,1)$.

If we now consider B as a sub-bundle of L, we find an expression for the forms W_a on L by maintaining the transformation law above but now for a general $GL(4,\mathbf{R})$ matrix. We obtain

$$W_a = \tfrac{1}{2}\sqrt{-g}\epsilon_{abcd}\theta^b \wedge \gamma^{cd}$$

where g is the determinant of the metric in the frame that corresponds to the bundle point that we are looking at. Calculating the exterior derivative of W_a we find that the Sparling form has to be defined as

$$\Gamma_a = \tfrac{1}{4}\sqrt{-g}\epsilon_{abcd}(\gamma^e{}_e \wedge \gamma^{dc} \wedge \theta^b + \gamma^b{}_e \wedge \gamma^{cd} \wedge \theta^e + \gamma^b{}_e \wedge \gamma^{de} \wedge \theta^c).$$

This can be written as $\Gamma_a = \gamma_{ab}^+ \wedge \gamma^{-b}{}_c \wedge \theta^c$, where $\gamma_{ab}^{\mp} = \frac{1}{2}(\gamma_{ab} \pm \frac{i}{2}\sqrt{-g}\epsilon_{ab}{}^{cd}\gamma_{cd})$ are the anti-self-dual and self-dual parts of γ_{ab}. Altogether we have an identity of the required form on the bundle of linear frames.

Let us now take a local coordinate section $s : U \to L$ on an open neighbourhood of M and determine the pullback of this identity on U. The local coordinate expression for W_a turns out to be

$$W_a = \frac{1}{2}(g^{b[c}\gamma_{ab}^{d]} + g^{be}\gamma_{be}^{[c}\delta_a^{d]} + \delta_a^{[c}g^{d]b}\gamma_{eb}^{e})\Sigma_{cd} = \frac{1}{4\sqrt{-g}}W_a{}^{cd}\Sigma_{cd}$$

where $\Sigma_{ab} = \frac{1}{2}\sqrt{-g}\epsilon_{abcd}du^c \wedge du^d$. If we define the so called 'Landau-Lifschitz superpotential' by $H^{abcd} = (-g)(g^{ac}g^{bd}-g^{ad}g^{bc})$ we can express the coefficients $W_a{}^{cd}$ of W_a as

$$W_a{}^{cd} = \frac{1}{\sqrt{-g}}g_{ae}H^{ebcd}{}_{,b}.$$

Using the fact, that in a coordinate frame $d(\frac{1}{\sqrt{-g}}\Sigma_{ab}) = 0$, we obtain

$$dW_a = \frac{1}{4\sqrt{-g}}W_a{}^{cd}{}_{,b}du^b \wedge \Sigma_{cd} = 2\Gamma_a{}^b\Sigma_b + G_a{}^b\Sigma^b.$$

Therefore we finally end up with the equation

$$W_a{}^{bd}{}_{,d} - 2\sqrt{-g}G_a{}^b = 4\sqrt{-g}\Gamma_a{}^b.$$

The left hand side of this equation defines the so called 'energy-momentum complex' first given by Einstein [7]. Hence, we find that this complex is up to a factor equal to the pullback of the Sparling form in a local coordinate frame. From this and the fact, that the Sparling form is a pseudotensorial form [14] it becomes clear, that the complex does not transform as a tensor under arbitrary coordinate transformations.

If instead of W_a we had considered the forms $W^a = g^{ab}W_b$, after going through the same procedure as above, we would have found the equation

$$H^{acbd}{}_{,cd} - 2(-g)G^{ab} = 4(g^{ac}g_{cd,e}H^{dfeb}{}_{,f} + g^{cd}g_{cd,f}H^{aebf}{}_{,e}) + 4(-g)\Gamma^{ab}.$$

As before, the lefthand side of the equation defines an 'energy-momentum complex', this time we obtain the Landau-Lifschitz pseudotensor [15] expressed with the components of the pullback of the Sparling form together with some additional terms.

8   Conclusions and Discussion

Unfortunately the key issues in this article remain unresolved; the proper derivation of the quasi-local mass construction, the detailed connections with supergravity and the relationship of these ideas with those of twistor theory. Indeed there are virtually no connections with the non-linear twistor constructions. There is however an intriguing analogy between Ashtekar's chiral description of general relativity and a pattern in twistor theory to do with the Sparling 3-form observed by Penrose [21].

First it is worth noting that there are chiral Lagrangians for all physical fields. These are related to the real Lagrangians by the addition of either total divergences or topological terms or quantities which are identically zero. For spin $\frac{1}{2}$ we have the chiral Lagrangian $i\overline{\psi}^{A'}\nabla_{AA'}\psi^A$ whose imaginary part is a total divergence, $\nabla_{AA'}(\overline{\psi}^A\psi^{A'})$. For Yang-Mills we have $iTr(F^- \wedge F^-)$ where F^- is the anti-self-dual part of the curvature of the Yang-Mills field. The imaginary part of this Lagrangian is the integrand for the Pontryagin class of the Yang-Mills bundle. For general relativity we have the imaginary part of the action used above which is $R_{[abcd]} = 0$, whose field equations are the condition that the connection be torsion free. In twistor theory there is an analogous but more detailed picture. Null geodesics correspond to points in twistor space with $Z \cdot \overline{Z} = 0$. On $Z \cdot \overline{Z} = 0$ we have the canonical 1-form $\theta = p_a dx^a = Re\{iZ^\alpha d\overline{Z}_\alpha\}$ and the form $Im\{iZ^\alpha d\overline{Z}_\alpha\}$ vanishes identically being $d(Z \cdot \overline{Z})$ restricted to $Z \cdot \overline{Z} = 0$. The structure θ and $d\theta$ give rise to the symplectic geometry of the space of null geodesics. If we now consider the Witten-Nester form $W^- = i\pi_{A'}d\overline{\pi}_A \wedge \theta^{AA'}$ we have that $Im\{W^-\} = id\theta$ is an exact divergence, and the real part is the only part that plays a rôle in the Hamiltonian considerations above. Finally we have $dW^- = \Gamma + \frac{1}{2}G^a_b\pi_{A'}\overline{\pi}_A\mathbf{X}^b$, so that $dW^- = \Gamma$ in vacuum. These ideas embody in a vague sense a Hamiltonian analogue of the ideas connected with the use of chiral Lagrangians above.

In order to make full contact with twistor theory, it will be necessary to reformulate the canonical formalism for general relativity in terms of structures on hypersurface twistor space. The remarkable simplifications discovered by Ashtekar and coworkers in this chiral version of the canonical formalism indicate that the chiral treatment necessitated by a twistor reformulation may be even simpler than the space-time formulation.

Acknowledgements

LJM would like to thank Abhay Ashtekar, Ted Jacobson, Roger Penrose, George Sparling and Paul Tod for discussions. JF thanks the Alexander Von Humboldt Foundation for financial support. LJM would like to thank the physics departments at the University of Pittsburgh and Syracuse University and the Maths Department at University of Washington for hospitality while this work was being written, and also the Esmee Fairbairn trust and the Science and Engineering Research Council for financial support. We would both like to thank E.T. Newman for suggesting this project.

References

[1] Ashtekar, A & Streubel, M. (1981) Symplectic geometry of radiative modes and conserved quantities at null infinity, Proc. Roy. Soc., A **376**, 595–607.

[2] Ashtekar, A. (1988) New perspectives in Canonical Gravity (Naples: Bibliopolis)

[3] Ashtekar, A., Jacobson, T. & Smolin, L. (1988) Comm. Math. Phys. **115** 631–648.

[4] Capovilla, R., Dell, J. & Jacobson, T. (1989) General relativity without the metric, Phys. Rev. Lett., **63**, 2325.

[5] Capovilla, R., Dell, J., Jacobson, T. & Mason, L. (1990) Self-dual 2-forms and general relativity, University of Maryland preprint.

[6] Dubois-Violette, M. & Madore, J. (1987) Conservation laws and integrability conditions for gravitational and Yang-Mills field equations, Comm. Math. Phys. **108**, 218–223.

[7] Einstein, A. (1916) Sitzungsberichte d. Preuss. Akad. Wiss. Phys. Math. Kl. **42** 1111.

[8] Frauendiener, J. (1989) Geometric description of energy momentum pseudotensors, University of Pittsburgh preprint.

[9] Gindikin, S., *Bundles of differential forms and the Einstein equation*, Nuclear Physics **36** (1982) 2(8), 537–548 (Russian).

[10] Goldberg, J.N. (1988) Triad approach to the Hamiltonian of general relativity, Phys. Rev. D **37**, 8 2116–2120.

[11] Jacobson, T. (1988) New variables for canonical super-gravity, Class. Quant. Grav. **5** 923–935.

[12] Jacobson, T. & Smolin, L. (1987) Phys. Lett. **196B** 39; (1988) Class. Quant. Grav. **5** 583.

[13] Kelly, R.M., Tod, K.P. & Woodhouse, N.M.J. (1986) Quasi-local mass for small surfaces, Class. & Quant. Grav. **3**, 1151–67.

[14] Kobayashi, S. & Nomizu, K. (1963) Foundations of Differential Geometry, Vol. 1 (Wiley: New York).

[15] Landau, L. K. & Lifschitz E. (1951) The Classical Theory of Fields (Addison-Wesley: Cambridge).

[16] Ludvigsen, M. & Vickers, J.A.G. (1982) A simple proof of the positivity of the Bondi mass, J. Math. Phys. **15**, L67–L70.

[17] Mason, L.J. (1989) A Hamiltonian interpretation of Penrose's quasi-local mass, Class. Quant. Grav **6** L7–L13.

[18] Mason L.J. & Newman E.T. (1989) A connection between the Einstein and Yang–Mills equations, Commun. Math. Physics **121**, 658–668.

[19] Nester, J.M. (1983) The gravitational Hamiltonian, Asymptotic Behaviour of Mass and Space-Time Geometry, (Lecture Notes in Physics 202) ed. F J Flaherty (Berlin: Springer) pp 155–63.

[20] Penrose, R. (1976) Non-linear gravitons and curved twistor theory, Gen. Rcl. Grav. **7**, 31, cf. also pp. 171.

[21] Penrose, R. (1983) The Sparling 3-form, Twistor Newsletter **16**.

[22] Penrose, R. (1982) Quasi-Local Mass and angular momentum in general relativity, Proc. R. Soc. London A **381** 53–63.

[23] Penrose, R. (1984) New improved quasi-local mass and the Schwarzchild solution, Twistor Newsletter **18**.

[24] Penrose, R. (1985) A suggested modification to the quasi-local mass formula, Twistor Newsletter **20**.

[25] Penrose, R. & Rindler, W. (1986) Spinors and Space-time, Vol. 2, C.U.P.

[26] Plebanski, J.F. (1977) On the separation of Einsteinian substructures, J. Math. Phys., **18**, 2511–2520.

[27] Regge, T. & Teitelboim, C. (1974) Rôle of surface integrals in the Hamiltonian formulation of general relativity, Ann. Phys., NY, **88**, 286.

[28] Reula, O. & Tod, K.P. (1984) Positivity of the Bondi energy, J. Math. Phys., **25**, 1004–8.

[29] Samuel, J. (1987) Pramana **28** L429.

[30] Sparling, G.A.J. (1983) Einstein's vacuum equations, University of Pittsburgh Preprint.

[31] Sparling, G.A.J. (1984) Twistor theory and the characterization of Fefferman conformal structures, University of Pittsburgh Preprint.

[32] Sparling, G.A.J. (1988) A development of the theory of classical supergravity, University of Pittsburgh Preprint.

[33] Tod, K.P. (1990) Penrose's quasi-local mass, in this volume.

[34] Wess, J. & Bagger, J. (1982) Supersymmetry and Supergravity, Princeton University Press.

[35] Witten, E. (1981) A new proof of the positive energy theorem, Commun. Math. Physics **80** 381–402.

Twistors and Strings

W.T. Shaw   L.P. Hughston

1 A Complex Analytic Approach

Twistor theory has provided many valuable insights into the nature of space-time geometry and the processes that take place within that geometry. In particular, the very special role played by complex analytic functions has been elucidated in several contexts [36, 37, 38, 39, 41, 43]. A feature common to many twistor constructions is the solution of linear or non-linear differential equations by consideration of suitable analytic objects and the construction of machinery for transforming such objects into the desired solutions.

As noted in [40], this particular approach was in use many years before the formal development of twistor theory began. Weierstrass' 1866 construction [53] of solutions of Plateau's problem in three dimensions, for example, may be regarded as an important model for the twistor solution of non-linear problems such as the non-linear graviton [39]. The programme of work described here is founded on the notion that there is much that might be learnt by exploring constructions of the Weierstrass type, especially if insight can also be gained into string theory and its relationship to gravity. That twistor theory might provide some clues regarding this relationship has been suspected for some time (see e.g. [17]).

There are several sources of motivation for considering possible links between twistor theory and string theory. Hitchin [18] explained how the Weierstrass parametrization is essentially the understanding of how arbitrary holomorphic curves in the twistor space for three dimensions correspond to null holomorphic curves in complex three-space. Given the general correspondence between minimal surfaces in real Euclidean n-space and null holomorphic curves in complex n-space, it is natural to investigate what form of correspondence exists between such objects and the appropriate twistor space. The correspondence in dimension four receives special attention in this review article, though we will explain briefly how some of these ideas generalize to other dimensions.

Our interest in this area spans 4 years, and began with two articles [21, 46]. Similar themes have been developed by other writers, for example, [10] and [49]. These approaches have as their common root the relationship between null curves in space-time, and spinor [42] or twistor [43] methods.

2   Historical Perspective

In the early part of the twentieth century, several writers found that complex
analytic null curves in Euclidean or Minkowskian 4-space could be parametrized in terms of two free functions of a complex variable. Montcheuil [34] and
Eisenhart [8, 9], by studying certain surfaces of translation, were led, rather
incidentally, to the following parametrization, for Minkowskian signature:

$$\begin{aligned}
t &= f' + g - \zeta g', \\
x &= g' + f - \zeta f', \\
iy &= g' - f + \zeta f', \\
z &= f' - g + \zeta g'.
\end{aligned}$$

In this system (x, y, z, t) are coordinates of 4-space, with flat metric

$$ds^2 = dt^2 - dx^2 - dy^2 - dz^2,$$

and $f(\zeta)$ and $g(\zeta)$ are analytic functions of the complex variable ζ. The
real part of such a null curve is a space-like minimal surface in Minkowskian
4-space.

It is straightforward to derive from this the result for three dimensions. If
we consider a curve confined to the hypersurface $t = 0$, then $f' = -g + \zeta g'$.
We may integrate this relation by writing $g = G'$, so that $f = \zeta G' - 2G$.
Then

$$\begin{aligned}
x &= (1 - \zeta^2)G'' + 2\zeta G' - 2G, \\
iy &= (1 + \zeta^2)G'' - 2\zeta G' + 2G, \\
z &= 2\zeta G'' - 2G'.
\end{aligned}$$

This formula has also been obtained directly from three-dimensional twistor
theory [18], and is one representation of the Weierstrass parametrization for
null curves in three dimensions [53].

We wish to elucidate the geometry underlying the Eisenhart formula, and
to establish corresponding formulae for real strings in four dimensions, and
to indicate how similar results for other dimensions may be established.

3   Complex Null Curves & Space-like Minimal Surfaces

There are several plausible starting points: In the context of minimal surface
theory a natural place to begin is with a variational formalism. In general
dimension d and for a general signature, one is interested in finding maps
$x^a : R^2 \to R^d$, represented as $x^a(u, v)$, such that the area of the immersed
2-surface is a minimum. In general this leads to non-linear differential equations. Matters simplify considerably if one uses the freedom to reparametrize

the surface to arrange that the 2-dimensional metric of the embedded surface has the form

$$ds^2 \propto du^2 + dv^2,$$

whence the equation of the surface becomes

$$(\partial_u^2 + \partial_v^2)x^a = 0$$

subject to the constraints

$$\partial_u x^a \partial_u x_a = \partial_v x^a \partial_v x_a,$$
$$\partial_u x^a \partial_v x_a = 0.$$

One can now introduce the complex variable $\zeta = u + iv$ and write

$$x^a = \mathrm{Re}\{X^a(\zeta)\}$$

where $X^a(\zeta)$ is analytic in ζ, and the constraints amount to requiring that this curve is null:

$$\partial_\zeta X^a \partial_\zeta X_a = 0.$$

This represents the correspondence between space-like minimal surfaces in any dimension greater than 2 (and essentially any signature), and complex analytic null curves. It should be emphasized that these relations are entirely local in nature.

4   Real Null Curves & Strings

In the case of a Minkowskian signature, one can in similar fashion look for 2-surfaces that are time-like, with signature $(+, -)$. In this case one chooses the coordinates, now labelled as τ, σ, such that the 2-dimensional metric of the immersed surface has the form

$$ds^2 \propto d\tau^2 - d\sigma^2,$$

whence the equation of the surface becomes

$$(\partial_\tau^2 - \partial_\sigma^2)x^a = 0$$

subject to the constraints

$$\partial_\tau x^a \partial_\tau x_a + \partial_\sigma x^a \partial_\sigma x_a = 0,$$
$$\partial_\tau x^a \partial_\sigma x_a = 0.$$

In this case, naturally enough, we are led to consider solutions of the 2-dimensional wave equation, rather than Laplace's equation.

The solution of the two-dimensional wave equation is, of course,

$$x^a(\sigma, \tau) = \xi^a(\tau - \sigma) + \eta^a(\tau + \sigma)$$

where $\xi^a(s)$ and $\eta^a(s)$ are real functions with continuous second derivatives. The constraints imply $\dot{\xi}^a\dot{\xi}_a = 0$ and $\dot{\eta}^a\dot{\eta}_a = 0$. (We use a dot to denote differentiation with respect to s). The fact that these curves must be null is made rather more transparent by the observation that the constraints are equivalent to the condition that each of $(\partial_\sigma \pm \partial_\tau)x^a$ should be null.

As for boundary conditions there are two primary elementary cases of general interest, depending on the topology of the resultant strings. These are:

Loops

In this case $\sigma \in [0, 2\pi]$. We require that the end-points must join, for each value of τ, in such a way that the tangent vectors agree:

$$\begin{aligned} x^a(2\pi, \tau) &= x^a(0, \tau), \\ \partial_\sigma x^a(2\pi, \tau) &= \partial_\sigma x^a(0, \tau) \end{aligned}$$

for all τ. From these conditions it follows that the functions ξ^a and η^a must satisfy the periodicity relations

$$\begin{aligned} \xi^a(s + 2\pi) - \xi^a(s) &= \tfrac{1}{2}P^a, \\ \eta^a(s + 2\pi) - \eta^a(s) &= \tfrac{1}{2}P^a. \end{aligned}$$

for some constant vector P^a.

Open Strings

Here $\sigma \in [0, \pi]$ with the edge condition $\partial_\sigma x^a = 0$ at $\sigma = 0, \pi$ for all τ. It follows that

$$x^a(\sigma, \tau) = \zeta^a(\tau - \sigma) + \zeta^a(\tau + \sigma)$$

for some real null curve $\zeta^a(s)$ which satisfies

$$\zeta^a(s + 2\pi) = \zeta^a(s) + P^a.$$

If units are chosen such that the 'string tension' is unity then the vector P^a is in each case the total momentum of the string, which is required to be future-pointing.

From the point of view of pure mathematics, as a matter of general principle one would be inclined to believe the case of the loop ('closed string') to be more fundamental and, in some respects, more natural than the case of the open string. However, in the present situation the reverse holds.

For further details of the underlying variational principle, see [13]. Schild [45] gives a lucid treatment of the geometric subtleties associated with the variational formulation. It is evident that the cases of open and closed strings do not cover all the possibilities for a time-like surface: but for a 'free' string theory, these periodicity relations are the conditions found generally to be applied in the physics literature.

5   Twistors and Null Curves

It is evident that the notion of a real or complex analytic *null* curve is fundamental to the description of strings or minimal surfaces. It is this property
that makes the introduction of twistor methods natural. It should be ap-
preciated that the detailed correspondence between surfaces of extremal area
and null curves which we have described is *directly* relevant to flat space-
times only. The twistor description always characterizes the null curves, and
it is only in flat space-time that these are so simply related to the minimal
surfaces. Bryant, however, [3] has constructed an example of a transform
appropriate for certain types of minimal immersions in the 4-sphere.

It is natural to treat the cases of space-like and time-like surfaces together.
In doing so it is convenient to adopt two levels of description: the first is
based on 2-component spinors, and is useful for generating explicit formulae,
while the second, based on the $\mathbb{CP}^5$ description [24], gives a more transparent
realization of the geometry.

We are interested in characterizing curves $\phi^a(s)$ such that $\dot{\phi}^a(s)$ is null.
In the case of real open or closed strings, further periodicity conditions are
to be imposed. The nullity of $\dot{\phi}^a(s)$ is equivalent to the existence of a spinor
field $\pi_{A'}(s)$ such that

$$\dot{\phi}^{AA'}(s)\pi_{A'}(s) = 0. \tag{1}$$

This spinor field is only defined projectively by this equation, which is invariant under replacements $\pi_{A'}(s) \longrightarrow \gamma(s)\pi_{A'}(s)$. We form a twistor field
$Z^\alpha(s) = \{\omega^A(s), \pi_{A'}(s)\}$ by imposing the incidence relation

$$\omega^A(s) = i\phi^{AA'}(s)\pi_{A'}(s) \tag{2}$$

at every point on the curve. It follows that

$$\dot{\omega}^A(s) = i\phi^{AA'}(s)\dot{\pi}_{A'}(s). \tag{3}$$

Evidently the twistor curve is just the field of α-planes that are tangent to
the null curve.

Having constructed the associated twistor curve, we can investigate how
to express the space-time curve in terms of it. At this point it is convenient to
make some assumptions that the curve is generic, or non-degenerate, in the
sense that $Z^\alpha(s)$ and $\dot{Z}^\alpha(s)$ are not proportional. We will consider shortly
what happens when this condition is not satisfied. Under this assumption,
we can express the null curve as

$$i\phi^{AA'}(s) = \frac{\omega^A(s)\dot{\pi}^{A'}(s) - \dot{\omega}^A(s)\pi^{A'}(s)}{\pi_{C'}(s)\dot{\pi}^{C'}(s)}. \tag{4}$$

It is straightforward to recover the Eisenhart formula from this equation.
Introduce a basis for spin space (o^A, ι^A), and define a new parameter ζ by

$$\zeta = \pi_{0'}/\pi_{1'}.$$

We parametrize the twistor curve in terms of ζ, by writing

$$\omega^A(\zeta) = \sqrt{2}i\pi_{1'}[f(\zeta), g(\zeta)]$$

The Eisenhart formula is then obtained by expressing equation (4) in standard (x, y, z, t) coordinates.

Some remarks are pertinent at this point. First, the relations between the space-time real or complex null curve are all invariant under reparametrizations of the curve. Second, and independently of this invariance, is the freedom to rescale the twistor curve by an s-dependent scaling. This latter freedom is an expression of the fact that the correspondence is between null curves and *projective* twistor space. It is possible therefore to normalize the twistor curve appropriately, to simplify calculations. Two choices have emerged as being particularly useful:

1. For real null curves, arrange that $\dot\phi^{AA'}(s) = \bar\pi^A \pi^{A'}$. This is the natural generalization to null curves of the conditions usually applied to the twistors representing null geodesics. As in that case, this normalization will lead to simple and natural expressions for the momentum, angular momentum and symplectic structure.

2. For non-degenerate real curves, normalize the spinor curve so that

$$\pi_{C'}(s)\dot\pi^{C'}(s) = 1.$$

   This normalization results in several simplifications, especially when explicit formulae for real null curves are being considered.

With any given choice of normalization, one can go further and choose a parametrization. We have already seen how the Eisenhart parameter emerges. Another choice is to require that, in addition to normalization 1 above, one component, say $\pi_{1'}$, is constant. This is equivalent to the so called 'light-cone-gauge' parameter choice, as described, e.g. in [12], and which was originally used to make first quantization possible.

The geometry of the correspondence becomes a trifle more transparent if one makes use of the $\mathbf{CP}^5$ notation [24]. Here we regard Z^α as a homogeneous coordinate for $\mathbf{CP}^3$ and the skew 2-index quantities $\phi^{\alpha\beta}$ as homogenous coordinates for $\mathbf{CP}^5$. Compactified and complexified Minkowski space is the quadric $\Omega \in \mathbf{CP}^5$ given by the set of points ϕ such that

$$\phi.\phi = \varepsilon_{\alpha\beta\gamma\delta}\phi^{\alpha\beta}\phi^{\gamma\delta} = 0.$$

Equivalently, $\phi^{\alpha\beta}$ is *simple*. With respect to the conformal metric on Ω, a simple curve $\phi^{\alpha\beta}(s)$ has a null tangent everywhere if and only if $\dot\phi^{\alpha\beta}$ is also simple.

Now consider a curve $Z^\alpha(s)$. The tangent to the curve is the space-time point represented projectively by

$$\phi^{\alpha\beta}(s) = Z^{[\alpha}\dot{Z}^{\beta]}$$

and the derivative of this curve is just

$$\phi^{\alpha\beta}(s) = Z^{[\alpha}\ddot{Z}^{\beta]}$$

so the space-time curve is necessarily null. This allows us to state a partial correspondence: *the field of tangents to a twistor curve is a null curve in space-time, and the twistor curve may be constructed as the field of α-planes tangent to the null curve.* The correspondence as thus stated is not yet an isomorphism, for there are some unaccounted for singular cases. To account for this fully, it is necessary to introduce dual twistors. This will be done in the next section. For the present it will suffice to explain that we can recover the 2-spinor description by picking a scaling for the curve as

$$\phi^{\alpha\beta}(s) = \frac{2 Z^{[\alpha}\dot{Z}^{\beta]}}{I_{\gamma\delta} Z^\gamma \dot{Z}^\delta} \tag{5}$$

and expanding $\phi^{\alpha\beta}$ in terms of spinors. This results in equation (4). We remark that at certain 'mildly' singular points, application (repeated if necessary) of l'Hôpital's rule to (4) or (5) results in a non-singular expression. This would not work, however, for a twistor curve degenerating to a single point. Such curves, which in space-time are confined to an α-plane, must be treated by the introduction of dual twistors.

6 In Search of Reality

The need to consider dual twistors is made evident by considering the case of a complex null curve confined to an α-plane. Such a curve has the form

$$\phi^{AA'}(s) = \phi_0^{AA'} + \Lambda^A(s)\pi^{A'},$$

where ϕ_0^{AA} and $\pi^{A'}$ are constant. Such a curve corresponds to a point in twistor space. The behaviour of this type of system can be described by dual twistors. Of course, not all curves can be described by dual twistors, as can be seen by considering the case of a curve confined to a β-plane. Thus for a general complex null curve, one must use both twistors and dual twistor curves to properly parametrize the entire space of such curves. In the case of real curves this is not the case: e.g., a real null curve confined to an α-plane is just a null geodesic, possibly with a non-affine parametrization, which should correspond precisely to a point in (null) twistor space.

To define the dual twistor curve, we consider the following incidence relations:

$$\dot{\phi}_{AA'}\lambda^A = 0,$$
$$\mu_{A'} = i\phi_{AA'}\lambda^A,$$
$$\dot{\mu}_{A'} = i\phi_{AA'}\dot{\lambda}^A.$$

It then emerges that the dual twistor curve $W_\alpha(s) = \{\lambda_A(s), \mu^{A'}(s)\}$ satisfies a set of relations with respect to the original twistor curve. A minimal set of such constraints can be written

$$Z.W = 0 = (\dot{Z}.W \text{ or } Z.\dot{W}) = \dot{Z}.\dot{W}.$$

We can also write these as

$$Z.W = 0 = \dot{Z}.W = \ddot{Z}.W,$$

or

$$Z.W = 0 = Z.\dot{W} = Z.\ddot{W}.$$

These constraints reveal the geometric relationship between the twistor curve and its dual: *the dual twistor curve is the family of planes that osculate the twistor curve, and vice versa.*

To account for the singular cases and obtain a one-one correspondence between the null curves on space-time and twistor structures, we introduce 'ambitwistor' space as

$$A = \{(Z^\alpha, W_\beta) \in \mathbf{CP}^3 \times \mathbf{CP}^3 | Z.W = 0\}.$$

We can state an isomorphism in terms of curves in A:

$$\{\phi^a(s) \in \mathbf{C}M | \dot{\phi}.\dot{\phi} = 0\} \equiv \{(Z(s), W(s) \in A | \dot{Z}.W = 0 = \dot{Z}.\dot{W}\}.$$

Generically, a null curve in space-time corresponds to an unconstrained curve in either twistor or dual twistor space. The ambitwistor description is subject to constraints, but encompasses the singular cases in a straightforward manner. By the introduction of so-called 'superambitwistors', e.g., as described by Ferber [11], it is possible to go further and construct supersymmetric null curves and define a twistor transform of a superstring. Although this can be done [47], we will not elaborate on this construction. When one considers the quantization of null geodesics, the choice of particular homogeneities for the quantum states leads naturally to states of half-odd-integer spin, without the introduction of fermionic variables. It is natural to expect that ultimately a similar twistor quanization of strings should contain this possibility also. The motivation for introducing fermionic variables at this stage is therefore unclear, to say the least.

When the space-time curve is real, in a space-time of Minkowskian signature, the dual twistor description becomes coincident with the description in terms of complex conjugate twistors. The twistor curve itself then becomes subject to reality constraints. These conditions are

$$Z.\bar{Z} = 0,$$
$$\dot{Z}.\bar{Z} = 0 = Z.\dot{\bar{Z}},$$
$$\dot{Z}.\dot{\bar{Z}} = 0.$$

The first condition is the familiar one which applies to real null geodesics, represented as points in PN. The second two equations represent additional conditions which must be applied in the case of strings. They are of central importance when one considers quantization. *Our purpose is to explain how to solve these constraints in terms of free functions, and thus obtain the analogue of the Eisenhart formula for real strings in Minkowski space-time.*

We remark, incidentally, that if one considers a space-time with signature $(+ + --)$ the reality conditions may be solved by considering curves on real twistor space $\mathbf{RP}^3$. The study of this 'ultra-hyperbolic' signature leads to a good deal of interesting mathematics [14], notwithstanding the difficulty of understanding the role of analyticity in the real case.

7   Explicit Formulae for Real Strings

The constraints on a twistor curve ensuring that it corresponds to a real null curve are singularly refractory: Nevertheless, two methods of solution have been found [27, 29], and both involve mixing the twistor curve and its complex conjugate in a strange, unusual fashion. The first method is a hands-on series of manipulations with the spinor variables, and has been found to be particularly useful in constructing and demonstrating string motions [29, 30]. This will be described in some detail. The second approach involves the construction of projection operators applied to a curve in PN. The reader is referred to [27] for details of this method.

Just as in the Eisenhart formula, it is necessary to assume that the curve is generic in a suitable sense. In this context we wish to impose the normalization $\pi_{A'}\dot{\pi}^{A'} = 1$. The null space-time curve then satisfies

$$\dot{\phi}^{AA'} = \psi\bar{\pi}^A\pi^{A'} \tag{6}$$

for some $\psi(s)$.

By differentiation of the normalization relation $\pi_{A'}\dot{\pi}^{A'} = 1$ we deduce that $\pi_{A'}\ddot{\pi}^{A'} = 0$ and hence that there exists a complex scalar $U(s)$ such that (the conjugate is introduced for consistency with [29])

$$\ddot{\pi}^{A'} = \bar{U}\pi^{A'}.$$

The significance of the 'potential' U, as will be evident shortly, is that for *real U* the associated curve lies in a *fixed time-like hyperplane*, whereas for *complex* U a non-planar null curve is produced. (The imaginary part of U is a fundamental invariant of the curve. Note, incidentally, that this equation is well known in the literature as 'Hill's Equation': see Magnus and Winkler [33] for a treatise on the subject; cf. also Hill [16].)

Thus open strings can be categorized simply according as to whether $\zeta^a(s)$ is planar or non-planar; whereas loops fall into several categories depending on whether $\xi^a(s)$ and $\eta^a(s)$ are planar—and, if they are both planar, whether the planes coincide.

The utility of this normalization on the spinor field $\pi_{A'}(s)$ is that it and its derivative form a normalized spinor basis, and by use of the relation between the second derivative and π, higher derivatives may be referred back to this basis. Furthermore, equation (4) becomes particularly simple to apply as the denominator is unity.

We begin by writing $\omega^A(s)$ in terms of the conjugate basis:

$$\omega^A(s) = a(s)\bar{\pi}^A(s) + b(s)\dot{\bar{\pi}}^A(s).$$

By differentiation, where we suppress the s-dependence, we have

$$\dot{\omega}^A = (\dot{a} + bU)\bar{\pi}^A + (a + \dot{b})\dot{\bar{\pi}}^A.$$

The constraints therefore take the form

$$Z.\bar{Z} = b + \bar{b} = 0,$$
$$\dot{Z}.\bar{Z} = \dot{b} + a - \bar{a} = 0,$$
$$\dot{Z}.\dot{\bar{Z}} = -\dot{a} - \dot{\bar{a}} - bU - \bar{b}\bar{U} = 0.$$

The first condition requires that b is imaginary, so we may write $b = i\Delta$, where Δ is real. The second condition requires that we write

$$a = \Omega - \frac{i}{2}\dot{\Delta},$$

where Ω is real. To treat the third condition, we write $U = \Sigma + i\Lambda$. The final condition is then just

$$\dot{\Omega} = \Delta\Lambda.$$

It is now evident that the reality of U, equivalent to the vanishing of Λ, plays a critical role. In this case Δ may be treated as a free function, and Ω is a freely determined constant, say ω. Also, because of the reality of $U(s)$ in the planar case, the vector $i(\bar{\pi}^A \dot{\pi}^{A'} - \dot{\bar{\pi}}^A \pi^{A'})$ is constant, and everywhere orthogonal to the tangent of the curve: it represents the space-like normal vector to the time-like hyperplane in which the curve resides.

Our expression for the general planar null curve is as follows. We now simply apply the formula (4), where the denominator is unity. Let $\pi^{A'}$ be normalized as above and chosen such that U is real. Let $\Delta(s)$ be a real function of s and ω a real constant. Then the general planar null curve can be put in the form

$$\phi^{AA'}(s) = \Delta\dot{\bar{\pi}}^A\dot{\pi}^{A'} - \tfrac{1}{2}\dot{\Delta}(\bar{\pi}^A\dot{\pi}^{A'} + \dot{\bar{\pi}}^A\pi^{A'}) - i\omega(\bar{\pi}^A\dot{\pi}^{A'} - \dot{\bar{\pi}}^A\pi^{A'}) + (\tfrac{1}{2}\ddot{\Delta} - U\Delta)\bar{\pi}^A\pi^{A'}.$$

A straightforward calculation shows that $\dot{\phi}^a(s)$ is given by equation (6) with

$$\psi = \tfrac{1}{2}\dddot{\Delta} - 2U\dot{\Delta} - \dot{U}\Delta.$$

We require $\phi^a(s)$ to satisfy the quasi-periodicity condition $\phi^a(s + 2\pi) - \phi^a(s) = \kappa P^a$ where $\kappa = 1$ for an open string and $\kappa = \tfrac{1}{2}$ for a loop: a necessary and sufficient condition for this to be the case is that the function $\Delta(s)$ should be of the form

$$\Delta(s) = \frac{\kappa s}{2\pi}P_{AA'}\bar{\pi}^A\pi^{A'} + \delta(s),$$

where $\delta(s + 2\pi) = \delta(s)$ but is otherwise arbitrary.

The general non-planar null curve can be constructed similarly. The curve can be expressed in the form

$$\phi^{AA'}(s) = \Delta\dot{\bar{\pi}}^A\dot{\pi}^{A'} - \tfrac{1}{2}\dot{\Delta}(\bar{\pi}^A\dot{\pi}^{A'} + \dot{\bar{\pi}}^A\pi^{A'}) - i\Omega(\bar{\pi}^A\dot{\pi}^{A'} - \dot{\bar{\pi}}^A\pi^{A'}) + (\tfrac{1}{2}\ddot{\Delta} - \Sigma\Delta)\bar{\pi}^A\pi^{A'},$$

where $\Omega(s)$ is given below in equation (7), and $\Delta = \dot{\Omega}/\Lambda$. Again a calculation shows that $\dot{\phi}^a$ is null, and is given by (6) with

$$\psi = \tfrac{1}{2}\dddot{\Delta} - 2\Sigma\dot{\Delta} - \dot{\Sigma}\Delta - 2\Lambda\Omega.$$

As for the periodicity condition, in this case we deduce that $\Omega(s)$ must be of the form

$$\Omega(s) = \frac{i\kappa s}{4\pi}P_{AA'}(\bar{\pi}^A\dot{\pi}^{A'} - \dot{\bar{\pi}}^A\pi^{A'}) + \gamma(s), \tag{7}$$

where $\gamma(s + 2\pi) = \gamma(s)$ but is otherwise freely specifiable.

It should be observed, incidentally, that the total momentum $P_{AA'}$ may be specified, in each case, independently of the remaining functions.

In the case of a planar null curve the free data are constituted by the quantities $P^a, \omega, \delta(s)$ and $\pi^{A'}(s)$. For a non-planar null curve the free data are $P^a, \gamma(s)$ and $\pi^{A'}(s)$. In each case the relevant functions are required to be periodic in s. In the case of δ or γ the period is 2π, whereas in the case of $\bar{\pi}^A$ we require one of the two cases $\bar{\pi}^A(s + 2\pi) = \pm\bar{\pi}^A(s)$.

As an elementary application of our procedure let us consider open strings ($\kappa = 1$) for which the potential U is a real constant. The various conditions lead to

$$U = -\tfrac{1}{4}n^2, \quad \bar{\pi}^A = \mathcal{N}[o^A\cos(ns/2) + \iota^A\sin(ns/2)], \quad o_A\iota^A = 1,$$

where n is a positive or negative integer and $\mathcal{N} = \sqrt{(2/n)}$. The curve is confined to the time-like three-plane orthogonal to the constant space-like vector

$$Z^{AA'} = \frac{i}{\sqrt{2}}[o^A\bar{\iota}^{A'} - \iota^A\bar{o}^{A'}].$$

If we set δ and ω to zero and set $P^a = MT^a$, where T^a is the unit time-like vector given by

$$T^{AA'} = \frac{1}{\sqrt{2}}(o^A \bar{o}^{A'} + \iota^A \bar{\iota}^{A'})$$

then the curve is given by

$$\zeta^a(s) = \frac{sM}{2\pi}T^a + \frac{M}{2\pi n}[\cos(ns)X^a + \sin(ns)Y^a],$$

a null curve that describes a right helix of helicity n in space-time. Here X^a and Y^a are constant unit space-like vectors orthogonal to T^a and Z^a as well as to one another.

The corresponding open string has as its edges a pair of such null curves, one displaced forward in time by half a period from the other, and is given explicitly (following some simplifications in the resulting formula) by:

$$X^a(\sigma,\tau) = \tau\frac{M}{\pi}T^a + \frac{M}{\pi n}\cos(n\sigma)[\cos(n\tau)X^a + \sin(n\tau)Y^a].$$

This is the so-called 'rigid rotator' solution of the string equation. It represents a string of length $2M/\pi n$ that rotates rigidly about its center.

The regular motion characteristic of the rigid rotator is in fact highly atypical of the generic string motion. To illustrate this let us consider, within the planar category, a family of 'perturbed' rotators, for which the fundamental spinor field $\bar{\pi}^A(s)$ is given by

$$\bar{\pi}^A(s) = \Gamma^A/(\Gamma_B \dot{\Gamma}^B)^{1/2},$$

where

$$\Gamma^A(s) = o^A \cos(s/2) + \iota^A[\sin(s/2) + \varepsilon\sin(3s/2)],$$

with $o_A\iota^A = 1$ as before.

Thus for very small values of the parameter ε we obtain a very slightly perturbed rotator; whereas for larger values of epsilon we expect nonlinearities to be more pronounced. For normalization we require ε to lie in the range $0 \leq \varepsilon < 2/3$.

The resulting analytical formulae for the string surfaces $x^a(\sigma,\tau)$ are exceedingly complicated, and do not convey much information. Nevertheless it a straightforward matter to obtain a good graphical representation of the corresponding string motions.

In Figures 1–4 we plot the motion of the string for several values of ε. Each string history is represented by a series of values of 'snap-shots' over one-half the period of the motion. The motion is presented in the rest-frame of the total momentum of the string, and within the plane in which the string resides. In the course of a half-turn the string returns, apparently, to its original position, but with the opposite orientation.

Figure 1: $\epsilon = 0.05$

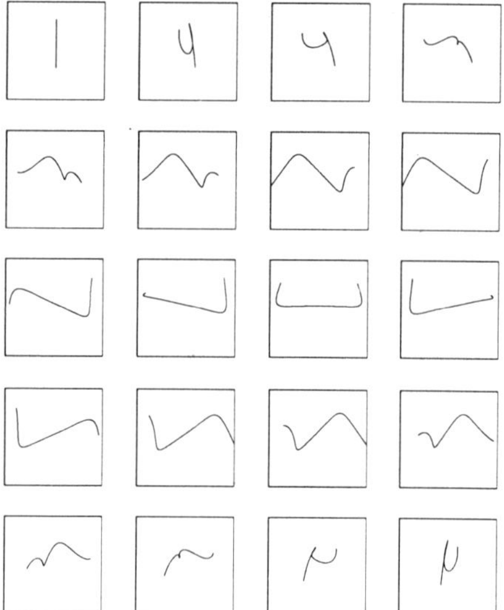

Figure 2: $\epsilon = 0.20$

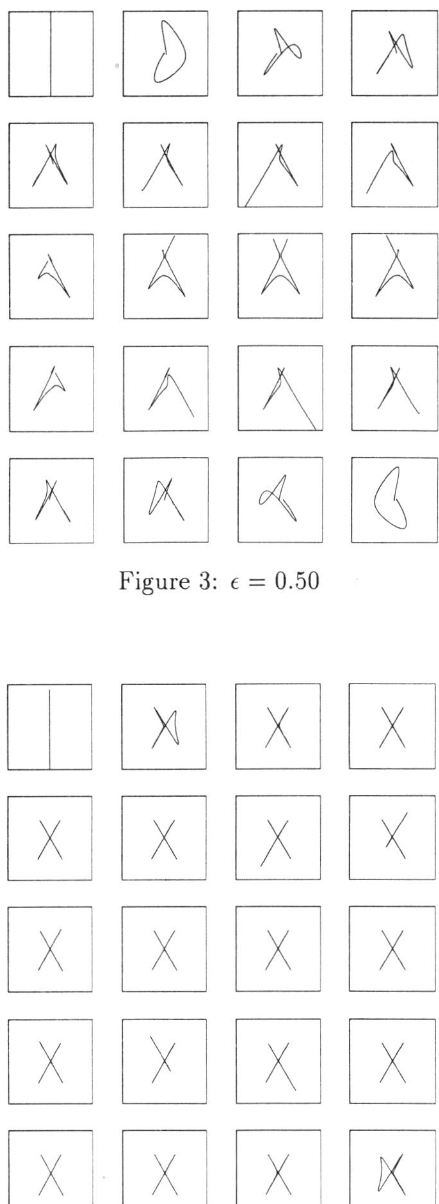

Figure 3: $\epsilon = 0.50$

Figure 4: $\epsilon = 0.65$

We consider motions for which $\varepsilon = 0.05, \varepsilon = 0.2, \varepsilon = 0.5$, and $\varepsilon = 0.65$. In the case $\varepsilon = 0.05$ (Figure 1), the motion is clearly indicative of a slight perturbation on the original rotator. In the case $\varepsilon = 0.2$ (Figure 2) the effects of the higher mode are manifested in distinctive non-linearities, while the 'ground-state' remains reasonably well-represented. At $\varepsilon = 0.5$ (Figure 3) the motion is highly non-linear, and exhibits a kind of 'wildness'; whereas at $\varepsilon = 0.65$ (Figure 4), near the critical value $\varepsilon = 2/3$, the string is very large compared to the ground state, and spends most its history in a tightly bound 'X' shaped configuration, unraveling itself briefly only once each period and half-period to a straight line.

In the case of non-planar strings the graphical representation is complicated by the need to introduce an appropriate device to convey the depth of the motion. Various approaches are under consideration, including representing the string as a thickened tube, with shading appropriate to certain light sources.

In the discussion so far, we have focused on geometrizing the Eisenhart formula, and providing an analogous formula for real strings in Minkowskian 4-space. Twistor techniques provide rather more than cute devices for solving the classical equations of motion, since it is possible to express other properties of the classical string quite naturally in terms of structures on twistor space. It will become evident that the string, in a very natural sense, represents the continuous analogue of the structures which were originally conceived as being the building blocks of the twistor particle programme [19, 23].

8   Classical Structures

In what follows we will be concerned with expressing the momentum, angular momentum, and symplectic structure of classical strings in terms of structures on twistor space. The discussion will make use of a particular normalization for the twistor curves: $\dot{\phi}^{AA'} = \bar{\pi}^A \pi^{A'}$. It follows that

$$P^{AA'} = \int_{-\pi}^{\pi} ds\, \bar{\pi}^A \pi^{A'}.$$

It will emerge that with this normalization the angular momentum can be expressed in a similar form, so that there are quite natural twistor expressions for quantities of interest. However, it is also important to understand how other views of the transform lead to expressions of a similar type. The role of the Eisenhart parameter ζ, and the extent to which the Eisenhart functions $f(\zeta)$, $g(\zeta)$ can be used as canonical variables, are not understood. Neither is the importance of the alternative normalization. This is important as, although the following series of manipulations leads to a highly suggestive formalism for quantization, the quantization has not yet been implemented, due to difficulties with the constraint algebra. It may be possible to resolve these difficulties with one of the alternative schemes.

The curve $\omega^A(s)$ is not periodic, but satisfies

$$\omega^A(s + 2\pi) = \omega^A(s) + iP^{AA'}\pi_{A'}(s).$$

Also, the translations do not act correctly on the twistor curve. Under a space-time translation of the string: $X^a \to X^a + V^a$ the null curve ϕ^a transforms according to

$$\phi^a \to \phi^a + \frac{1}{2}V^a,$$

and so

$$\omega^A(s) \to \omega^A(s) + \frac{i}{2}V^{AA'}\pi_{A'}(s).$$

This is *half* the displacement in twistor space that one expects. One may define a new twistor curve which is periodic and transforms suitably under translations as follows. Define

$$\gamma_s^a = \frac{1}{2\pi}\int_{-\pi}^{\pi} ds \ \phi^a(s),$$

this quantity being the average $< \phi^a >_s$ with respect to the given parametrization. Now define $\Omega^A(s)$ by

$$\omega^A(s) = i\left[\frac{s}{2\pi}P^{AA'} - \gamma_s^{AA'}\right]\pi_{A'}(s) + \Omega^A(s).$$

The new curve $P^\alpha(s) = \{\Omega^A(s), \ \pi_{A'}(s)\}$ has several useful properties. Since $\Omega^A(s)$ is periodic, $P^\alpha(s)$ defines a *loop* on twistor space. Also, under translations Ω^A transforms correctly:

$$\Omega^A(s) \to \Omega^A(s) + iV^{AA'}\pi_{A'}(s).$$

The constraints are modified when expressed in terms of $P^\alpha(s)$. An independent set can be taken to be

$$P^\alpha \bar{P}_\alpha = 0,$$
$$\dot{P}^\alpha \bar{P}_\alpha + \frac{i}{2\pi}P^{AA'}\bar{\pi}_A\pi_{A'} = 0,$$
$$\dot{P}^\alpha \dot{\bar{P}}_\alpha + \frac{i}{2\pi}P^{AA'}[\dot{\bar{\pi}}_A\pi_{A'} - \bar{\pi}_A\dot{\pi}_{A'}] = 0,$$

where the reality of γ^a has been used.

Consider now the symplectic structure. From any Lagrangian field theory one can construct an associated 2-form (see e.g. [56]). Applying these ideas to string theory, suppose one has a string $X^a(\tau,\sigma)$ and two nearby strings $X^a + V_1^a$, $X^a + V_2^a$. The symplectic form is described by an integral:

$$2\omega(V_1, V_2) = \int_0^\pi d\sigma\{V_2^a V_{1a,\tau} - V_1^a V_{2a,\tau}\}.$$

The integral can be taken over the $\tau = 0$ cross-section, for the equations of motion and boundary condtions ensure that ω is independent of τ. This can be written in terms of variations in the corresponding null curve as

$$2\omega(V_1, V_2) = \int_{-\pi}^{\pi} ds[\delta_2\phi^a\delta_1\dot\phi_a - \delta_2\dot\phi_a\delta_1\phi^a] + \delta_1\phi^a(\pi)\delta_2\phi_a(-\pi) - \delta_2\phi^a(\pi)\delta_1\phi_a(-\pi)$$

Now let V_i^a correspond to variations $\delta_i P^\alpha$ in P^α. Then some calculation leads to:

$$\omega(V_1, V_2) = \frac{i}{2}\int_{-\pi}^{\pi} ds \ \{\delta_1 P^\alpha\delta_2\bar{P}_\alpha - \delta_2 P^\alpha\delta_1\bar{P}_\alpha\}. \tag{8}$$

Thus if the open strings in space-time are represented by loops in twistor space then the symplectic structure on the space of open strings is just the integral around the loop of the standard twistor symplectic structure. It should be noted that there are no s-derivatives in this expression: it is not what might have been guessed on the grounds of invariance under reparametrizations. The twistor fields are, however, with the given normalization, weighted quantities with respect to reparametrizations. It may be verified from the normalization conditions, and the change of variables to the twistor loop variables, that this expression is reparametrization invariant. This is also true of the angular momentum expression, given below. It would be useful to make this invariance more manifest.

The twistor expression for the energy-momentum of the string has already been given in terms of the loop variables. One may write down a corresponding expression for the angular momentum tensor M^{ab}. This is given in the space-time by

$$M^{ab} = \int_0^{\pi} d\sigma \ \{X^a X^{,b}_{,\tau} - X^b X^{,a}_{,\tau}\},$$

the integral being taken over any constant-τ cross-section. One may express this in terms of the null curve as

$$M^{ab} = \int_{-\pi}^{\pi} ds\{\phi^a\dot\phi^b - \phi^b\dot\phi^a\} + \phi^a(-\pi)\phi^b(\pi) - \phi^a(\pi)\phi^b(-\pi).$$

This tensor may be expressed in spinor form as

$$M^{ab} = \mu^{AB}\varepsilon^{A'B'} + \bar\mu^{A'B'}\varepsilon^{AB},$$

where μ^{AB} is symmetric. In terms of the twistor loop coordinates one finds that

$$\mu^{AB} = \int_{-\pi}^{\pi} ds \ i\Omega^{(A}\bar\pi^{B)}.$$

These expressions may be combined into a simple expression for the angular momentum twistor as an integral over the twistor loop variables:

$$A^{\alpha\beta} = \int_{-\pi}^{\pi} ds \ 2P^{(\alpha}I^{\beta)\gamma}\bar{P}_\gamma.$$

When expressed in terms of the loop variables the structures describing
the string are just the integrals around a loop in twistor space of the structures
appropriate to the space of null geodesics. The connection with the twistor
description of massive particles is now evident. If we were to make a Fourier
expansion of the twistor loop, these expressions would become an infinite sum
of the simple expressions appropriate for a massless particle. Such finite sums
are the conventional twistor description of massive particles.

9 Quantization

Given the form of the structures described in the previous Section, one can
write down what at first seems to be a simple quantization rule. One expands
the twistor loops as a Fourier series, and quantizes each mode. For each ex-
cited mode, there is a contribution to the momentum and angular momentum
structure.

Just as in the case of first quantization of null geodesics, one must impose
the constraints on the quantum states. In the case of null geodesics, the
reality condition at the quantum level leads to states of definite homogeneity.
In the present case, there are three constraints to be imposed. However, with
our 'obvious' quantization scheme, these lead to difficulties. Conditions (8)
and (8) present no difficulties, and lead to a semi-direct sum of the Virasoro
algebra with a commutative algebra, which is closed under commutation.
However, condition (8), turns out to be a second class constraint, as can be
checked at the classical Poisson bracket level, or at the quantum level.

In view of this, one may be led to consider the utility of a twistor description at all. The constraints in the space-time picture are first class, so
what is the point of making the quantization more awkward? The quantization of a system containing second class constraints is rather more intricate.
The procedure is described by [15], following Dirac's exposition of how it
should be done [7]. It would be an interesting exercise to follow this through
in the present context. However, other routes may be more fruitful. The
point of introducing twistors in the first place was to *eliminate* the awkward
constraints. It just happens that early attempts to write down canonical
variables based on a simple expression for the appropriate 2-form have led
to the re-introduction of reality constraints which are more complex than
the space-time nullity constraints they were intended to replace. Therefore
it is more appropriate to re-consider which variables are the most appropriate canonical variables. This has not yet been followed through successfully.
The attempt to write down a suitable symplectic structure (8) is plagued by
significant degeneracies, corresponding to the presence of symmetries. These
are the reparametrization invariance, already present at the space-time level,
and the local phase symmetry $P^\alpha(s) \to e^{i\phi(s)}P^\alpha$. On the constraint surface,
the 2-form ω vanishes evaluated on a pair of twistor curve variations where
at least one variation arises from a local phase change or reparametrization.

One can re-write the twistor correspondence in such a way that the constraint algebra is modified. In [44] the real space-time curve is first complexified in such a way as to preserve the first two constraints but eliminate the third and awkward one. However, this is done by defining the twistor curve to be incident with the curve

$$y^a(s) = \phi^a(s) + i\ddot{\phi}^a(s)$$

and it is difficult to justify the treatment of the resulting spinor variables as canonical variables. A more promising approach is to try to use the Eisenhart functions, or the appropriate structures for real strings, as canonical objects. This is currently under investigation.

It is rather intruiging that the magnitude of the difficulties concerning the reality constraints is linked to the dimension and signature of the space-time. In dimensions and signatures where real twistors may be defined, matters are rather more straighforward. The details of the quantization scheme in dimension three, with signature $(+,-,-)$ have been worked out [48]. In this case the inner product on twistor space is real symplectic rather than complex unitary. By choosing to work with real twistors, the phase degeneracy may be elimnated completely, and the analogues of the constraints (8) and (8) are satisfied *identically*. This leaves the analogue of (8). Through a series of fairly obvious definitions, one is led to a first quantized theory where the quantum string states are 'functions' (ultimately in a suitable cohomological sense) of a real spinor and a series of complex twistors, holomorphic in each of their twistor arguments, and subject to constraints on their homogeneity. The constraints define a Virasoro [52] algebra in the form:

$$[V_m,\ V_n] = (m-n)V_{m+n} + \frac{d}{12}(m^3 - m)\delta_{m,-n}\ ,$$

and the string states Ψ satisfy

$$[V_0 - h]\Psi = 0, \quad V_m\Psi = 0, \quad m \geq 1.$$

The parameters of the representation are given by $d = 2$ and $h = \frac{1}{4}$. It is possible to find exact solutions of the constraints, which are realised as differential equations on the states, and these correspond, via appropriate contour integrals, to solutions of the *massive* wave equation on space-time, with quantized mass and spin. Further details are given in [48].

10   Other Dimensions: Pure Spinors

This article has focused on four dimensions thus far, though we have noted the simplifications which arise in the quantization when one considers dimension three. In 'traditional' string theory, dimensional constraints arise for quite different reasons, as discussed by [13]. There are two reasons for considering

higher dimensional analogues of the results described above. The first is
the considerable interest, amongst many physicists, in higher dimensional
systems. The second, which we regard as being at least as important, is the
desire to investigate the generality of the ideas which allowed us to geometrize
the Eisenhart result. This might be regarded as a similar line of reasoning to
that which led to six [6] and ten [55] dimensional analogues of the transforms
for four-dimensional Yang-Mills theory, as described by [31, 54].

It is not appropriate to claim that the methods we will outline here are
the only sensible approach to the problems in general dimension. Another
interesting approach is described by [50]. A particular feature of the geometrical techniques we have found to be useful is the consistent role played by
spinors and twistors for higher dimensions. In particular, spinors that are
pure in the sense of Cartan [5] play a key role. The appropriate notion of
twistor space for space-time M_d dimension d appears to be the space of pure
(reduced) spinors for $O(d + 2)$. If d is even then this space corresponds to
the space of null $\frac{d}{2}$ planes in the complexification of M_d. In what follows we
will explain how the purity conditions arise when considering null curves in
dimensions 6 and 10. The purity conditions for dimension 10 have also been
considered by Nillson [35].

For a discussion of the properties of spinors in general dimension, the work
of Brauer & Weyl [2] is recommended. More up-to-date views are provided
by Penrose & Rindler [43], and Kugo et al [32] provide a complementary
discussion which spells out the reality conditions in some detail, for various
dimensions and signatures. The purity conditions are also discussed in [4].
Properties of the twistor equation in general dimension are described in [1],
while the application of twistor methods to higher dimensional wave equations, and related results are described by [20, 22, 25]. Further details of the
results for null curves in six and ten dimensions can be found in [26] and [28]
respectively.

The results for six dimensions may be most straightforwardly given by
use of the $\mathbf{CP}^5$ notation used in Section 5. Here, however, $X^{\alpha\beta}$ is treated
non-projectively as a coordinate system for complex 6-space. We seek to
parametrize curves $X^{\alpha\beta}(s)$ such that $\dot{X}^{\alpha\beta}$ is null (or simple). This is equivalent to being able to choose a spinor field $P^\beta(s)$ such that

$$\dot{X}_{\alpha\beta}P^\beta = 0.$$

This is the six-dimensional analogue of equation (1). We now write down the
incidence relations that are the analogues of equations (2) and (3):

$$Q_\alpha + X_{\alpha\beta}P^\beta = 0, \tag{9}$$

$$\dot{Q}_\alpha + X_{\alpha\beta}\dot{P}^\beta = 0. \tag{10}$$

We may now observe the role of purity in this case. The relation (9) has no
solution for $X_{\alpha\beta}$ unless $P.Q = 0$. The existence of such a constraint required

that we consider a subspace of all spinor pairs, the subspace of *pure spinors*. When this constraint is satisfied the relation (9) has solutions for $X^{\alpha\beta}$ which define a null 3-plane.

The spinor curve defined by equations (9) and (10) also satisfies $\dot{P}.\dot{Q} = 0$. We call such curves *pure curves*. We can express the curve $X^{\alpha\beta}(s)$ in terms of the pure curve in an analogous form to (4), as

$$X^{\alpha\beta} = \frac{\{\varepsilon^{\alpha\beta\gamma\delta}Q_\gamma\dot{Q}_\delta + fP^{[\alpha}\dot{P}^{\beta]}\}}{\dot{P}.Q},\tag{11}$$

where $f = \frac{1}{2}X^{\alpha\beta}X_{\alpha\beta}$.

There are other features of the construction in dimension six that have no parallel in four dimensions. First, the spinor P is not unique, even projectively, and the different choices define a null or pure line in projective twistor space. This reflects the fact that we have in fact defined a curve on twistor space that has the same properties as the original space-time curve. This may be made evident by temporarily regarding the pure spinors as projective coordinates for compactified and complexified six-dimensional space-time. The remaining condition $\dot{P}.\dot{Q} = 0$ is then just a condition that a curve in this 6-space be null or pure! A corresponding curve may be obtained in the dual twistor space, and the set of three curves thus obtained defines three isomorphic structures. This is one manifestation of the triality principle for $O(8)$ spinors.

However, we have not just obtained a re-write of the space-time nullity constraints in terms of twistor space purity constraints, for the machinery developed allows us to parametrize the curve in terms of free functions, just as in dimensions three and four. For suitably non-degenerate curves $(\dot{P}.Q \neq 0)$, we have the following:

Let $X^{\alpha\beta}(s)$ be a non-degenerate analytic null curve in C^6. By a suitable choice of parametrization, unique up to bilinear transformation, $X^{\alpha\beta}(s)$ can be expressed as

$$X^{\alpha\beta}(s) = Z^{[\alpha}\dot{Z}^{\beta]} + \phi(s)\ddot{Z}^{[\alpha}\dddot{Z}^{\beta]},$$

for some choice of $\phi(s)$ and $Z^\alpha(s)$. Conversely, any curve thus expressed is necessarily null.

In dimension six, and other dimensions above four, the possible degeneracies are more complicated and there is, of course, greater variety in the possible reality conditions one might consider. These are discussed in [26], and will not be elaborated on here. The reader should note that the form of the expression (11) for six dimensions is a sum of two curves which correspond to null curves (when viewed projectively) in four dimensions. This appears to be a general property which makes it possible to represent null curves in dimension $d + 2$ in terms of pairs of null curves for dimension d. We will see this again in the result for ten dimensions, where a null curve in eight dimensions arises as a similar special case, where one wears projective 'glasses'.

In dimension ten one must introduce an appropriate set of Van der Waerden symbols or reduced gamma matrices γ_i^{AB} and γ_{iAB}. These are symmetric in A and B and satisfy the fundamental Clifford algebra identity

$$\gamma_{AB}^{(i}\gamma^{j)BC} = \delta_A^C g^{ij} \ .$$

Furthermore, we have a set of important quadratic relations

$$\gamma^{i(AB}\gamma_i^{C)D} = 0, \quad \gamma_{i(AB}\gamma_{C)D}^i = 0$$

which we shall call 'purity syzygies': these relations are responsible for many of the special geometrical features of ten dimensional space-time. We renormalize the gamma matrices, defining $\Gamma_{AB}^i = \frac{1}{4}\gamma_{AB}^i$. Vectors V^i may be converted to and from their spinor forms V^{AB} (and corresponding dual objects) by the relations

$$V^{AB} = \Gamma_i^{AB}V^i, \quad V^i = \Gamma_{AB}^i V^{AB},$$

and corresponding dual relations.

A pure spinor β_B for ten dimensions is one that satisfies

$$\Gamma_i^{AB}\beta_A\beta_B = 0$$

or the corresponding dual equation with index positions reversed. The appropriate twistors for ten dimensions consist of a pair $\{\alpha^A,\ \beta_B\}$ satisfying the purity conditions for $O(12)$:

$$\Gamma_i^{AB}\beta_A\beta_B = 0, \quad \Gamma_{iAB}\alpha^A\alpha^B = 0, \quad \alpha.\beta = 0. \tag{12}$$

The incidence relations may be used, as in dimensions four and six, to define a twistor curve. Given a curve $X^{AB}(s)$, where $\dot{X}^{AB}$ is null, one picks a pure spinor field $\beta_B(s)$ satisfying

$$\dot{X}^{AB}\beta_B = 0.$$

A twistor curve is then defined by the pair of conditions

$$\alpha^A = X^{AB}\beta_B,$$

$$\dot{\alpha}^A = X^{AB}\dot{\beta}_B.$$

The twistor curve satisfies the purity conditions (12), but in this case no such constraints are implied or required on the derivatives. The incidence relations may be inverted to yield

$$X^i = \frac{8\Gamma_{AB}^i\dot{\alpha}^A\dot{\alpha}^B + \frac{1}{2}X^2\Gamma^{iAB}\dot{\beta}_A\dot{\beta}_B}{\dot{\alpha}.\dot{\beta}}.$$

It will be evident from this result, and the corresponding results and incidence relations for dimensions four and six, that a quite general pattern exists. The details of the calculation vary depending on the dimension, and in ten dimensions explicit formulae require some detailed manipulations with the Clifford algebra and the purity syzygies.

Just as in six dimensions, the expression for the space-time curve may be simplified somewhat, provided we make suitable genericity assumptions. The use of some rescalings and changes of spinor variable leads to the following:

A generic curve $X^i(s)$ in ten dimensions is null if and only if there exists a pair of pure spinor curves $\xi^A(s)$, $\eta^A(s)$ such that

$$X^i(s) = \Gamma^i_{AB}\dot{\xi}^A\dot{\xi}^B + \Gamma^{iAB}\dot{\zeta}_A\zeta_B,$$

where $\zeta_A = \Gamma_{iAB}\Gamma^i_{CD}\eta^B\dot{\xi}^C\ddot{\xi}^D$.

We now state the result for eight dimensions, where X^i, subject to $X.X = 0$ is now a projective coordinate system for complex 8-space:

A generic curve $X^i(s)$ in eight dimensions is null if and only if there exists a curve of pure spinors $\xi^A(s)$ such that

$$X^i = \Gamma^i_{AB}\dot{\xi}^A\dot{\xi}^B.$$

Note that the specialization from ten to eight dimensions follows the same pattern as the specialization from six to four. In a natural way, a null curve in dimension $d + 2$ may be reduced to a pair of null curves in dimension d. This may be regarded as a natural form of 'dimensional reduction'. In some ways, the construction for ten dimensions is more straightforward than that for six dimensions: we only require that the spinor curves satisfy the algebraic constraints of purity, with no restrictions on the derivatives.

11 Issues

The ideas we have discussed here represent a foundation for considerable further development. There are a great many open questions related to this general circle of ideas. Some of the more immediate issues are the following:

Conformal Aspects We have been concerned mainly with properties of strings as objects in flat space-time. The underlying structure is a conformal one: the null curve. Structures such as the momentum of the string arise through the breakage of conformal invariance by the identification of a definite 'pitch' to the null curves. However, the conformal properties are of interest in themselves, as discussed in [51]. A related question concerns the solution of the equations of string motion in conformally flat space-time, such as might be considered an appropriate background for 'cosmic strings'. The boundary conditions for a string in such a background may be characterized in terms of null curves, but is is less clear how to generate the 2-surface from the null curves.

Quantization We have already considered some of the issues related to quantization. This must represent the greatest challenge to the twistor description of string theory. In view of the natural extension to higher dimensions which exist, it is not unreasonable to expect that it is possible to interpret (or possibly even circumvent) the dimensional constraints on quantum string theory within a twistor description. The obstructions we have thus far encountered in trying to quantize in dimension four seem to be rather more of a technical nature than a manifestation of a fundamental obstruction.

Gravity A more ambitious proposal is to try to understand the relationship between string theory and gravity in twistor terms. At a superficial level, the potential point of unity is the key role played in string theory, and also in non-linear graviton constructions, by curves in twistor space. It is appropriate to investigate whether a more substantive link exists. One possible route involves trying to construct dual twistor (i.e. googly) descriptions of non-linear gravitons by dualizing the appropriate holomorphic curves. Dualization is in some respects more natural on the spaces of twistor curves than it is on the point sets, and it may be useful to proceed by considering first some specific examples. This programme must be regarded as considerably more speculative.

12 Acknowledgements

This work has profited from discussions with numerous individuals during its evolution. Lectures on 3-dimensional twistor theory given by N.J. Hitchin at the Oxford Mathematical Institute, and some suggestions by M. Perry were instrumental in starting off this programme. Since then, discussions with M. Awada, C.R. LeBrun, D. Fairlie, G.W. Gibbons, R. Penrose, V.W. Guillemin have all been of great assistance of the development of these ideas. LPH has benefited much from conversations with colleagues as the Scuola Internazionale Superiore di Studi Avanzati, Trieste, especially P. Budinich and L. Dabrowski. One of us (WTS), has been supported in part by a Research Fellowship from Clare College, Cambridge and by NSF grant no. DMS 8603523. The numerical and graphical work described here was carried out while one of us (LPH) was a visitor to the National Center for Supercomputer Applications at the University of Illinois, Urbana-Champaign, where D. Bernstein and L.J. Wicker gave very useful assistance and advice; L.L. Smarr is thanked for his generous hospitality.

References

[1] Awada, M.A., Gibbons, G.W. & Shaw, W.T. 1986, Conformal Supergravity, Twistors and the Super-BMS Group, *Annals of Physics* **171**, 52.

[2] Brauer, R. & Weyl, H. 1935, Spinors in n dimensions, *Am. J. Math.* **57**, 425.

[3] Bryant, R.L. 1982, Conformal and Minimal Immersions of compact surfaces into the 4-sphere, *J. Diff. Geo.* **17**, 455.

[4] Budinich, P. & Trautman, A. 1986, Remarks on Pure Spinors, *Lett. Math. Phys.* **11**, 315.

[5] Cartan, E. 1937, *Leçons sur la théorie des spineurs*; trans. 1966, *The theory of spinors,* Hermann, Paris; 1981, New York, Dover.

[6] Devchand, C. 1986, Integrability on lightlike lines in six dimensional superspace *Z. Phys.* **C 32**, 233.

[7] Dirac, P.A.M., 1964, *Lectures on Quantum Mechanics* Belfer Graduate School of Science Monograph series, **2**, Yeshiva Univ. Press, New York.

[8] Eisenhart, L.P. 1911, A Fundamental Parametric Representation of Space Curves, *Ann. Math.* (Ser. II) **XIII**, 17.

[9] Eisenhart, L.P. 1912, Minimal Surfaces in Euclidean Four-space, *Amer. J. Math.* **34**, 215.

[10] Fairlie, D.B. & Manogue, C.A. 1986, Lorentz Invariance and the Composite String, *Phys. Rev.* **D34**, 1832.

[11] Ferber, A. 1978, Supertwistors and Conformal Supersymmetry, *Nucl. Phys.* **B 132**, 55.

[12] Goddard, P., Goldstone, J., Rebbi, C. & Thorne, C.B. 1973, Quantum Dynamics of a Massless Relativistic String, *Nucl. Phys.* **B 56**, 109.

[13] Green, M., Schwarz, J. & Witten, E. 1987, *Superstring theory, Vols. 1 & 2.* Cambridge University Press.

[14] Guillemin, V.W. & Sternberg, S. 1986, An Ultra-Hyperbolic Analogue of the Robinson-Kerr Theorem, *Lett. Math. Phys.* **12**, 1.

[15] Hanson, A., Regge, T. & Teitelboim, C., 1976, *Constrained Hamiltonian Systems,* Accademia Nazionale dei Lincei, Rome.

[16] Hill, G.W., 1886, On the Part of the Motion of the Lunar Perigee which is a Function of the Mean Motions of the Sun and the Moon, *Acta Mathematica* **8**, 1.

[17] Horowitz, G.T. 1986, Introduction to String Theories. In *Topological Properties and Global Structure of Spacetime* (ed. Bergmann, P. & DeSabatta, V.) NATO ASI series.

[18] Hitchin, N.J. 1982, Monopoles and Geodesics, *Commun. Math. Phys.* **83**, 579.

[19] Hughston, L.P. 1979, *Twistors and Particles.* Springer Lecture Notes in Physics, **97**.

[20] Hughston, L.P. 1987, Applications of $SO(8)$ Spinors. In *Gravitation and Geometry, a volume in honour of I. Robinson* (ed. Rindler, W. & Trautman, A.). Bibliopolis, Naples.

[21] Hughston, L.P. 1986, Supertwistors and superstrings, *Nature,* **321**, 381.

[22] Hughston, L.P. 1988, Applications of Cartan Spinors to Differential Geometry in n Dimensions. In *Spinors in Physics and Geometry* (ed. Furlan, G. & Trautman, A.), p 226. World Scientific Press.

[23] Hughston, L.P. & Hurd, T.R. 1981, A cohomological description of massive fields, *Proc. R. Soc. Lond.* **A 378**, 141.

[24] Hughston, L.P. & Hurd, T.R. 1983, A $\mathbb{CP}^5$ calculus for space-time fields. *Phys. Repts.* **100**, 273.

[25] Hughston, L.P. & Mason, L.J. 1988, A Generalized Kerr-Robinson Theorem *Class. Quantum Grav.* **5**, 275.

[26] Hughston, L.P. and Shaw, W.T. 1987, Minimal Curves in Six Dimensions, *Class. Quantum Grav.* **4**, 869.

[27] Hughston, L.P. & Shaw, W.T. 1987, Real Classical Strings, *Proc. Roy. Soc. Lond.* **A414**, 415.

[28] Hughston, L.P. & Shaw, W.T. 1987, Classical Strings in Ten Dimensions, *Proc. R. Soc. Lond.* **A414**, 423.

[29] Hughston, L.P. & Shaw, W.T. 1988, Constraint-free analysis of relativistic strings, *Class. Quantum Grav.* **5**, L69.

[30] Hughston, L.P. & Shaw, W.T. 1989, Spinor parametrizations of minimal surfaces, in *Mathematics of Surfaces* III, Institute of Mathematics and its Applications conference proceedings.

[31] Isenberg, J., Yasskin, P.B. & Green, P.S. 1978, Non-self-dual gauge fields, *Phys. Lett.* **78B**, 462.

[32] Kugo, T. & Townsend, P. 1983, Supersymmetry and the Division Algebras, *Nucl. Phys.* **B 221**, 357.

[33] Magnus, W. & Winkler, S., 1966, *Hill's Equation*, Wiley (Dover) New York, London, 1966.

[34] Montcheuil, M. 1905, Résolution de léquation $ds^2 = dx^2 + dy^2 + dz^2$, *Bull. Soc. Math. France* **33**, 170.

[35] Nillson, B.E.W. 1986, Pure spinors as auxiliary fields in the ten-dimensional supersymmetric Yang-Mills theory, *Class. Quantum Grav.* **3**, L41.

[36] Penrose, R. 1967, Twistor Algebra, *J. Math. Phys.* **8**, 345.

[37] Penrose, R. 1968, Twistor quantization and curved space-time, *Int. J. Theor. Phys.* **1**, 61.

[38] Penrose, R. 1975, Twistor Theory: its aims and achievements, in *Quantum Gravity, an Oxford Symposium* (ed. Isham, C.J., Penrose, R. & Sciama, D.W.) Oxford University Press.

[39] Penrose, R. 1976, Nonlinear gravitons and curved twistor theory, *Gen. Rel. Grav.* **7**, 31-52; The non-linear graviton, *ibid.*, 171.

[40] Penrose, R., 1987, On the Origins of Twistor Theory. In *Gravitation and Geometry, a volume in honour of I. Robinson* (ed. Rindler, W. & Trautman, A.). Bibliopolis, Naples.

[41] Penrose, R. & MacCallum, M.A.H. 1972, *Phys. Rep.* **6C**, 241.

[42] Penrose, R. & Rindler, W. 1984, *Spinors and space-time Vol. 1: Two-spinor Calculus and Relativistic Fields.* Cambridge University Press.

[43] Penrose, R. & Rindler, W. 1986, *Spinors and space-time Vol. 2: Spinor and Twistor Methods in Space-time Geometry.* Cambridge University Press.

[44] Perjés, Z., 1988, *Twistor String Models.* In *Quantum Gravity 4* (eds. Markov, M.A., Berezin, V. & V.P. Frolov), World Scientific Press.

[45] Schild, A. 1977, Classical Null Strings, *Phys. Rev.* D **16**, 1722.

[46] Shaw, W.T. 1985, Twistors, minimal surfaces and strings, *Class. Quantum Grav.* **2**, L113.

[47] Shaw, W.T. 1986, An ambitwistor description of bosonic or supersymmetric minimal surfaces and strings in four dimensions, *Class. Quantum Grav.* **3**, 753.

[48] Shaw, W.T. 1987, Twistor Quantization of Open Strings in Three Dimensions. *Class. Quantum Grav.* **3**, 753.

[49] Urbantke, H., 1988, Classical Strings and Minimal Surfaces. In *Spinors in Physics and Geometry* (ed. Furlan, G. & Trautman, A.), p 258. World Scientific Press.

[50] Urbantke, H. 1988, On Pinl's Representation of Null Curves in n Dimensions. In *Relativity Today: Proceedings of the 2nd Hungarian Relativity Workshop*, (ed. Z. Perjés) World Scientific Press.

[51] Urbantke, H. 1989, Local Differential Geometry, of Null Curves in Conformally Flat Space-time, University of Vienna preprint UWThPh-1989-1.

[52] Virasoro, M.A. 1970, *Phys. Rev.* D 1, 2933.

[53] Weierstrass, K. 1866, *Monats. Berliner Akad.* 612.

[54] Witten, E. 1978, An Interpretation of Classical Yang-Mills Theory, *Phys. Lett.* 77B, 394.

[55] Witten, E. 1986, Twistor-like transform in ten dimensions, *Nucl. Phys.* **B 266**, 245.

[56] Woodhouse, N.M.J. 1980, *Geometric Quantization*. Oxford: Clarendon Press.

Integrable Systems in Twistor Theory

R.S. Ward

1 Introduction

An equation is 'integrable' (or 'solvable') if, roughly speaking, all its solutions are well-behaved and can (at least in principle) be constructed explicitly. This is a very stringent requirement: almost all non-linear equations are *not* integrable in this sense. For example, the Einstein vacuum equations and the Yang-Mills equations are certainly non-integrable, since one knows that they admit solutions which behave chaotically. So why bother about the (very few) equations which *are* integrable? Partly because it is something we can do: we search for the proverbial lost key under the lamp-post, since we have little hope of finding it anywhere else. Partly because the subject involves a great deal of very beautiful mathematics. And partly because integrable equations *are* relevant to the real world, in describing real phenomena such as solitons, or in serving as a first approximation to a more accurate model.

The twistor description has turned out to be particularly appropriate for many (if not all) classical integrable systems. (It has not, as yet, had much impact in the area of integrable quantum systems.) This review will attempt to show how the plethora of known integrable systems are related to one another, and how they fit into the twistor framework.

2 What is Integrability?

Let us begin by considering ordinary differential equations. In classical mechanics, the standard 'Liouville' definition of integrability is that there should exist a sufficient number of constants of motion, enabling one to reduce the equations of motion to quadratures (see, for example, [2]). If the phase space is $2m$-dimensional, then one wants m independent integrals of the motion, in involution with one another. It is crucial that these should exist *globally*, i.e. that they be smooth, single-valued functions on phase space. And one can add various other technical requirements. One particularly strong version is that of 'algebraic complete integrability' (see [23]). This involves complexifying the system, so that the phase-space variables become holomorphic functions of 'complex time'. And integrability then means that the time-evolution of the system consists of linear flow on m-dimensional abelian varieties (complex algebraic tori) in complexified phase space (these tori are the level sets of the m constants of motion).

There is a well-known way of generating systems of ordinary differential equations having some constants of motion. Let L be an $n \times n$ matrix of functions of t; and let P be another matrix, the elements of which are given by some local expressions in terms of the elements of L. Consider the equation

$$L_t = [L, P], \tag{1}$$

where the subscript denotes differentiation with respect to t. Equation (1) is a set of n^2 coupled, nonlinear, ordinary differential equations for the n^2 functions that make up L. Furthermore, the spectrum of L is invariant: its eigenvalues, or equivalently the quantities

$$I_k = \mathrm{tr}(L^k), \quad k = 1, 2, \ldots, n, \tag{2}$$

are constant in time. (Here 'tr' denotes trace).

But the existence of these n constants of motion is not in general enough to ensure integrability. One needs to specify more precisely how the elements of P are defined, and maybe also reduce the number of independent functions appearing in L. There is a systematic way of doing this which uses Lie algebra theory and which is associated with the names of Adler, Kostant and Symes. But there is an even more effective procedure, which involves introducing a parameter ζ (sometimes called a spectral parameter). See, for example, [1, 10]. The idea is to let L and P be Laurent polynomials in ζ, i.e. of the form

$$\begin{aligned}
L(\zeta) &= \sum_{k=-p}^{q} L_k \zeta^k, \\
P(\zeta) &= \sum_{k=-p}^{q} P_k \zeta^k,
\end{aligned} \tag{3}$$

for some fixed value of p and q. The equation (1) is required to hold for all ζ, which implies that we get more equations than before; and the quantities (2) become polynomials in ζ and ζ^{-1}, each of the coefficients of which is a conserved quantity. If things are set up correctly, then P is determined in terms of L by the equations, and there are enough conserved quantities to ensure integrability. Some examples will be mentioned in §4.

The spectrum of $L(\zeta)$ depends on ζ, but is independent of t. As ζ varies, the eigenvalues of $L(\zeta)$ trace out a Riemann surface S, which is an n-fold branched covering of the Riemann sphere (the ζ-space). If we fix t for the moment, then to each point of S is associated a complex line $\mathbb{C}$, namely the eigenspace of $L(\zeta)$ corresponding to that particular point of S. In other words, we get a holomorphic line bundle over S. Now the line bundles over S (of a given topological type) are parametrized by the Jacobian $J(S)$, which is a complex torus of dimension equal to the genus of S. So for our fixed value of t, we get a point H_t on the torus $J(S)$. If we now allow t to vary, then S remains unchanged, since the eigenvalues are constants of motion; but the eigenspaces change, and hence the line bundle H_t changes. Thus we get a flow on the torus $J(S)$. Under certain circumstances, this flow is linear, and the system is integrable (cf. [10]).

What seems to lie at the heart of integrable systems of ordinary differential equations, therefore, is the 'Lax equation' (1) with spectral parameter. It can be rewritten as

$$\left[\frac{d}{dt} + P(\zeta), L(\zeta)\right] = 0; \tag{4}$$

in other words, as the vanishing of the commutator of two linear operators depending on ζ.

Let us move on to consider systems of partial differential equations, and list some suggestions as to how one might define what it means for such a system to be integrable.

1. The phase space is now infinite-dimensional, so one might require that there exist an infinite number of conserved currents (the analogue of the Liouville definition). But this is certainly no good as a definition, because it does not guarantee that there will be enough conserved quantities: the word 'infinite' is too imprecise. It is easy to construct examples of systems having infinitely many conserved quantities, but where the behaviour of solutions is so wild that one would not wish to call these systems integrable.

2. Integrability should imply that the solutions of the equations are well-behaved, in some appropriate sense. One could try to formulate this as a definition, for example in the form of the 'Painlevé property' (PP). Certainly the PP is useful as a test for integrability in classical mechanics (cf. [23]); in this case, PP is simply the statement that all solutions are meromorphic functions of complex time (i.e. that they have no singularities worse than poles). This can be generalized to partial differential equations in various different ways (cf. [16, 28]), and seems to be useful as an indicator of integrability. But as a definition it is not very satisfactory, since it is sensitive to the choice of variables that one makes.

3. In some special cases, such as the Korteweg-de Vries equation, the inverse scattering transform provides, in effect, a canonical transformation to action-angle variables: there is a close analogy with classical mechanics. But one wants also to be able to deal with elliptic equations where there is no 'time' variable as such, and a mechanical analogy is impossible.

4. The most useful approach seems to be to generalize equation (4), and to obtain integrable equations by requiring that certain linear, first-

order partial differential operators, depending on a spectral parameter ζ, commute. This approach is central to the twistor description.

3 Self-Dual Gauge Fields

How can one generalize the Lax equation (4)? The simplest possibility is to require that the two linear operators in question depend *linearly* on ζ, and then the most general such operators have the form

$$\begin{aligned}
\Delta_0 &= \partial_u + \zeta\partial_v + \Phi_u + \zeta\Phi_v, \\
\Delta_1 &= \partial_y + \zeta\partial_z + \Phi_y + \zeta\Phi_z.
\end{aligned} \tag{5}$$

Here u, v, y, z are four independent variables, and $\partial_u = \partial/\partial u$ etc; the Φ's are four $n \times n$ matrices of functions of these coordinates. (Actually, there is another restriction in force as well: we are only considering operators in which the derivative terms ∂_a have coefficients independent of u, v, etc.. Relaxing this constraint allows curved-space-time equations; cf. §6.)

Before going any further, let us convert to coordinates which express the underlying symmetry. Namely, ζ is taken to be the ratio between the two components of a primed spinor:

$$\zeta = \pi^{1'}/\pi^{0'}; \tag{6}$$

and the four coordinates u, v, y, z are renamed as $x^{AA'}$, where

$$\begin{pmatrix} x^{00'} & x^{01'} \\ x^{10'} & x^{11'} \end{pmatrix} = \begin{pmatrix} u & v \\ y & z \end{pmatrix}. \tag{7}$$

Then (5) can be rewritten as

$$\Delta_A = \pi^{A'}\nabla_{AA'} := \pi^{A'}(\partial_{AA'} + \Phi_{AA'}) \tag{8}$$

(ignoring an overall factor of $\pi^{0'}$, which is irrelevant).

Clearly what is involved here is a trivial vector bundle, of rank n, over a four-dimensional space (which may be either $\mathbf{R}^4$ or $\mathbf{C}^4$, depending on whether the coordinates x^a are taken to be real or complex). The gauge potential Φ_a defines a connection on this vector bundle; the covariant derivative is ∇_a, and the curvature 2-form (gauge field) is

$$F_{ab} = [\nabla_a, \nabla_b] = 2\partial_{[a}\Phi_{b]} + [\Phi_a, \Phi_b]. \tag{9}$$

The condition that Δ_0 and Δ_1 commute, for all $\pi^{A'}$, is

$$\epsilon^{AB}F_{AA'BB'} = 0, \tag{10}$$

which are the anti-self-dual Yang-Mills (ASDYM) equations. They are a set of coupled, first-order, nonlinear partial differential equations for Φ_a, and provide the prototype of an integrable system [24, 7].

Usually one wishes to work on $\mathbf{R}^4$ rather than $\mathbf{C}^4$, and to impose a reality condition on the fields Φ_a. For example, we may require that the Φ_a, as $n \times n$ matrices, are anti-hermitian and trace-free: this corresponds to the gauge group being $SU(n)$. Now the metric on the underlying space $\mathbf{R}^4$ is given by the usual expression

$$ds^2 = \epsilon_{AB}\epsilon_{A'B'}dx^{AA'}dx^{BB'}, \tag{11}$$

and its signature depends on precisely what reality condition is imposed on the $x^{AA'}$. There are three possible choices: $+ + ++$ (Euclidean), $+ - --$ (Lorentzian), or $+ + --$. In the Lorentzian case, the reality condition on Φ_a is inconsistent with the equations (10): in other words, non-trivial solutions of (10) in Minkowski space-time are necessarily complex. Such complex solutions have been studied (see, for example, [21, 14, 9]); but from now on I shall restrict to the cases $+ + ++$ and $+ + --$.

It is well-known that solutions of (10) correspond to holomorphic vector bundles over (regions of) twistor space $\mathbf{C}P^3$ [24, 5, 3, 37]. The integrability of the equations is tied up with this correspondence. Rather than describing the correspondence in detail, I shall mention three of the methods for constructing explicit solutions which arise from it.

1. The Ansätze A_k. In the case $n = 1$, i.e. abelian gauge group, everything becomes linear and easy to deal with explicitly. In particular, the equations (10) with $n = 1$ are linear, and solutions of them correspond to holomorphic line bundles over twistor space. One can build vector bundles of rank two (i.e. $n = 2$) as extensions of one line bundle by another, and this leads to a sequence of ansätze $A_0, A_1, A_2, \ldots$ for solutions of (10) with $n = 2$. For details, see [6, 26, 39, 40, 37]. Basically, these ansätze convert solutions of linear equations into solutions of non-linear equations (10) with $n = 2$. The ideas generalize to $n > 2$, and give ansätze there as well, but these have not been developed very far.

2. The Method of Simple Poles. Since the two operators Δ_0 and Δ_1 commute, one can solve the simultaneous equations

$$\Delta_A \psi = 0, \tag{12}$$

where $\psi = \psi(x^a, \zeta)$ is an $n \times n$ matrix. For ζ in some domain, ψ is required to be holomorphic in ζ, and nonsingular. Outside of this domain, ψ could have nasty singularities. The method of simple poles (also referred to as the Zakharov-Shabat transformation, or dressing method, or Riemann problem with zeros), consists of requiring that ψ and ψ^{-1} have only simple poles in the ζ space (including at $\zeta = \infty$). For each arrangement of poles, one gets an explicit family of gauge potentials Φ_a. The simplest example arises when ψ and ψ^{-1} each has exactly one pole. See, for example, [7]; Tafel [22] has described how this method is related to the ansätze mentioned previously. A slightly more general variant of the method is to begin with a known solution $\Phi_a^{(0)}$, with associated $\psi^{(0)}$, and then to construct a new ψ of the form $\psi = \chi\psi^{(0)}$, where χ only has simple poles in ζ.

3. The ADHM Construction (Method of Monads). The idea here is to construct vector bundles over twistor space in the following way. Let U, V and W be complex vector spaces and let

$$f(Z^\alpha) : U \to V, \quad g(Z^\alpha) : V \to W$$

be linear maps between them, with f and g depending linearly on the complex parameters Z^α, and such that $g(Z^\alpha)f(Z^\alpha) = 0$ for all Z^α. Then one defines a holomorphic vector bundle E over twistor space by saying that the fibre of E over a point Z^α in twistor space is

$$E_Z = \frac{\operatorname{Ker} g(Z^\alpha)}{\operatorname{Im} f(Z^\alpha)}.$$

The gauge potential Φ_a can be constructed directly from the data U, V, W, f, g, without going via twistor space. For more details, see [4, 3, 18].

All three of these methods were used in the construction of instantons, i.e. real, finite-action solutions of (10) on Euclidean four-space $\mathbb{R}^4$. The ADHM construction is the best-adapted to this particular problem, and considerable progress was made in using it to construct instanton solutions.

Geometrically, and physically, the connection is regarded as being specified by Φ_a modulo the gauge freedom

$$\Phi_a \mapsto \Lambda^{-1}\Phi_a\Lambda + \Lambda^{-1}\partial_a\Lambda, \tag{13}$$

where $\Lambda(x^a)$ takes values in the gauge group (a subgroup of $GL(n, \mathbb{C})$). The ASDYM equations (10) are invariant under such transformations (13). There is an alternative form of (10) involving a gauge-invariant field J taking values in the gauge group; in this, the equations look like those of a modified chiral model. But this can only be achieved at the cost of breaking the rotational symmetry in space-time: one has to choose two constant primed spinors. If we choose these to be the basis spinors with components (0,1) and (1,0), then we get the following.

Using the gauge freedom (13) and the ASDYM eqns (10), we can find a matrix J such that

$$\Phi_{A1'} = 0, \quad \Phi_{A0'} = J^{-1}\partial_{A0'}J. \tag{14}$$

And then J satisfies

$$\epsilon^{AB}\partial_{A1'}(J^{-1}\partial_{B0'}J) = 0. \tag{15}$$

This is an alternative form of the ASDYM equations (for example, de Vega [9] uses this form). See [33] for a discussion of a family of equations to which (15) belongs, namely

$$G^{ab}\partial_a(J^{-1}\partial_b J) = 0,$$

where G^{ab} is a constant tensor.

4 Reductions of the ASDYM Equations

The purpose of this section is to demonstrate how most well-known (and not-so-well-known) integrable systems arise as reductions of the anti-self-dual Yang-Mills (ASDYM) equations, and so fit into the twistor framework. By 'reduction' I mean the following.

1. First, one can reduce the number of independent variables (to fewer than four) by factoring out by a subgroup of the Poincaré group.

2. Secondly, one can reduce the number of dependent variables by imposing algebraic constraints on the Φ_a or on J (but this has to be done in a way which is consistent with the equations).

These procedures are illustrated by the examples which follow. One obvious example of (2) which might be mentioned immediately is that of reducing the gauge group to a subgroup: for example, the gauge group $U(n)$ is reduced to $SU(n)$ by imposing the condition $\det J = 1$, or $\operatorname{tr}\Phi_a = 0$.

4.1 Reductions to Three Dimensions

The only case which will be considered here is that of factoring out by a translation in $\mathbf{R}^4$. As a first example, take the positive-definite case, in which $x^{AA'}$ has the form

$$x^{AA'} = \begin{pmatrix} x^1 + ix^2 & x^3 + ix^4 \\ -x^3 + ix^4 & x^1 - ix^2 \end{pmatrix} \tag{16}$$

with the x^a all real. We reduce to $\mathbf{R}^3$ by requiring the gauge potentials Φ_a to be independent of x^4, i.e. functions only of $x^\alpha = (x^1, x^2, x^3)$. Then the ASDYM eqns (10) become

$$\nabla_\alpha \Phi = \tfrac{1}{2}\epsilon_{\alpha\beta\gamma}F^{\beta\gamma} \tag{17}$$

where $\Phi := \Phi_4$, and $\epsilon_{\alpha\beta\gamma}$ is the alternating tensor with $\epsilon_{123} = 1$. These eqns (17) are the Bogomolny equations, describing static Yang-Mills-Higgs monopoles in $\mathbf{R}^3$. Each of the three construction methods listed in the previous section can, and has, been used to construct multi-monopole solutions. (In this case the ADHM construction is called ADHMN, since it was first applied to this problem by W. Nahm; the vector spaces U, V and W are infinite-dimensional here.) For details, see [11, 12, 8].

As a second example, begin in $\mathbf{R}^4$ with the metric of signature $+ + --$; this is achieved by taking

$$x^{AA'} = \begin{pmatrix} x + s & t - y \\ t + y & x - s \end{pmatrix} \tag{18}$$

with x, y, t, s all real. If we reduce by requiring the fields to be independent of the second 'time' coordinate s, we end up with a hyperbolic equation in $(2+1)$-dimensional flat space-time (the hyperbolic version of the Bogomolny equations (17)). There is some advantage in using the J-description (15) in this case, because then one has a nice energy density functional. For details, see [34], where soliton solutions of this equation are constructed using the 'method of simple zeros'.

These reductions can be implemented very neatly in the twistor picture: one factors out by a holomorphic vector field on twistor space, and obtains a two-dimensional complex manifold sometimes known as 'mini-twistor space'. In other words, there is a 'reduced' Penrose correspondence between $\mathbf{R}^3$ (with $+ + +$ or $+ + -$ metric) and the mini-twistor space (which, as a complex manifold, is isomorphic to the holomorphic tangent bundle of the Riemann sphere $\mathbb{C}P^1$). For more details, see [11, 13, 35, 36].

4.2 *Reductions to Two Dimensions*

Above, we factored out by a constant non-null Killing vector. Let us now factor out, in addition, by another Killing vector field, thereby reducing to two dimensions. There are several possibilities which lead to interesting equations; let us concentrate on reductions of the $+ + --$ space (with coordinates as in (18) and with s already factored out).

One could choose a constant null Killing vector, such as $V = \partial_t - \partial_y$ (i.e. all fields are assumed constant along V). The reduced equations are then parabolic (with $t + y$ interpreted as 'time'), and in the $n = 2$ case boil down to either the nonlinear Schrödinger or the Korteweg-de Vries equations [15]. Another possibility is to factor out by a boost $V = y\partial_t + t\partial_y$, and use the J-description (15). This essentially leads to Einstein's vacuum equation for cylindrically-symmetric space-time [42]. (If one begins in the $+ + +$ signature and factors out by a rotation, one gets the 'Ernst' equation, i.e. Einstein's vacuum equation for stationary axisymmetric space-times; see [27, 41, 43]).

A third possibility is to factor out by a constant spacelike Killing vector. For example, if we factor out by $V = \partial_y$, then the J-equation (15) reduces to

$$\partial_t(J^{-1}\partial_t J) - \partial_x(J^{-1}\partial_x J) = 0, \tag{19}$$

which is the chiral field equation in $1 + 1$ dimensions. Special cases of this include well-known integrable systems such as nonlinear sigma models and

$\mathbb{C}P^n$ models. Proceeding in a slightly different way leads to the Toda lattice equations and their generalizations, including the sine-Gordon equation (cf. [38, 17, 31, 36]).

4.3 Reductions to One Dimension

This is where we began, since the linear system (5) gets reduced to the Lax equation (4). However, there are at least two different versions: we can reduce to a non-null coordinate (such as x^4), or to a null coordinate (such as $x^{00'}$). To illustrate the first of these possibilities, let us start with the ASDYM equations in positive-definite 4-space, and reduce by requiring the gauge potentials Φ_a to depend on x^4 only. Choose a gauge such that $\Phi_4 = 0$, and rename x^4 as t. Then the reduced equations are

$$d\Phi_\alpha/dt = -\tfrac{1}{2}\epsilon_{\alpha\beta\gamma}[\Phi_\beta, \Phi_\gamma]. \tag{20}$$

This set of ordinary differential equations, for the three matrices (Φ_1, Φ_2, Φ_3), is called the Nahm equation; it is involved in the ADHMN construction for monopoles referred to earlier. In a sense, the Nahm equation (20) is 'reciprocal' to the Bogomolny equation (17) (cf. [12, 8]). If the gauge group is $SU(2)$, then (20) is essentially just Euler's equation for a spinning top (a famous integrable dynamical system), while for larger gauge groups, (20) reduces to the Toda molecule equation (cf. [30, 32]). It is worth noting that the Lax pair (cf. § 2) for (20) is given by

$$\begin{aligned}
L &= -\Delta_1 + \zeta^{-1}\Delta_0 = 2\Phi_3 - \zeta(\Phi_1 + i\Phi_2) + \zeta^{-1}(\Phi_1 - i\Phi_2),\\
P &= -i\Phi_3 + i\zeta(\Phi_1 + i\Phi_2).
\end{aligned}$$

The equation $L_t = [L, P]$ is equivalent to (20).

The other possibility is to begin in the $+ + --$ signature, and reduce down to the null coordinate $u = x^{00'}$. In this case, the Lax pair obtained from (5) is

$$\begin{aligned}
L &= \Phi_y + \zeta\Phi_z,\\
P &= \Phi_u + \zeta\Phi_v,
\end{aligned}$$

which again is associated with well-known integrable dynamical systems (cf. [1, 23, 10, 31]).

4.4 Summary

Many of the best-known integrable differential equations arise as reductions of the ASDYM equations. In general, however, one needs to go to higher-dimensional generalizations of ASDYM, and these are discussed in the next section. And there are some integrable equations, such as the Kadomtsev-Petviashvili (KP) and Davey-Stewartson equations in $2+1$ dimensions, which do not, at present, appear to fit into this scheme at all.

5 Higher-Dimensional Generalizations of the ASDYM Equations

The ASDYM equations in four dimensions are associated with the linear system (5) (or 8). One can generalize the ASDYM equations by generalizing this linear system, which can be done in several different ways. Note that (5) consists of two equations, each linear in the spectral parameter ζ. So one could:

1. increase the number of equations; or

2. allow each equation to be a polynomial in ζ of degree ≥ 2; or

3. increase the number of spectral parameters;

or any combination of these. Each of these has the effect of increasing the number of space-time dimensions. One has a gauge potential Φ_a in the higher-dimensional space, and the consistency condition for the linear system is a set of linear relations on the corresponding gauge field F_{ab}. These relations, which are differential equations for Φ_a, are generalized ASDYM equations.

The various possibilities are discussed in [29]. Equations of type (3) (i.e. involving more than one spectral parameter) have not, as yet, found much application. Those of type (2) are, in effect, a restriction of type (1). So type (1) is the most natural; it is what underlies the 'hyperkähler structure' mentioned in the next section. Let us examine type (1) in slightly more detail.

The system (8) of linear operators is replaced by

$$\Delta_\alpha = \pi^{A'} \nabla_{\alpha A'} \qquad (21)$$

where the index α runs from 0 to $n - 1$. So the 'space-time' has dimension $2n$: its coordinates are $x^{\alpha A'}$. And the projective twistor space is $\mathbb{C}P^{n+1}$, with homogeneous coordinates $(\omega^\alpha, \pi_{A'})$. (In [29], this structure is denoted A_n. The 'standard' case is, of course, $n = 2$.) Holomorphic objects such as cohomology classes or vector bundles on $\mathbb{C}P^{n+1}$, correspond to solutions of systems of differential equations on $\mathbb{C}^{2n}$ (or $\mathbb{R}^{2n}$).

Mason and Sparling [15] discuss an interesting reduction of this system, obtained by factoring out the vector fields

$$V_\alpha = \partial_{\alpha 1'} - \partial_{(\alpha+1)0'} \text{ for } 0 \leq \alpha \leq n - 2. \qquad (22)$$

Since there are $n - 1$ of these vectors, the quotient space has dimension $2n - (n - 1) = n + 1$. And the twistor space becomes two-dimensional; indeed, the reduced twistor space is simply the holomorphic line bundle of Chern number n over $\mathbb{C}P^1$ (if $n = 2$, it is the mini-twistor space mentioned in §4.1). Holomorphic vector bundles over this space then correspond to a system of generalized Bogomolny equations on $\mathbb{R}^{n+1}$; and factoring out $x^{00'}$ leads to the well-known nonlinear Schrödinger and Korteweg-de Vries hierarchies of integrable equations.

6   Self-Dual Einstein Metrics

In generalizing the Lax equation (4) to the ASDYM system (5), we assumed that the derivative terms had constant coefficients. A more general possibility would be

$$\begin{aligned}
\Delta_0 &= U + \zeta V, \\
\Delta_1 &= Y + \zeta Z
\end{aligned} \tag{23}$$

where U, V, Y and Z are four independent vector fields in a four-dimensional space. One could also add in Φ_a terms as in (5), which would give a gauge field on the curved space-time background; to keep things simple these will be omitted.

In effect, the vector fields U, V, Y, Z define a null tetrad, and hence also a metric, on the four-dimensional space (cf. [20]). And the commutator $[\Delta_0, \Delta_1]$ vanishes for all ζ if and only if the conformal curvature of this metric is anti-self-dual (see for example, [19, 25, 37]). One gets a 'curved twistor space', which is the space of all integral surfaces of the distributions defined by (23). With some additional structure on the twistor space, one can solve the (anti-self-dual) Einstein equations, with or without a cosmological constant.

As in §5, one can also generalize to higher dimensions. The best-known case of this corresponds to adding more linear operators to the two of (23), i.e. what in §5 was called A_n. Here, one gets quaternionic and hyperkähler spaces of dimension $2n$ (for n even), which generalize the $n = 2$ case of ASD Einstein and ASD vacuum spaces mentioned above.

References

[1] Adler, M. & van Moerbeke, P. (1980) *Completely integrable systems, Euclidean Lie algebras, and curves. Linearization of Hamiltonian systems, Jacobi varieties and representation theory.* Adv. Math **38**, 267–379

[2] Arnold, V.I. (1978) *Mathematical Methods of Classical Mechanics* (Springer-Verlag, New York)

[3] Atiyah, M.F. (1979) *Geometry of Yang-Mills Fields* (Scuola Normale Superiore, Pisa)

[4] Atiyah, M.F., Drinfeld, V.G., Hitchin, N.J. & Manin, Yu. I. (1978) *Construction of instantons*, Phys. Lett. **A 65**, 185–187

[5] Atiyah, M.F., Hitchin, N.J. & Singer, I.M. (1978) *Self-duality in four-dimensional Riemannian geometry.* Proc. R. Soc. Lond. **A 362**, 425–461

[6] Atiyah, M.F., & Ward, R.S. (1977) *Instantons and algebraic geometry.* Commun. Math. Phys. **55**, 117–124

[7] Belavin, A.A. & Zakharov, V.E. (1978) *Yang-Mills equations as inverse scattering problem.* Phys. Lett. **B 73**, 53–57

[8] Corrigan, E. & Goddard, P. (1984) *Construction of instanton and monopole solutions and reciprocity.* Ann. Phys. **154**, 253–279

[9] de Vega, H.J. (1988) *Non-linear multi-plane wave solutions of self-dual Yang-Mills theory.* Commun. Math. Phys. **116**, 659–674

[10] Griffiths, P.A. (1985) *Linearizing flows and a cohomological interpretation of Lax equations.* Amer. J. Math **107**, 1445–1483

[11] Hitchin, N.J. (1982) *Monopoles and geodesics.* Commun. Math. Phys. **83**, 579–602

[12] Hitchin, N.J. (1983) *On the construction of monopoles.* Commun. Math. Phys. **89**, 145–190

[13] Jones, P.E. & Tod, K.P. (1985) *Minitiwistor spaces and Einstein-Weyl spaces.* Class. Quantum Grav. **2**, 565–577

[14] Kovacs, E. & Lo, S.Y. (1979) *Self-dual propagating wave solutions in Yang-Mills gauge theory.* Phys. Rev. **D 19**, 3649–3652

[15] Mason, L.J. & Sparling, G.A.J. (1989) *Nonlinear Schrödinger and Korteweg-de Vries are reductions of self-dual Yang-Mills.* Phys. Lett. **A 137**, 29–33

[16] Newell, A.C., Tabor, M. & Zeng, Y.B. (1987) *A unified approach to Painlevé expansions,* Physica **29D**, 1–68

[17] Olive, D & Turok, N. (1983) *The symmetries of Dynkin diagrams and the reduction of Toda field equations.* Nucl. Phys. **B 215**, 470–494

[18] Osborn, H. (1982) *On the Atiyah-Drinfeld-Hitchin-Manin construction for self-dual gauge fields.* Commun. Math. Phys. **86**, 195–219

[19] Penrose, R. (1976) *Nonlinear gravitons and curved twistor theory.* Gen. Rel. Grav. **7**, 31–52

[20] Penrose, R & Rindler, W. (1984) *Spinors and Space-Time.* (Cambridge University Press)

[21] Rebbi, C. (1978) *Self-dual Yang-Mills fields in Minkowski space-time.* Phys. Rev. **D 17**, 483–485

[22] Tafel, J. (1989) *A comparison of solution generating techniques for the self-dual Yang-Mills equations.* J. Math. Phys. **30**, 706–710

[23] van Moerbeke, P. (1985) *Algebraic geometrical methods in Hamiltonian mechanics.* Phil. Trans. R. Soc. Lond. **A 315**, 379–390

[24] Ward, R.S. (1977) *On self-dual gauge fields.* Phys. Lett **A 61**, 81–82

[25] Ward, R.S. (1980) *Self-dual space-times with cosmological constant.* Commun. Math. Phys. **78,** 1–17

[26] Ward, R.S. (1981) *Ansätze for self-dual Yang-Mills fields.* Commun. Math. Phys **80**, 563–574

[27] Ward, R.S. (1983) *Stationary axisymmetric space-times; a new approach.* Gen. Rel. Grav. **15**, 105–109

[28] Ward, R.S. (1984) *The Painlevé property for the self-dual gauge-field equations.* Phys. Lett **A 102**, 279–282

[29] Ward, R.S. (1984) *Completely solvable gauge-field equations in dimension greater than four.* Nucl. Phys **B 236**, 381–396

[30] Ward, R.S. (1985) *Generalized Nahm equations and classical Yang-Baxter equations.* Phys. Lett. **A 112**, 3–5

[31] Ward, R.S. (1987) *Multi-dimensional integrable systems. In: Field Theory, Quantum Gravity and Strings II*, ed. H.J. de Vega & N. Sanchez (Springer Lecture Notes in Physics, **Vol 280**)

[32] Ward, R.S. (1987) *The Nahm equations, finite-gap potentials and Lamé functions.* J. Phys. **A 20**, 2679–2683

[33] Ward, R.S. (1988) *Integrability of the chiral equations with torsion term.* Nonlinearity **1**, 671–679

[34] Ward, R.S. (1988) *Soliton solutions in an integrable chiral model in 2+1 dimensions.* J. Math. Phys **29**, 386–389

[35] Ward, R.S. (1989) *Classical solutions of the chiral model, unitons, and holomorphic vector bundles.* Commun. Math. Phys., to appear

[36] Ward, R.S. (1989) *Twistors in 2+1 dimensions.* J. Math. Phys., **30**, 2246–2251

[37] Ward, R.S. & Wells, R.O. (1989) *Twistor Geometry and Field Theory* (Cambridge University Press)

[38] Wilson, G. (1981) *The modified Lax and two-dimensional Toda lattice equations associated with simple Lie algebras.* Ergod. Th. & Dynam. Sys. **1**, 361–380

[39] Woodhouse, N.M.J. (1983) *On self-dual gauge fields arising from twistor theory.* Phys. Lett. **A 94**, 269–270

[40] Woodhouse N.M.J. (1985) *Real methods in twistor theory.* Class. Quantum Grav. **2**, 257–291

[41] Woodhouse, N.M.J.(1987) *Twistor description of the symmetries of Einstein's equations for stationary axisymmetric spacetimes.* Class. Quantum Grav. **4**, 799–814

[42] Woodhouse, N.M.J. (1989) *Cylindrical gravitational waves.* Class. Quantum Grav. **6**, 933–943

[43] Woodhouse, N.M.J. & Mason, L.J. (1988) *The Geroch group and non-Hausdorff twistor spaces.* Nonlinearity **1**, 73–114

Twistor Characterization of Stationary Axisymmetric Solutions of Einstein's Equations

J. Fletcher   N.M.J. Woodhouse

1   Introduction

One of the nonlinear systems that can be 'solved' by twistor methods is a reduced form of Einstein's vacuum equations for gravitational fields with two commuting Killing vectors. It is an intriguing example because it remains one of the central aims of twistor theory to tackle the full Einstein equations without any special assumptions about symmetry or self-duality. The solution of the reduced problem is a step towards achieving this. However, unlike the nonlinear graviton [11], which was based on a direct generalization of the familiar geometry of flat twistor space, the construction in this case is indirect and ungeometric. It was developed by Ward [18], who followed up Witten's observation [19] that the Ernst equation [3] is equivalent to a reduction of the self-dual Yang-Mills equations.

Ward's twistor analysis of stationary axisymmetric fields has been extended in two papers. The first (Woodhouse and Mason [21]) describes the connection with the solution generation techniques in relativity and uses twistor theory to 'explain' the occurrence of Riemann-Hilbert problems in the construction of exact solutions (the solution generation techniques are reviewed by Cosgrove [2] and in a collection of articles edited by Hoenselaers and Dietz [5]). The second (Woodhouse [20]) applies Ward's construction to gravitational waves with cylindrical symmetry.

The connection between the twistor geometry and the space-time geometry remains obscure and it is still possible that the construction reflects no more than an accidental correspondence between equations. There are, however, some indications that it is more than this in the remarkable way in which it is possible to read off local and global features of the space-time geometry directly from its twistor description without writing down the metric explicitly.

In this article, we shall review the basic construction, concentrating on the case of solutions which are generated by Ward's ansätze, but giving rather more details than are available in his original rather brief paper [18]. The twistor spaces of these special solutions are somewhat simpler than those of the general stationary axisymmetric solution. We shall then indicate how the twistor description encodes some of the global geometry of the maximally extended space-time, mostly by reference to particular examples where the geometry is already well known and well understood.

2   Space-times with Two Commuting Killing Vectors

We shall look at vacuum space-times with two commuting Killing vectors X_1 and X_2 which generate an orthogonally transitive group action. This means that the two-plane elements orthogonal to X_1 and X_2 are surface-forming, a condition that is less restrictive than it might appear since it is equivalent to the vanishing of the two *twist scalars*

$$e_{abcd} X_1^a X_2^b \nabla^c X_1^d, \qquad e_{abcd} X_1^a X_2^b \nabla^c X_2^d.$$

In a vacuum space-time, the twist scalars are constant ([9], [8, p 193]) and so they vanish identically if any combination of X_1 and X_2 has a zero.

Put $J_{ij} = X_i^a X_{ja}$ $(i, j = 1, 2)$. We shall assume that the symmetric matrix J is non-singular almost everywhere, which excludes the case in which the group orbits are everywhere null, as they are, for example, when X_1 and X_2 are orthogonal translation vectors in Minkowski space and $X_1^a X_{1a} = 0$.

Away from the singularities of J, the metric takes the form

$$g_{ab} = J^{ij} X_{ia} X_{jb} + h_{ab}$$

where $J^{ij} J_{jk} = \delta_k^i$ and h_{ab} is orthogonal to the Killing vectors. We shall identify h_{ab} with the induced metric on the space Σ of orbits.

Since the J_{ij} are constant on the orbits, we can regard J as a matrix-valued function on Σ. The vacuum field equation $R_{ab} = 0$ in the original space-time is then equivalent to the two equations

$$D_a(r J^{-1} D^a J) = 0 \tag{1}$$

$$\mathrm{Tr}\left(2 J^{-1} D_a D_b J + (D_a J^{-1})(D_b J)\right) = -4 R_{ab}^{(2)} \tag{2}$$

where $a, b, \dots$ are now two-dimensional tensor indices, $r^2 = -\det J$, D_a is the intrinsic covariant derivative on Σ, and $R_{ab}^{(2)}$ is the intrinsic Ricci tensor. The first equation is invariant under conformal transformations on Σ. Its trace gives $D_a D^a r = 0$.

The equations behave in different ways according to the signature of h_{ab} and the nature of $D_a(\det J)$. The various possibilities are illustrated in Table 1. The first three cases are the most straightforward and they are the ones that have been dealt with by twistor methods. In these, the problem is solved by first constructing a solution of eqn (1) on a two-dimensional conformal manifold of the appropriate signature and then by using equation

Case	h_{ab}	$D_a(\det J)$	Examples in Minkowski space
1	$(--)$	spacelike	$X_1 = x\partial_y - y\partial_x, X_2 = \partial_t$
2	$(+-)$	spacelike	$X_1 = x\partial_y - y\partial_x, X_2 = \partial_z$
3	$(+-)$	timelike	$X_1 = x\partial_t + t\partial_x \quad (t^2 > x^2), \; X_2 = \partial_z$
4	$(+-)$	null	$X_1 = y(\partial_z - \partial_t) - (z+t)\partial_y, X_2 = \partial_x$
5	any	zero	$X_1 = \partial_x, X_2 = \partial_y$

Table 1: The cases in which J is non-degenerate

(2) to determine a particular metric h_{ab} within the conformal class (up to a multiplicative constant). That such a conformal factor exists can be seen by exploiting the fact that $D_a D^a r = 0$ to write the metric on Σ in the form $ds^2 = -\Omega^2(dr^2 + dz^2)$ where z is the harmonic conjugate of r (r is imaginary in cases (2) and (3) and Ω is imaginary in case (3); z is the same as the Minkowski z coordinate in case (1), but not in the other cases). If we put $w = z + ir, \tilde{w} = z - ir$, then eqn (2) becomes

$$2\mathrm{i}\frac{\partial}{\partial w}\left(\log r\Omega^2\right) = r\,\mathrm{Tr}\left(\frac{\partial J^{-1}}{\partial w}\frac{\partial J}{\partial w}\right) \tag{3}$$

together with the conjugate equation in which w is replaced by $\tilde{w}$, and i by $-$i. These are automatically integrable if J satisfies eqn (1).

In the first three cases, therefore, the essential problem is to solve eqn (1) on Σ. In case (1), which includes the stationary axisymmetric solutions, eqn (1) is elliptic and its solutions are analytic (which is also an easy consequence of the twistor theory). In cases (2) and (3), which include, respectively, cylindrical gravitational waves and interior black hole solutions, the equation is hyperbolic and its solutions need not be analytic.

We shall consider below the extent to which it is possible to use twistor methods not only to solve eqn (1), but also to patch together local solutions of different types to build up global, maximally extended vacuum space-times. One difficulty is that we only know how to use twistor methods to solve eqn (1): the conformal factor must still be found indirectly by integrating eqn (3). As far as the J matrix and the twistor theory are concerned, there is no distinction between cases (2) and (3): one can pass between them simply by changing the sign of Ω^2. The global behaviour illustrated by the two flat examples in the table is, however, very different. In case (2), $r = 0$ is a two-dimensional timelike plane, while in case (3) it is the pair of null hyperplanes $t = \pm x$, across which the solution can be extended to a region of the first type. But Σ and J are the same in the two examples, so they cannot in themselves determine the global structure of the maximally extended space-time. Additional information is needed and it is not clear how to encode it into the twistor construction.

Cases (4) (translation and null rotation) and (5) (two translations) seem less straightforward and, so far, have defied twistor analysis (with an appropriate choice for the two Killing vectors, case (4) includes gravitational plane waves). In neither of these cases is it necessarily possible, given a general solution of eqn (1), to find a conformal factor on Σ such that eqn (2) holds.

3 The Ernst Potential

We shall now restrict attention to cases (1) to (3) and concentrate on eqn (1). Our strategy will be to begin by picking a (real or imaginary) solution of the equation $D_a D^a r = 0$ and then to solve eqn (1) subject to the constraint $\det J = -r^2$ (in addition to the requirements that J should be real, symmetric, and of the appropriate signature).

Let z be the conjugate function to r on Σ (z is determined by r up to the addition of a constant). Then the metric on Σ is conformal to $dr^2 + dz^2$ and eqn (1) takes the local form

$$r^{-1}\partial_r(rJ^{-1}\partial_r J) + \partial_z(J^{-1}\partial_z J) = 0. \tag{4}$$

Ward's twistor approach exploits the fact that eqn (4) is the same as the reduction of Yang's equation [22] in Minkowski space-time under the group generated by a boost or rotation together with a commuting translation. Yang's equation is itself a form of the anti-self-dual (ASD) Yang-Mills equations and can be solved by a splitting construction.

An alternative approach is to put $r = uv$ and $z = \frac{1}{2}(u^2 - v^2)$. Then eqn (4) becomes

$$u^{-1}\partial_u(uJ^{-1}\partial_u J) + v^{-1}\partial_v(vJ^{-1}\partial_v J) = 0. \tag{5}$$

This is also a reduction of Yang's equation under the group generated by a boost and an orthogonal rotation—that is by ∂_θ and ∂_ψ when the Minkowski metric is written in the form

$$ds^2 = du^2 - u^2 d\psi^2 - dv^2 - v^2 d\theta^2. \tag{6}$$

In both cases the twistor transform of the four-dimensional Yang-Mills field can be reduced by factoring out the two-parameter symmetry group to give a direct twistor solution of eqn (4) or eqn (5). The geometry of the group orbits in **PT** is rather different in the two cases, so it less obvious than it may appear (but nonetheless true) that the two starting points lead to the same reduced twistor construction.

The first approach is useful for examining the behaviour of J near symmetry axes or horizons, where one of the Killing vectors vanishes or becomes null. The second is more useful for understanding what happens near events where both Killing vectors vanish (as ∂_θ and ∂_ψ do at the origin in Minkowski space).

Eqn (4) has the obvious symmetry $J \mapsto A^t J A$ where A is a constant matrix. This corresponds to the linear transformation

$$(X_1 \; X_2) \longmapsto (X_1 \; X_2)A$$

of the Killing vectors in the original space-time. It is also invariant under $J \mapsto J^t$ and $J \mapsto J^{-1}$ as well as under the following less obvious 'hidden symmetry'.

Assume that J is symmetric and write

$$J = \begin{pmatrix} f\omega^2 + \hat{f}^{-1} & -f\omega \\ -f\omega & f \end{pmatrix}.$$

This amounts to writing the metric in Weyl coordinates in the canonical form

$$ds^2 = f(dt - \omega d\theta)^2 + \hat{f}^{-1}d\theta^2 - \Omega^2(dr^2 + dz^2).$$

The field equation (4) then becomes

$$\nabla^2 \log f = f\hat{f}(\partial_r\omega)^2 + f\hat{f}(\partial_z\omega)^2 = \nabla^2 \log \hat{f}$$

$$\partial_r(rf\hat{f}\partial_r\omega) + \partial_z(rf\hat{f}\partial_z\omega) = 0 \qquad (7)$$

where $\nabla^2 = r^{-1}\partial_r(r\partial_r) + \partial_z^2$ (when r is real, this is the axisymmetric form of the three-dimensional Laplacian in cylindrical polar coordinates).

The hidden symmetry transforms J to

$$J' = -\frac{1}{r^2\hat{f}} \begin{pmatrix} 1 & -\psi \\ -\psi & \psi^2 - r^2 f\hat{f} \end{pmatrix}$$

where

$$\partial_z\psi = rf\hat{f}\partial_r\omega, \qquad \partial_r\psi = -rf\hat{f}\partial_z\omega,$$

for which the integrability condition is eqn (7).

The quantity $\mathcal{E} = f + i\psi$ is the *Ernst potential* [3] of J. It is often taken as the basic variable in the analysis of stationary axisymmetric fields. We shall also refer to J' as the Ernst potential. If $\det J = -r^2$, then J' satisfies the simpler constraint $\det J' = 1$.

Note that since ψ is only determined up to a constant, J' is only determined up to

$$J' \longmapsto \begin{pmatrix} 1 & 0 \\ \gamma & 1 \end{pmatrix} J' \begin{pmatrix} 1 & \gamma \\ 0 & 1 \end{pmatrix}$$

where γ is constant. Also the construction of J' is not covariant with respect to general linear transformations in the space of Killing vectors, although under the restricted class of lower triangular transformations

$$(X_1 \; X_2) \longmapsto (X_1 \; X_2)\begin{pmatrix} \alpha & 0 \\ \beta & \delta \end{pmatrix},$$

J' transforms according to

$$ J' \longmapsto \begin{pmatrix} \alpha & 0 \\ 0 & \delta \end{pmatrix} J' \begin{pmatrix} \alpha & 0 \\ 0 & \delta \end{pmatrix}. \tag{8} $$

Thus the construction of J' involves the singling out of the direction of X_2 in the space of Killing vectors.

We can attach boundary points to Σ at which $r = 0$. These correspond to group orbits which lie on a symmetry axis—where some combination of the Killing vectors vanishes in the same way as a rotation in Minkowski space—or a Killing horizon—where some combination becomes null in the same way as a boost in Minkowski space. By examining the behaviour of J in geodesic normal coordinates near an axis or horizon event at which the space-time metric is nonsingular, one can see that provided that $X_2^a X_{2a}$ does not itself vanish, J' remains smooth at $r = 0$. In the elliptic case, J' is positive definite on an axis and negative definite on a horizon; in the hyperbolic case, J' is indefinite at both types of event.

The transformations of J' induced by upper triangular transformations in the space of Killing vectors

$$ (X_1 \ X_2) \longmapsto (X_1 \ X_2) \begin{pmatrix} 1 & \beta \\ 0 & 1 \end{pmatrix}, $$

which do alter the direction of X_2, are complicated; but on $r = 0$ in a space-time with a regular axis or horizon, they reduce to

$$ J' \longmapsto \begin{pmatrix} 1 & 0 \\ 2\beta z & 1 \end{pmatrix} J' \begin{pmatrix} 1 & 2\beta z \\ 0 & 1 \end{pmatrix}. \tag{9} $$

Thus, to within the freedom allowed by the transformations (8) and (9), the *boundary values* of J' are independent of the choice of X_1 and X_2, provided only that $X_2^a X_{2a} \nrightarrow 0$ as $r \to 0$.

The Ernst formalism is not well adapted to global problems because one may want to make incompatible choices for X_2^a to make $\mathcal{E}$ well behaved at infinity and on the various components of the symmetry axis and Killing horizons. The potential J' suffers from the additional drawback that it may be singular at events where the space-time metric is well behaved (if $X_2 = \partial_t$ in the Kerr solution, for example, then J' blows up on the ergosphere).

Note that J' certainly behaves badly at events where both $X_1^a X_{1a}$ and $X_2^a X_{2a}$ vanish—for example, at the intersection of a horizon cross-over and an axis, where the metric looks like (6) to within terms that are quadratic in the Minkowski position vector.

4 Yang's Equation

The basis of the twistor approach is the correspondence between eqn (4) and a reduction of Yang's equation. It arises as follows.

Suppose that Φ_a is the $\mathrm{sl}(2,\mathbb{C})$-valued connection 1-form of an ASD Yang-Mills field in complex Minkowski space. For any constant spinor field $o^{A'}$, the equation

$$o^{A'}H^{-1}\nabla_{AA'}H = o^{A'}\Phi_{AA'}$$

is integrable; H takes values in $\mathrm{SL}(2,\mathbb{C})$ and the equation is the condition that the columns of H^{-1} should be covariantly constant with respect to Φ over the α-planes tangent to $o^{A'}$. Let $\iota^{A'}$ be a second constant spinor field such that $o_{A'}\iota^{A'} = 1$. Choose $\hat{H}$ (also with values in $\mathrm{SL}(2,\mathbb{C})$) such that $\iota^{A'}\Phi_{AA'} = \iota^{A'}\hat{H}^{-1}\nabla_{AA'}\hat{H}$ and put $J = H\hat{H}^{-1}$. Then J determines Φ_a up to gauge since Φ_a is gauge equivalent to

$$-\iota_{A'}o^{B'}J^{-1}\nabla_{AB'}J.$$

The ASD condition on Φ_a is equivalent to $\iota^{A'}o^{B'}\nabla^{A}_{A'}(J^{-1}\nabla_{AB'}J) = 0$, which is Yang's equation.

Now suppose that Φ_a is invariant under Lie propagation along $X_1 = \partial_\theta$ and $X_2 = \partial_t$ in cylindrical polar coordinates t, r, θ, z. Take $o^{A'}$ and $\iota^{A'}$ along the two real null directions in the t, z plane and choose H and $\hat{H}$ so that they are functions of r and z alone. Then J is also a function of r and z alone and it is determined by the Yang-Mills field up to $J \to AJB$ where A and B are constant matrices. Moreover Yang's equation reduces to eqn (4).

All that is needed to solve the reduced Einstein equations, therefore, is to encode the reality conditions and other constraints on the space-time J or its Ernst potential into the construction of the holomorphic bundle corresponding to Φ_a. The resulting connection between the geometry of the gravitational field and the structure of a bundle over flat twistor space is, however, rather indirect and obscure. To try to shed some light on it, we look next at a straightforward generalization in which the ASD Yang-Mills field is defined on an ASD vacuum space-time. If the gravitational field is then taken to be the complex field represented by the ASD space-time itself, we could expect the connection to be more geometrically transparent (see §8.6).

5   Yang's Equation in ASD Space-time

Suppose that M is an ASD vacuum space-time (that is, $\phi_{ABA'B'} = 0$, $\Lambda = 0$, and $\psi_{A'B'C'D'} = 0$, so that M admits covariantly constant spinors $\pi_{A'}$). It makes no difference whether M is thought of as a real four-dimensional Riemannian manifold (with a negative-definite metric), or as a complex space-time, or, in the flat case, as real Minkowski space.

Suppose also that there are two independent commuting Killing vectors on M and that their action is orthogonally transitive. By Lie propagation on constant primed spinors, these generate an abelian subalgebra of $\mathrm{sl}(2,\mathbb{C})$, which can be at most one-dimensional. We shall assume that it *is* one-dimensional and that it does not consist only of null rotations. Then, without loss of generality,

$$\nabla_A^{A'} X_2^{AB'} = 0, \qquad \nabla_A^{A'} X_1^{AB'} = 2io^{(A'}\iota^{B')} \tag{10}$$

where $o^{A'}$ and $\iota^{A'}$ are constant, with $o_{A'}\iota^{A'} = 1$.

It follows from the anti-self-duality of X_2 that there exists a Killing spinor $\omega^{A'B'}$ such that $\nabla_{AA'}\omega_{B'C'} = X_{2A(B'}\epsilon_{C')A'}$ (see [16]). It is not hard to show that

$$2X_2^{A(A'}X_{1A}{}^{B')} = ire^{-i\theta}o^{A'}o^{B'} + ire^{i\theta}\iota^{A'}\iota^{B'},$$

$$2\omega^{A'B'} = -re^{-i\theta}o^{A'}o^{B'} + 2zo^{(A'}\iota^{B')} + re^{i\theta}\iota^{A'}\iota^{B'}, \tag{11}$$

where r, z, θ are functions on M constant along X_2, with r and z also constant along X_1. Moreover, the metric on Σ—as before, the space of orbits in M—is

$$ds^2 = -(X_2^a X_{2a})^{-1}(dr^2 + dz^2).$$

If M is Minkowski space and X_1 and X_2 are as in the previous section, then r, z and θ are cylindrical polar coordinates.

Armed with these relations, we can carry out the same reduction of the ASD Yang-Mills equations as in Minkowski space. Given an ASD connection 1-form which is invariant under Lie propagation along X_1 and X_2, we define $H, \hat{H}$ and J as before, but now using the two spinors $o^{A'}$ and $\iota^{A'}$ in eqn (10). We can choose them so that J is constant along X_1 and X_2. Then, since

$$o^{A'}\iota^{B'}\nabla_{A(A'}r\nabla^A_{B')}z = 0$$

as a consequence of eqn (11) and the Killing spinor equation on $\omega^{A'B'}$, the ASD condition on the Yang-Mills field reduces to $\nabla_a(J^{-1}\nabla^a J) = 0$ and hence to eqn (4). Thus the reduced Einstein equations are locally equivalent to the reduced ASD Yang-Mills equations in *any* ASD space-time with the appropriate Killing vectors.

6 A Twistor Construction for Yang's Equation

We shall ignore reality conditions for the moment and think of M as a complex ASD vacuum space-time. The corresponding twistor space PT is the reduction of $\mathsf{F} = M \times \{\pi_{A'}\}$ by the foliation spanned by the horizontal vectors $\pi^{A'}\nabla_{AA'}$ and the vertical field $\Upsilon = \pi_{A'}\partial/\partial\pi_{A'}$.

Ward's correspondence associates ASD Yang-Mills fields on M with holomorphic vector bundles over PT in such a way that the local holomorphic sections of the bundle corresponding to the connection 1-form Φ_a are identified with the solutions of

$$\pi^{A'}(\nabla_{AA'} + \Phi_{AA'})s = 0, \qquad \Upsilon(s) = 0 \tag{12}$$

on F, [17, 13]. We shall look at the reduced construction in which all the objects involved are invariant under the group generated by X_1 and X_2.

The Lie derivative lifts X_1 and X_2 to $\mathbf{F}$. Since X_2 is ASD, its lift is horizontal, but the lift of X_1 has a nonzero vertical component. The lifted vector fields project onto $\mathbf{PT}$ where they span a foliation by complex surfaces (in the flat case, a pencil of quadrics). The space of leaves is a non-Hausdorff Riemann surface, which we denote by R and call the *reduced twistor space*.

When Φ_a is invariant along X_1 and X_2, we can add to eqns (12) the conditions $X_1(s) = 0$ and $X_2(s) = 0$ and so associate Φ_a with a holomorphic bundle over R. We shall not consider the full details of this reduction, which proceed along lines that are already familiar in twistor theory and which are complicated only by the fact that R is non-Hausdorff. All that we need to understand the reduced construction directly without reference to the four-dimensional version is, first, that the foliation of $\mathbf{PT}$ is given by

$$w = \frac{\omega_{A'B'}\pi^{A'}\pi^{B'}}{o_{C'}\pi^{C'}\iota_{D'}\pi^{D'}} = \frac{r}{2}(\zeta^{-1} - \zeta) + z = \text{constant}$$

where $\pi^{A'} = e^{i\theta}\zeta\iota^{A'} - o^{A'}$; and second, that eqns (12) together with the supplementary conditions $X_1(s) = 0$ and $X_2(s) = 0$ are equivalent to $s = s(z, r, \zeta)$ together with

$$(\partial_r + \zeta\partial_z + r^{-1}\zeta\partial_\zeta)s + (J^{-1}\partial_r J)s = 0 \tag{13}$$

$$(-\zeta\partial_r + \partial_z + r^{-1}\zeta^2\partial_\zeta)s + (J^{-1}\partial_z J)s = 0. \tag{14}$$

The starting point for the direct construction is not M, but instead a two-dimensional complex conformal manifold Σ on which there is given a holomorphic solution r of Laplace's equation. As before, z will denote the conjugate function.

Let $F = \Sigma \times X$, where X is the ζ Riemann sphere. The (reduced) twistor space R associated with Σ and r is constructed from F by identifying (σ, ζ) with (σ', ζ') whenever they both lie on the same connected component of one of the surfaces given by

$$r\zeta^2 - 2(z - w)\zeta - r = 0 \tag{15}$$

for some value of w. We can use w as a local holomorphic coordinate on R, which is a non-Hausdorff Riemann surface.

Each value of w corresponds to one point of R if one can continuously change one root of eqn (15) into the other by varying $\sigma \in \Sigma$ with w fixed, and to two points otherwise.

Let S denote the w Riemann sphere and let $V \subset S$ denote the (open) subset of values of w for which there is just one point of R. If Σ is simply connected, then
$$V = \{z(\sigma) \pm ir(\sigma) : \sigma \in \Sigma\}.$$
In general, V is not connected.

When $w = \infty$, either $\zeta = 0$ or $\zeta = \infty$ whatever the values of r and z, so $w = \infty$ is never in V. We shall denote the two corresponding points of R by ∞_0 (where $\zeta = 0$) and ∞_1 (where $\zeta = \infty$).

The two vector fields $Z_1 = \partial_r + \zeta \partial_z + r^{-1}\zeta \partial_\zeta$ and $Z_2 = -\zeta \partial_r + \partial_z + r^{-1}\zeta^2 \partial_\zeta$ are tangent to the surfaces of constant w in F. Given a solution J of eqn (4), we construct a holomorphic bundle $E \to R$ by taking the fibre of E over a point of R to be the space of solutions of eqns (13) and (14) on the corresponding connected surface in F. The integrability condition is precisely (4) (see also [1]).

Conversely, given $E \to R$, we can recover J to within $J \mapsto AJB$, where A and B are constant matrices, as follows. Pick $\sigma \in \Sigma$ and let $\pi : X \to R$ be the map defined by restricting the projection $F \to R$ to $\{\sigma\} \times X$. We have to assume that $\pi^*(E)$ is a trivial holomorphic bundle on X (this is not as restrictive as it appears—if it is true at σ, then it is true in a neighbourhood of σ). Suppose that E is given by patching matrices $\{P_{\alpha\beta}(w)\}$ relative to an open cover $\{R_\alpha\}$ of R such that $\infty_0 \in R_0$ and $\infty_1 \in R_1$. Then $\pi^*(E)$ is given by patching matrices

$$P_{\alpha\beta}\left(r(\sigma)\tfrac{1}{2}(\zeta^{-1} - \zeta) + z(\sigma)\right) \tag{16}$$

relative to the open cover $\pi^{-1}(R_\alpha)$ of X, so by our triviality assumption there exist splitting matrices $K_\alpha(\zeta)$ such that

$$P_{\alpha\beta}\left(r(\sigma)\tfrac{1}{2}(\zeta^{-1} - \zeta) + z(\sigma)\right) = K_\alpha(\zeta)K_\beta(\zeta)^{-1}.$$

The value of J at σ is $K_0(0)K_1(\infty)^{-1}$. Given the $P_{\alpha\beta}$, J is independent of the choice of splitting matrices since the only freedom in $\{K_\alpha\}$ is to replace each K_α by $K_\alpha C$ where C is a constant matrix. This leaves J unchanged. Moreover, the K_α and J depend smoothly on r and z as σ varies, although J may have singularities at points where the triviality condition does not hold.

We can exploit the freedom in the choice of splitting matrices to set $K_1(\infty) = 1$. Then $J = K_0(0)$. By differentiating eqn (16) and using the fact that w is constant along Z_1 and Z_2, we deduce that

$$K_\alpha^{-1}Z_i(K_\alpha) = K_\beta^{-1}Z_i(K_\beta); \quad i = 1, 2$$

and hence from the Liouville theorem that both sides are independent of ζ. It follows by putting $\zeta = 0$ that the columns of K_0^{-1} are solutions of eqns (13) and (14) and therefore that J is a solution of eqn (4).

The only other freedom in the construction of J from E is to change the local trivializations, for example by replacing $P_{\alpha\beta}$ by $Q_\alpha P_{\alpha\beta}Q_\beta^{-1}$ where each $Q_\alpha : R_\alpha \to GL(2, \mathbb{C})$ is holomorphic. This has the effect of replacing J by AJB^{-1} where $A = Q_0(\infty_0)$ and $B = Q_1(\infty_1)$. So the freedom to multiply J by constant matrices corresponds to the freedom to make linear transformations in the fibres of E over ∞_0 and ∞_1. It is not hard to present

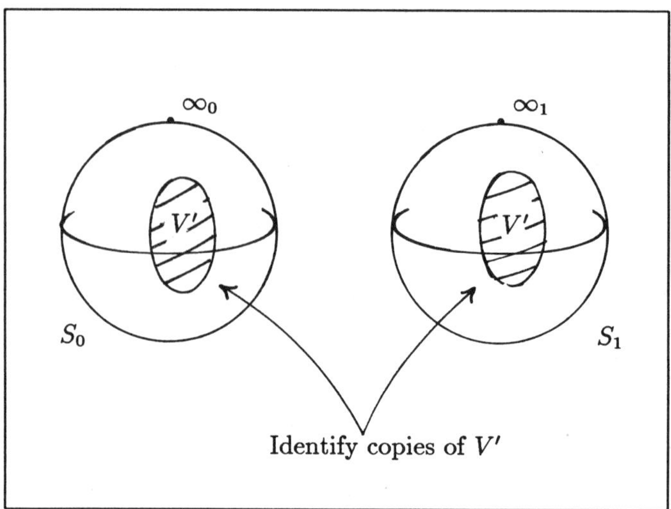

Figure 1: The construction of the non-Hausdorff Riemann surface R'

the construction of J from E in a more abstract form in which J emerges not as a matrix but as a linear map $J : E_{\infty_1} \to E_{\infty_0}$. Thus if J is to be interpreted as a solution of the reduced Einstein equation, then we must identify E_{∞_1} with the space of Killing vectors in the space-time and E_{∞_0} with its dual space.

The general conditions on $E \to R$ that ensure that it corresponds to a real, symmetric J satisfying the constraint $\det J = -r^2$ are quite complicated. They are explained by Woodhouse and Mason [21]. If, however, we only look at Js derived from space-times with a regular symmetry axis or Killing horizon, then the construction simplifies and the constraints are more straightforward.

The simplification involves the enlargement of V so that it becomes a connected open subset $V' \subset S$ by making further identifications between double points of R. The resulting space, which we denote R', consists of two copies S_0 and S_1 of the w Riemann sphere which are identified at points of V'. We shall choose the labelling of the copies so that $\infty_0 \in S_0$ and $\infty_1 \in S_1$ (see figure 1).

Whatever further identifications we make, we shall still have a projection $F \to R'$ and so we shall still be able to construct a solution J from a holomorphic bundle $E \to R'$; but not every solution will arise in this way since, given a general bundle $E \to R$, it may not be possible to extend the identifications to the fibres of E over the double points of R. Those that do we shall call *regular solutions*—they are the solutions given by the various ansätze by Ward [18].

A holomorphic bundle $E \to R'$ is easy to characterise. We take a four-set open cover $\{U_0, U_1, U_2, U_3\}$ of R' such that U_0 and U_2 cover S_0, with $V' \subset U_2$ and $\infty_0 \in U_0$; and such that U_1 and U_3 cover S_1, with $V' \subset U_3$ and $\infty_1 \in U_1$. Since the restrictions of E to S_0 and S_1 must be standard bundles on the Riemann sphere, we can choose local trivializations for E such that

$$P_{02} = \begin{pmatrix} (2w)^p & 0 \\ 0 & (2w)^q \end{pmatrix}, \quad P_{13} = \begin{pmatrix} (2w)^{p'} & 0 \\ 0 & (2w)^{q'} \end{pmatrix} \tag{17}$$

(for simplicity, we are assuming that $w = 0$ is in V'; this can always be arranged by adding a real constant to w). That is $E|_{S_0} = L^p \oplus L^q$ and $E|_{S_1} = L^{p'} \oplus L^{q'}$ where, contrary to the statement of Woodhouse and Mason [21], $L \to S$ is the tautological bundle, whose local sections are homogeneous functions of degree -1. If the pull-back of E to F is to be trivial on the copies of X, then $p' = -p$ and $q' = -q$, so the regular solutions are determined by two integers p and q and by a single holomorphic patching matrix $P(w) = P_{23}$, defined for $w \in V'$.

The procedure for reconstructing J from R' and the data P, p, q reduces to the following. Pick r, z and let π be the map from X onto R' given by

$$\zeta \longmapsto w = \tfrac{1}{2}r(\zeta^{-1} - \zeta) + z$$

where X is the corresponding line in F. This expression does not, in itself, determine π uniquely since, outside V', each value of w corresponds to two points of R': we must add a rule for deciding which of these is to be $\pi(\zeta)$. It is clearly not possible to make π surjective unless both the branch points $w = z \pm ir$ lie in V', so the construction will only determine $J(z, r)$ where this is satisfied.

When V' is simply connected, π is fixed by requiring that $\zeta = 0$ should be mapped to S_0 and $\zeta = \infty$ to S_1. But in the examples we shall look at, V' is the complement of a finite set $\{\infty, w_1, w_2, ..., w_n\}$, made up of the point at infinity and a collection of isolated singularities of P. Each w_i corresponds to two points of X, which are the two roots of

$$r\zeta^2 - 2(z - w_i)\zeta - r = 0. \tag{18}$$

The additional information needed to fix π in this case is the labelling of one root as ζ_i^0 (which is mapped to S_0), and the other as ζ_i^1 (which is mapped to S_1).

To evaluate $J(z, r)$, choose a cover V_0, V_1 of X such that $\{0, \zeta_1^0, ..., \zeta_n^0\} \subset V_0$ and $\{\infty, \zeta_1^1, ..., \zeta_n^1\} \subset V_1$ and look for splitting matrices $Q_0(\zeta)$ and $Q_1(\zeta)$ such that

$$\begin{pmatrix} r^p\zeta^{-p} & 0 \\ 0 & r^q\zeta^{-q} \end{pmatrix} P\left(\tfrac{1}{2}r(\zeta^{-1} - \zeta) + z\right) \begin{pmatrix} (-r\zeta)^p & 0 \\ 0 & (-r\zeta)^q \end{pmatrix} = Q_0 Q_1^{-1} \tag{19}$$

with Q_0 holomorphic in ζ and nonsingular for $\zeta \in S_0$ and Q_1 holomorphic in ζ and nonsingular for $\zeta \in V_1$. Finally, we put $J(z,r) = Q_0(0)Q_1(\infty)^{-1}$ to obtain a solution of eqn (4). In fact we get a number of apparently different solutions by making different choices for the labelling of the roots of eqn (18). In general these are analytic continuations of each other. For real z,r, they correspond to different parts of the Penrose diagram of the maximal analytic extension of the space-time metric.

7 The Patching Matrix

Most of the following are stated by, or implicit in, Ward [18] and are derived in [21].

1. If P is symmetric then so is $J(z,r)$. Conversely if $J(z,r)$ is symmetric, then $E|_{S_0}$ is dual to $E|_{S_1}$. If the representation of $E|_{S_0}$ and $E|_{S_1}$ respects this duality, then P is symmetric.

2. If $\det P = 1$ then $\det J = (-r^2)^{p+q}$.

3. $J(z,r)$ is real in each of the cases

   (a) $\overline{P(w)} = P(\overline{w})$, z,r real;

   (b) $\overline{P(w)} = P(\overline{w})$, z real, r imaginary;

   (c) $\overline{P(w)} = P(-\overline{w})$, z imaginary, r real.

4. If J is obtained from a space-time with a regular axis or horizon, then $p = 1$, $q = 0$ and $P(z) = J'(z,0)$ on the axis or horizon, where J' is the Ernst potential. That is, the patching matrix is the analytic continuation of the boundary values of the Ernst potential. In case (a) above, which corresponds to case (1) in Table 1 (see §1), the boundary values of J' are positive definite on an axis and negative definite on a horizon.

5. If J comes from an asymptotically flat space-time (in the sense that its Ernst potential has the same asymptotic form as the Ernst potential of Minkowski space with rotation and translation Killing vectors), then $P(\infty) = 1$, and conversely.

The fourth statement conceals a number of issues concerning the uniqueness of P. Suppose that V' is the complement of $\{\infty, w_1, ..., w_n\}$ and that we are given $E \to R'$, with $p = 1$, $q = 0$. Then there are two sources of freedom in the representation of E as a pair of standard bundles over the spheres S_0 and S_1, glued together by a patching matrix $P(w)$.

(I) There is the discrete freedom in the representation of R' as $S_0 \cup S_1$. If we fix, once and for all, the labelling of the two copies of $w = \infty$ and denote the sphere containing ∞_0 as S_0 and the sphere containing ∞_1 as S_1, then we can still choose which of the copies of $w_1, ..., w_n$ to assign to S_0 and which to assign to S_1. Different choices lead to completely different patching matrices and even to different values of the integers p and q (see the example below, where this phenomenon 'explains' the different natures of the singularities in the Schwarzschild and Kerr solutions).

As $r \to 0$, one root of eqn (18) goes to zero and the other goes to infinity (provided that $z \neq w_i$). Statement **4** is true provided that the representation of R' as $S_0 \cup S_1$ and the definition of π are such that $\zeta_i^0 \to 0$ and $\zeta_i^1 \to \infty$ for each i (given π, we shall describe such a representation as being *adapted* to the relevant portion of the axis or horizon). The fifth statement is true only if the representation is adapted to part of the axis that extends to infinity.

(II) There is further (continuous) freedom in the representation of $E|_{S_0}$ and $E|_{S_1}$ as direct sums of line bundles. Suppose, for simplicity, that P and J are symmetric. Then, without upsetting eqn (17) or the symmetry of P, we can change the trivializations of E on U_0, U_1, U_2, U_3 so that

$$P_{02} \mapsto AP_{02}B^{-1}, \quad P_{13} \mapsto (A^t)^{-1}P_{13}B^t, \quad P = P_{23} \mapsto BPB^t$$

where

$$A = \begin{pmatrix} \alpha & 0 \\ \beta + (2w)^{-1}\gamma & \delta \end{pmatrix}, \quad B = \begin{pmatrix} \alpha & 0 \\ \gamma + 2w\beta & \delta \end{pmatrix}$$

for constant $\alpha,\beta,\gamma,\delta$.

Under the identification of E_{∞_1} with the space of Killing vectors in space-time, this corresponds to the freedom to make linear transformations of the form

$$X_1 \mapsto \alpha X_1, \quad X_2 \mapsto \delta X_2 + \beta X_1$$

where X_1 is the Killing vector that vanishes or becomes null at $r = 0$. The corresponding effect on the boundary values of the Ernst potential is

$$J' \mapsto \begin{pmatrix} \alpha & 0 \\ \gamma + 2\beta z & \delta \end{pmatrix} J' \begin{pmatrix} \alpha & \gamma + 2\beta z \\ 0 & \delta \end{pmatrix}$$

(see §2).

Note that the direction in E_{∞_1} of X_1 (the Killing vector that vanishes or becomes null) is singled out by the holomorphic structure of $E|_{S_1}$, but that the direction of X_2 is not. The one-dimensional space spanned by X_1 is the image of $L_{\infty_1}^{-1}$ under

$$L^{-1} \longrightarrow E|_{S_1} = L^{-1} \oplus L \longrightarrow L.$$

Finally, to return to our list of properties of the patching matrix, the proof of the following is given by Fletcher [4].

6. Suppose that J arises from a space-time which has an axis and a horizon (with a cross-over); and that there is a nonsingular event at which $r = z = 0$ and at which the axis and horizon cross-over intersect. Then, with an appropriate choice of basis in the space of Killing vectors,

$$P = \begin{pmatrix} (2w)^{-1}a & b \\ b & 2wc \end{pmatrix} \tag{20}$$

where a, b and c are holomorphic at $w = 0$ and $ac - b^2 \neq 0$. Conversely, if P satisfies (20), then J behaves in the correct way as $r, z \to 0$ for $r = z = 0$ to be a nonsingular space-time event.

8 Examples

8.1 *Minkowski Space with Translation and Rotation*

In cylindrical polar coordinates, the metric is $ds^2 = dt^2 - r^2 d\theta^2 - dr^2 - dz^2$. If we take $X_1 = \partial_\theta$ and $X_2 = \partial_t$, then

$$J = \begin{pmatrix} -r^2 & 0 \\ 0 & 1 \end{pmatrix}, \qquad J' = \begin{pmatrix} 1 & 0 \\ 0 & 1 \end{pmatrix}.$$

The metric is generated by the data $p = 1$, $q = 0$, $P = 1$, with V' the entire complex plane.

8.2 *Minkowski Space with Boost and Rotation*

Here the metric is given by equation (6). If we take $X_1 = \partial_\theta$ and $X_2 = \partial_\psi$, then

$$J = \begin{pmatrix} z - \sqrt{z^2 + r^2} & 0 \\ 0 & z + \sqrt{z^2 + r^2} \end{pmatrix},$$

$$J' = \begin{pmatrix} (z + \sqrt{z^2 + r^2})^{-1} & 0 \\ 0 & z + \sqrt{z^2 + r^2} \end{pmatrix}.$$

By putting $r = 0$ in the second of these, we find that the solution is generated by

$$P = \begin{pmatrix} (2w)^{-1} & 0 \\ 0 & 2w \end{pmatrix}$$

with $p = 1$, $q = 0$, as can be checked directly by using eqn (19)—the splitting is straightforward.

This example illustrates a number of 'global' phenomena that also occur in less trivial metrics. If u and v in eqn (6) are real and non-negative, then the coordinate patch covers only one of the two regions of Minkowski space in which the boost Killing vector ∂_ψ is timelike. The Weyl coordinates z and r, which in this case are *not* cylindrical polar coordinates, correspondingly range over the real half-plane $r > 0$. On the symmetry axis, $r = 0$ and $z > 0$; and on the horizon, where ∂_ψ becomes null, $r = 0$ and $z < 0$. Both z and r vanish at the intersection of the axis and horizon 'cross-over', which is the origin in Minkowski space.

Note that $J'(z,0) = P(z)$ only on the positive z axis. On the negative z axis, the Ernst potential J' is singular, but $P(z)$ is still well behaved. In the terminology of the previous section, the description is adapted to the axis rather than to the horizon.

The two copies of $w = 0$ in R' are the images under $\pi : X \to R'$ of

$$\zeta^0 = r^{-1}(z - \sqrt{z^2 + r^2}), \quad \zeta^1 = r^{-1}(z + \sqrt{z^2 + r^2})$$

(taking the positive square root in both cases). The open sets U_0 and U_1 are copies of the Riemannn sphere, less $w = 0$; and the open sets U_2 and U_3 are copies of the w plane, with U_2 containing the image of ζ^0 and U_3 the image of ζ^1. As $r \to 0$, $\zeta^0 \to 0$ and $\zeta^1 \to \infty$ only for $z > 0$.

To get a description adapted to the horizon, we must represent R' as a union of two spheres in a different way by interchanging the two copies of $w = 0$ between S_0 and S_1. In other words, we put $R' = \hat{S}_0 \cup \hat{S}_1$ where $\hat{S}_0 = U_0 \cup U_3$ and $\hat{S}_1 = U_1 \cup U_2$. The restrictions of E to $\hat{S}_0$ and $\hat{S}_1$ are given by

$$P_{03} = P_{02}P_{23} = \begin{pmatrix} 1 & 0 \\ 0 & 2w \end{pmatrix}, \quad P_{12} = P_{13}P_{23}^{-1} = \begin{pmatrix} 1 & 0 \\ 0 & (2w)^{-1} \end{pmatrix}$$

and the new patching matrix is

$$\hat{P}(w) = P_{32} = P(w)^{-1} = \begin{pmatrix} 2w & 0 \\ 0 & (2w)^{-1} \end{pmatrix}$$

The bundles on $\hat{S}_0$ and $\hat{S}_1$ are in standard form, but now with $p = 0$, $q = 1$. To recover $p = 1$, $q = 0$, we interchange the two Killing vectors.

This is a general phenomenon. If $V' = \{\infty, w_1, ..., w_n\}$, then we have 2^n choices for S_1 by making different assignments of the double points $w_1, ..., w_n$. In each case, the decomposition of $E|_{S_1}$ into $L^{-1} \oplus \mathbb{C}$ picks out a different one-dimensional subspace $L_{\infty_1}^{-1}$ of E_{∞_1}. These one-dimensional spaces contain the Killing vectors which vanish or become null on an axis or horizon somewhere in the maximal analytic extension of the solution.

The other global feature that this flat example shares with more general metrics is the way in which the analytic continuation to the 'interior', where the boost ∂_ψ becomes spacelike, is encoded in the same bundle $E \to R'$ as the exterior.

In the 'exterior', where ∂_ψ is timelike, the two branch points of $\pi : X \to R'$ are at $= z + ir$ and $w = z - ir$, where z and r are real. In the 'interior', r is imaginary and both branch points lie on the positive real axis. Given two such branch points, we can fix the discrete freedom in π by mapping ζ^0 to S_0

and ζ^1 to S_1. The corresponding space-time events then lie outside the light cone of the origin in Minkowski space, but inside the 'interior' region. The events inside the light cone are obtained by mapping ζ^1 to S_0 and ζ^0 to S_1.

The matrix J remains well behaved as one branch point approaches $z = 0$; and, as a consequence of statement (6) in the previous section, the space-time metric remains nonsingular when both branch points approach the origin the w plane. The fact that it is possible to piece together the various portions of the space-time across the light cone of the origin (where one branch point is at $w = 0$) and the horizon (where both branch points coincide), and to include the origin as a regular space-time event, is a general consequence of the fact that P has a pole at $w = 0$ of the form of that in eqn (20) and is not special to this simple geometry, except that the form of the pole does not exclude NUT-like behaviour (see below).

Finally note that P in this example is not holomorphic at $w = \infty$ since J is not asymptotically flat in the sense of the definition in §6 because the boost Killing vector blows up at infinity. It is not clear how to characterise in terms of the twistor construction space-times which are asymptotically flat, but which have Killing vectors which are not tangent to $\mathcal{I}$. It is also not clear how to recognise directly that the patching matrices in these first two examples generate the same space-time, but with different Killing vectors.

8.3   Conical Singularities and Weyl Solutions

It follows from eqn (3) that $f\Omega^2$ is constant over each connected component of a regular symmetry axis or Killing horizon, but it may take different constant values on different components. Such behaviour gives rise to conical singularities in some of the well known Weyl solutions.

In the Weyl solutions, J is diagonal, $\log f$ is an axisymmetric solution of Laplace's equation, and $\hat{f} = -f/r^2$. Penrose [12] discusses the case

$$\log f = -\frac{1}{\sqrt{(r^2 + (z + 1)^2)}} - \frac{1}{\sqrt{r^2 + (z - 1)^2}}$$

(see also [15, p 312]). Here

$$P = \exp\begin{pmatrix} \frac{1}{w+1} + \frac{1}{w-1} & 0 \\ 0 & -\frac{1}{w+1} - \frac{1}{w-1} \end{pmatrix}.$$

The symmetry axis has three components, $z > 0$, $-1 < z < 1$, and $z < -1$, on all of which P is positive definite. The value of $f\Omega^2$ differs on the middle component by a factor of $e^{\frac{1}{2}}$ from its value on the two outer components. To make the space-time non-singular on the axis, we must be able to scale X_1 so that its orbits are periodic with period 2π, which requires that $f\Omega^2 = 1$; so there must be conical singularities on either the inner or outer portions of the axis.

It is not clear how to predict this behaviour directly from the twistor theory, but it appears to be related to the existence of essential singularities in the patching matrix.

8.4 The Kerr Solution

The Kerr solution (with $a < m$) is generated by the patching matrix

$$P = \frac{1}{w^2 - \sigma^2} \begin{pmatrix} (w+m)^2 + a^2 & 2am \\ 2am & (w-m)^2 + a^2 \end{pmatrix}$$

where $\sigma = \sqrt{m^2 - a^2}$. The open set V' is the complement of $\{\infty, w_1, w_2\}$ where $w_1 = \sigma$ and $w_2 = -\sigma$.

In Boyer-Lindquist coordinates, the metric is

$$ds^2 = dt^2 - 2mR\rho^{-2}(dt - a\sin^2\theta d\phi)^2 - (R^2 + a^2)\sin^2\theta d\phi^2 - \rho^2(\Delta^{-1}dR^2 + d\theta^2)$$

where $\Delta = R^2 + a^2 - 2mR$ and $\rho^2 = R^2 + a^2\cos^2\theta$. The Boyer-Lindquist coordinates are related to the Weyl coordinates by $r = \sqrt{\Delta}\sin\theta$, $z = (R - m)\cos\theta$. The Killing vectors are $X_1 = \partial_\phi$ and $X_2 = \partial_t$.

To recover the metric from P, we must first decide how to treat the double points of R' under $\pi : X \to R'$. The two copies of $w_1 = \sigma$ are the images of

$$\zeta_1^0 = r^{-1}\left((z - \sigma) - \sqrt{(z - \sigma)^2 + r^2}\right), \quad \zeta_1^1 = r^{-1}\left((z - \sigma) + \sqrt{(z - \sigma)^2 + r^2}\right)$$

and the two copies of $w_2 = -\sigma$ are the images of

$$\zeta_2^0 = r^{-1}\left((z + \sigma) - \sqrt{(z + \sigma)^2 + r^2}\right), \quad \zeta_2^1 = r^{-1}\left((z + \sigma) + \sqrt{(z + \sigma)^2 + r^2}\right)$$

(taking the positive square roots in all four cases). The various regions of the maximal extension are obtained as follows.

- Outside the black hole, π maps ζ_1^0 and ζ_2^0 to S_0 and has branch points at $w = z \pm ir$, where z and r are real.

- Between the horizons, π has branch points in the interval $[-\sigma, \sigma]$ on the real axis in the w plane. The different ways of mapping ζ_i^0 and ζ_i^1 to w_i ($i = 1, 2$) correspond to different regions separated by the light cones of the events where both Killing vectors vanish (the intersections of the horizon cross-overs with the symmetry axes).

- In the negative mass region beyond the ring singularity, π maps ζ_1^0 and ζ_2^0 to S_1 and has branch points at $w = z \pm ir$, where z and r are real.

There are other possibilities for π which also give real values of J, but they result in metrics with the wrong signature.

The patching matrix has the right singular behaviour at $w = \pm\sigma$ for the metric to be regular at $z = \pm\sigma, r = 0$ (i.e. at the intersections of the axes and horizon cross-overs).

As presented, the description is adapted to the part of the axis on which $\theta = 0$ (since $\zeta_1^0 \to 0$ and $\zeta_2^0 \to 0$ as $r \to 0$ only if $z > \sigma$). As in the flat example, to adapt it to the other part of the axis or to one of the horizons, we must interchange double points of R' between S_0 and S_1, as follows.

- Outer horizon: interchange the two copies of w_1.

- Inner horizon: interchange the two copies of w_2.

- Axis with $\theta = \frac{\pi}{2}$: interchange the copies of both w_1 and w_2.

Consider, for example, the outer horizon. Let $\hat{S}_1 \subset R'$ be the sphere that contains ∞_1, the copy of w_2 on S_1, and the copy of w_1 on S_0. Then $E|_{\hat{S}_1}$ has patching matrices

$$P_{13} = \begin{pmatrix} (2w)^{-1} & 0 \\ 0 & 1 \end{pmatrix}, \quad P_{23} = P$$

relative to the three set open cover $\{U_1, U_2, U_3\}$ of $\hat{S}_1$, where U_1 is a neighbourhood of $w = \infty$, U_2 is a neighbourhood of $w = \sigma$, and $U_3 = \{\infty, \sigma\}^c$. Put

$$P_+ = \frac{1}{w-\sigma} \begin{pmatrix} w + \sigma + m & -\lambda m \\ m\lambda^{-1} & w + \sigma - m \end{pmatrix},$$

$$P_- = \frac{1}{w+\sigma} \begin{pmatrix} w - \sigma + m & \lambda m \\ -m\lambda^{-1} & w - \sigma - m \end{pmatrix}$$

where $\lambda = a/(\sigma + m)$. Then $P = P_+ P_-$. Moreover, P_- is nonsingular on $U_1 \cup U_3$ and P_+ is nonsingular on U_2. Thus if we put

$$Q_1 = \begin{pmatrix} (2w)^{-1} & 0 \\ 0 & 1 \end{pmatrix} P_- \begin{pmatrix} 2w & 0 \\ 0 & 1 \end{pmatrix}, \quad Q_2 = P_+^{-1}, \quad Q_3 = P_-$$

then

$$Q_1 P_{13} Q_3^{-1} = \begin{pmatrix} (2w)^{-1} & 0 \\ 0 & 1 \end{pmatrix}, \quad Q_2 P_{23} Q_3^{-1} = 1.$$

By using Q_1, Q_2, Q_3 to change the trivializations on U_1, U_2, and U_3, therefore, we can reduce $E|_{\hat{S}_1}$ to a standard sum of line bundles $L^{-1} \oplus \mathbf{C}$. The change of trivialization over the neighbourhood of ∞_1 results in a change of basis in E_{∞_1}:

$$(X_1 \ X_2) \mapsto (X_1 \ X_2) Q_1(\infty)^{-1} = (X_1 \ X_2) \begin{pmatrix} 1 & 0 \\ 2m\lambda^{-1} & 1 \end{pmatrix}$$

(recall the identification of the space of Killing vectors with E_{∞_1}). Thus the Killing vectors that become null on the outer horizon are proportional to

$$X_1 + 2m\lambda^{-1}X_2 = \frac{\partial}{\partial\phi} + 2ma^{-1}R_+\frac{\partial}{\partial t},$$

as can be seen directly from the metric (R_+ is the value of R on the outer horizon).

In this example, and in general, picking out the Killing vectors that vanish or become null elsewhere in the space-time involves the solution of a Riemann-Hilbert problem of the form $P = P_+P_-$ around the poles of P.

The effect of interchanging both w_1 and w_2 follows from a more general result. Suppose that $V' = \{\infty, w_1, ..., w_n\}$, $p = 1$, $q = 0$, and that $P(\infty) = 1$. Let $\hat{S}_1 \subset R'$ be the sphere containing ∞_1 and the copies of $w_1, ..., w_n$ originally on S_0. Then $E|_{\hat{S}_1} = L^{-1} \oplus \mathbf{C}$ and a basis $\{\hat{X}_1, \hat{X}_2\}$ for E_{∞_1} such that $\hat{X}_1$ spans $L^{-1}_{\infty_1}$ is given by $(\hat{X}_1\ \hat{X}_2) = (X_1\ X_2)F$ where F^{-1} is the value at $w = \infty$ of

$$\begin{pmatrix} (2w)^{-1} & 0 \\ 0 & 1 \end{pmatrix} P \begin{pmatrix} 2w & 0 \\ 0 & 1 \end{pmatrix}. \tag{21}$$

In the case of the Kerr solution, the off-diagonal entries in P fall off like w^{-2} and $F = 1$. Therefore the Killing vector that vanishes on the part of the axis given by $\theta = \frac{\pi}{2}$ is the same as the one that vanishes at $\theta = 0$. This is not true in the Taub-NUT solution.

We remark finally that it is not easy to see from the twistor description that the Kerr solution is *not* geodesically complete. The events on the ring singularity are given by taking the branch points of $\pi : X \to R'$ at $w = \pm ia$ and treating the double points in the way approprite to the negative mass region. A little algebra shows that $\pi^*(E) = L^{-2} \oplus L^2$. This is not trivial, so the construction of J fails.

8.5 The Schwarzschild and Taub-NUT Solutions

The patching matrices of these two metrics are very similar to that of the Kerr solution. The reasons for their radically different global structures are quite subtle.

The Schwarzschild solution is generated by

$$P = \frac{1}{w^2 - m^2}\begin{pmatrix} (w+m)^2 & 0 \\ 0 & (w-m)^2 \end{pmatrix}$$

(with $p = 1$, $q = 0$). The Taub-NUT solution is generated by

$$P = \frac{1}{w^2 - \sigma^2}\begin{pmatrix} (w+m)^2 + l^2 & 2lw \\ 2lw & (w-m)^2 + l^2 \end{pmatrix},$$

where $\sigma^2 = l^2 + m^2$. Both have the same reduced twistor space as the Kerr solution, and both have patching matrices with simple poles at two points on the real axis.

The Schwarzschild singularity appears when one tries to repeat the analysis of the Kerr solution and adapt the twistor description to the (non-existent) inner horizon by restricting E to the sphere $\hat{S}_1$ containing ∞_1, the copy of $w_1 = m$ originally on S_1 and the copy of $w_2 = -m$ originally on S_0. The result is not $L^{-1} \oplus \mathbb{C}$, but $L^{-2} \oplus L$, which gives rise to a singular space-time metric as $r \to 0$.

In the Taub-NUT solution, the different global structure arises from the fact that the off-diagonal terms in P fall off like w^{-1}, so the matrix F is not the identity, but

$$F = \begin{pmatrix} 1 & 0 \\ -4l & 1 \end{pmatrix}$$

(see eqn(21)). Thus the Killing vector that vanishes on the component $\theta = \frac{\pi}{2}$ of the axis is not X_1, but $X_1 - 4lX_2$. Identifications must be made in the space-time to make the orbits of both X_1 and $X_1 - 4lX_2$ periodic with period 2π and this leads to non-Hausdorff behaviour in the maximal extension (see [10, 7]).

8.6   *ASD Solutions*

If the space-time is ASD, then some combination of X_1 and X_2 (X_2 say) must be ASD (see §4). It follows from the results of Tod and Ward [16] that if the metric admits a positive definite real slice and if $\det J$ is not constant, then

$$J = \begin{pmatrix} r^2 f^{-1} + f\omega^2 & -f\omega \\ -f\omega & f \end{pmatrix}, \quad \Omega^2 = f^{-1} \tag{22}$$

where ω and f are any functions of r and z satisfying

$$\omega_r = -rf^{-2}f_z, \quad \omega_z = rf^{-2}f_r.$$

These imply that $V = f^{-1}$ is an axisymmetric solution of Laplace's equation. In the 'Euclidean Taub-NUT' solution, for example, [6]

$$V = 1 + \frac{2m}{\sqrt{r^2 + z^2}}.$$

If the solution is regular, then the corresponding Ernst potential and patching matrix are

$$J' = \begin{pmatrix} V & 1 \\ 1 & 0 \end{pmatrix}, \quad P(w) = \begin{pmatrix} g(w) & 1 \\ 1 & 0 \end{pmatrix}.$$

where $g(z) = V(z,0)$, so that $g(w)$ is the twistor function of V. Note that $\hat{f} = f/r^2$ in this case.

We remarked in §4 that a solution $J(z,r)$ of the reduced form of Yang's equation can be associated with an ASD Yang-Mills field in *any* ASD space-time with the appropriate Killing vectors. In particular, the J in eqn (22) generates an ASD field in its *own* backgound. The corresponding connection is the Levi-Civita connection acting on the bundle of unprimed spinors ω^A. This is generated by the holomorphic bundle over projective twistor space whose fibre at each point is the tangent space to the surface of constant $\pi_{A'}$, twisted by $\mathcal{O}(1)$ (recall that the twistor space of an ASD vacuum solution is foliated by surfaces of constant $\pi_{A'}$ [11]); see Salamon [14].

Acknowledgement

We thank Richard Ward for some suggestions concerning ASD solutions.

References

[1] Belinsky V A and Zakharov V E (1979) Stationary gravitational solitons with axial symmetry *Sov. Phys.—JETP* **50** 1-9

[2] Cosgrove C M (1980) Relationships between group-theoretic and soliton–theoretic techniques for generating stationary axisymmetric gravitational solutions *J. Math. Phys.* **21** 2417-47

[3] Ernst F J (1968) New formulation of the stationary axially symmetric gravitational field problem *Phys. Rev.* **167** 1175-8

[4] Fletcher J (1990) D.Phil thesis (Oxford)

[5] Hoenselaers C and Dietz W (1984) Solution of Einstein's equations: techniques and results *Lecture Notes in Physics* **205** (Berlin: Springer)

[6] Hawking S W (1977) Gravitational instantons *Phys. Lett.* **A60** 81-3

[7] Hawking S W and Ellis G F R (1973) *The large scale structure of space-time* (Cambridge: Cambridge University Press)

[8] Kramer D, Stephani H, MacCallum M, and Herlt E (1980) *Exact solutions of Einstein's equations* (Cambridge: Cambridge University Press)

[9] Kundt W and Trümper M (1966) Orthogonal decomposition of axisymmetric stationary space-times *Z. Phys.* **192** 419-422

[10] Misner C W (1963) The flatter regions of Newman, Unti and Tamborino's generalised Schwarzschild space *J. Math. Phys.* **4** 924-37

[11] Penrose R (1976) Nonlinear gravitons and curved twistor theory *Gen. Rel. Grav.* **7** 31-52

[12] Penrose R (1986) Hermann Weyl, space-time and conformal geometry. In: *Hermann Weyl 1885-1985* ed. K Chandrasekharan (Berlin: Springer)

[13] Penrose R and Rindler W (1986) *Spinors and space-time. Volume 2: spinor and twistor methods in space-time geometry* (Cambridge: Cambridge University Press)

[14] Salamon S M (1983) Topics in four-dimensional Riemannian geometry. In: *Geometry seminar 'Luigi Bianchi'* ed E Vesentini *Lecture Notes in Mathematics* **1022** (Berlin: Springer)

[15] Synge J L (1960) *Relativity: the general theory* (Amsterdam: North-Holland)

[16] Tod K P and Ward R S (1979) Self-dual metrics with self-dual Killing vectors *Proc. Roy. Soc. Lond.* **A368** 411-27

[17] Ward R S (1977) On self-dual gauge fields *Phys. Lett.* **61A** 81-2

[18] Ward R S (1983) Stationary axisymmetric space-times: a new approach *Gen. Rel. Grav.* **15** 105-9

[19] Witten L (1979) Static axially symmetric solutions of self-dual SU(2) gauge fields in Euclidean four-dimensional space *Phys. Rev.* **D19** 718-20

[20] Woodhouse N M J (1989) Cylindrical gravitational waves *Class. Quant. Grav.* **6** 933-43

[21] Woodhouse N M J and Mason L J (1988) The Geroch group and non-Hausdorff twistor spaces *Nonlinearity* **1** 73-114

[22] Yang C N (1977) Condition of self-duality for SU(2) gauge fields on Euclidean four-dimensional space *Phys. Rev. Lett.* **38** 1377-9

A Two-surface Encoding of Radiative Space-times

C.N. Kozameh   C.J. Cutler   E.T. Newman

1 Introduction

The purpose of this work it to represent an alternative approach or description of the geometry of asymptotically simple space-times [1]. The essential idea behind this approach is the attempt to analyze the gravitational field at any given interior point of a space-time using the imprint left at null infinity by light originating from that point. To be more specific, given a point x^μ and its future (or past) light cone N_x, the light cone cuts of null infinity are defined as $C_x = N_x \cap \mathcal{I}$, where $\mathcal{I}$, null infinity, is the null boundary attached to an asymptotically simple space-time [2]. One then shows that knowledge of the geometry of the cut C_x and the neighboring cuts, for an arbitrary point x^μ, is equivalent to knowledge of the conformal metric at that point [1]. (These cuts have actually been explicitly calculated and the conformal metrics reconstructed in the case of the Schwarzschild, Kerr and charged Kerr metrics [1].)

Though the goals of this formalism have been outlind elsewhere [3] one can say that the main areas of research in the program are: a) kinematical description of the cuts, b) dynamical description of the cuts, and c) quantum description of the cuts.

The main purpose of the present work is to present some new results concerning part a) of the program, though towards the end we will make comments on part b). A new geometrical characterization of the cuts is given and a 'global' coordinate system (or more accurately an atlas) for the space-time, which depends on the cuts, is constructed.

The point of view that we would like to adopt is that null infinity, i.e. $\mathcal{I}$, which has topology $S^2 \times R$, is to be the basic object, with all other objects arising as structures on $\mathcal{I}$. The asymptotically flat space-time manifold would then be *defined* by an appropriately chosen four-dimensional set of cuts, with the local space-time geometry arising from the geometry of these cuts. Local coordinate systems for the space time, are then to be obtained or defined from the cuts themselves. This is what we referred to in the previous paragraph, as the atlas for the space-time.

In §2 we define the cuts as a certain class of local sections on $\mathcal{I}$. We describe the embedding properties of these two-dimensional surfaces using the available geometry of $\mathcal{I}$. We also give an alternative description using either the the cotangent or contact bundle over $\mathcal{I}$. In §3 we discuss several theorems on the behaviour of null cone congruences in order to show that these cuts provide a 'global' coordinate system for the space-time. This 'global' coordinate system is naturally adapted to the discussion of radiation problems. Finally, in §4 we discuss the connection of these results with our programme and outline how the Einstein equations can be encoded in this new point of view.

We will use x^μ to denote coordinates on the four dimensional space-time, x^a as the coordinates on the three-dimensional $\mathcal{I}$, and x^i as coordinates on S^2.

2   Geometry of the Light Cone Cuts

We will consider that we have been given $\mathcal{I}$, with its intrinsic properties (described below). We will then consider cuts of $\mathcal{I}$ intrinsically, i.e. independently of the underlying geometry of the space-time, though we will extract several properties of the cuts from their space-time origin.

The null cone cuts (or cuts for short) are constructed by the intersection of the future light cone of a point x^μ with the null boundary $\mathcal{I}$. Locally, they are two-surfaces of $\mathcal{I}$, though globally they will in general have cusps and self-intersections. The local description of the cuts is given in §2.1. It is also possible to give a global description of the cuts (in §2.2) as an embedded submanifold of the cotangent or contact bundle of $\mathcal{I}$.

Since the cuts are objects defined at $\mathcal{I}$, in order to give their description it is necessary to first review the geometry of null infinity, $\mathcal{I}$ [4, 5]. $\mathcal{I}$ is a 3-dimensional manifold with topology $S^2 \times R$. It comes equipped with a pair (q_{ab}, n^c) where n^c is a vector field whose integral curves foliate $\mathcal{I}$ and whose orbit space is diffeomorphic to S^2, (the orbits themselves being the fibers over S^2 defining the bundle) q_{ab} is a degenerate metric ($q_{ab}n^b = 0$) with signature $(0, +, +)$ which is Lie propagated by n. We consider the q and n to be given up to the equivalence $(q_{ab}, n^c) \sim (\Omega^2 q_{ab}, \Omega^{-1} n^c)$ for any nowhere-vanishing Ω. These fields, (q_{ab}, n^c), provide the kinematic description of $\mathcal{I}$. Any asymptotically flat space-time will yield the same pair (q, n).

Null infinity comes also equipped with a collection of connections D_a (one connection for each radiative space-time) satisfying

$$D_a q_{bc} = 0; \quad D_a n^b = 0 \qquad (1)$$

(equation (1) fails to determine D completely since q_{ab} is degenerate). The connections D_a have a kinematic or universal part (the same for every space-time) and a dynamical part that represents the radiation field of each space-time.

The kinematic part of the connection is obtained by operating D_a on forms w_a such that $w_a n^a = 0$, i.e.

$$D_a w_b =: \partial_{[a} w_{b]} + \mathcal{L}_w q_{ab}, \tag{2}$$

where the first term on the r.h.s. of (2) is the exterior derivative of w and the second term is the Lie derivative of q with respect to $w^a = q^{ab} w_b$. Although q^{ab} is not well defined (since q_{ab} is degenerate, q^{ab} has an ambiguity of the addition of a term $v^a n^b$ for an arbitrary v) one can easily show that the second term in (2) is unambiguous. The universal part of the connection is the pull back of the metric connection of S^2.

The dynamical part of D_a is obtained by operating on those forms l_a such that $l_a n^a = 1$, namely

$$D_a l_b =: S_{ab}.$$

There is however a gauge freedom in this part, so that $D'_a \sim D_a$ if

$$(D'_a - D_a) l_b = f q_{ab}. \tag{3}$$

The radiation fields at $\mathcal{I}$ are represented by the dynamical part of the equivalence class $\{D_a\} =: D_a/\sim$, with $\sim$ given by (3). They can be given freely as the symmetric, trace free tensors

$$\Sigma_{ab} = D_a l_b - \tfrac{1}{2} q_{ab} q^{cd} D_c l_d. \tag{4}$$

With an appropriate choice of l_a, Σ_{ab} becomes, what is known as, the Bondi shear, the free radiation data.

(Note that for simplicity of presentation we have chosen a *particular* conformal factor satisfying the Bondi gauge condition and not used the conformal freedom $(q_{ab} n^c) \sim (\Omega^2 q_{ab}, \Omega^{-1} n^c)$.)

This exhausts the usual geometry assigned to $\mathcal{I}$.

2.1 *Local Geometry of the Cuts.*

As we mentioned earlier, the light cone cuts, or simply 'cuts', are the intersection of the future light cone N_x of a point x^μ, with null infinity, $\mathcal{I}$, i.e. $C_x = N_x \cap \mathcal{I}$. That is, the cuts are defined as the image of the smooth mapping that takes the sphere of null directions at x^μ into $\mathcal{I}$ via the null geodesics. That this mapping is smooth follows from the theorem on the differentiability of the solutions of a differential equation with respect to the initial conditions. (Note that although the mapping is smooth, the image of the mapping, the cut, need not be an immersed submanifold of $\mathcal{I}$; i.e. the rank of the map can be less than two at some points and hence the cut can have 'cusps'. For example, consider the map from $\mathbb{R}$ into $\mathbb{R}^2$ given by $x = t^2$ and $y = t^3$. It is clearly smooth but has rank zero at $t = 0$; the curve in $\mathbb{R}^2$ has a cusp at $x = y = 0$.)

One can locally describe the cuts as a class of (local) sections $Z : S^2 \to \mathcal{I}$. By the construction of C_x, a vector tangent to $\mathcal{I}$ lies in C_x iff it is orthogonal to the null geodesics from x at that point of $\mathcal{I}$. Two independent vectors tangent to a cut are thus given by (in local coordinates $x^a = (u, x^i)$, with u the local fiber coordinate given by $n^a \partial/\partial x^a = \partial/\partial u$)

$$E_i = e_i + \partial Z/\partial x^i \cdot \partial/\partial u, \quad \text{with } e_i = \partial/\partial x^i,$$

(Note that we are using here (u, x^i), for the local coordinates of $\mathcal{I}$. The u is defined (as above) by the tangent vectors to the generators of $\mathcal{I}$; the x^i label the generators. The conventional and very useful choice for these coordinates has been the complex stereographic coordinates, $x^i = (\zeta, \bar{\zeta})$. See §2.2.)

Although the one form $K = K_a dx^a$ which defines a section, (so $K_a E_i^a = 0$) does not determine K uniquely one can obtain a canonical form by demanding

$$K_a n^a = 1. \tag{5}$$

Assuming choice (5) has been made, one can write the form K in local coordinates as

$$K = du - (\partial Z/\partial x^i) \cdot dx^i \tag{6}$$

or

$$K_a = \nabla_a F, \quad F = u - Z(x^i),$$

with $F = 0$ at the cut.

We are now ready to introduce the first and second fundamental forms of the cuts. In the usual situation of a smooth ambient metric space, the cuts describe the embedding properties of a hypersurface [6]. However, since in our particular case the metric of $\mathcal{I}$ is degenerate and the connection has unphysical degrees of freedom, certain objects become trivial and others are not well defined.

In particular, the first fundamental form q (the restriction of the degenerate metric to the cuts) contain no new information. It is defined by

$$q_{ij} =: q(E_i, E_j) = q(e_i, e_j)$$

i.e., it does not depend on the cut since the metric q_{ab} annihilates the component of E_i in the n^a direction.

The non-trivial information is provided by the second fundamental form $I\!I$ whose components $I\!I_{ij}$ are given by

$$I\!I_{ij} =: E_i^a E_j^b D_a K_b. \tag{7}$$

This symmetric tensor, also called extrinsic curvature, gives the local description of the cuts. However, there is an unphysical degree of freedom in this tensor induced by the gauge freedom of the connection D. To see this explicitly, consider the behavior of the mean curvature of the cut (we are extending the definition of mean curvature to our degenerate case)

$$R =: q^{ij} II_{ij} \tag{8}$$

where q^{ij} is the inverse to q_{ij}. Under a gauge transformation of the connection, R transforms as

$$R' = q^{ij} E_i^a E_j^b D'_a K_b = q^{ij} E_i^a E_j^b D_a K_b + f = R + f$$

(R will later play an important role in the construction of the global coordinate system). The geometry of the embedding is thus characterized by the trace-free part of II_{ij}, namely,

$$\sigma_{ij} =: II_{ij} - \tfrac{1}{2} q_{ij} q^{mn} II_{mn}. \tag{9}$$

One can easily check that σ_{ij} is invariant under a gauge transformation of the connection.

(One can show that if the σ_{ij} were known for all the light cone cuts, i.e. as a function of the cuts, then all the conformal information of the space-time could be reconstructed from the σ_{ij}, i.e. *The metric is a functional of σ_{ij}* [1].)

2.2 *Global Description of the Cuts.*

In Minkowski space, the light cone cuts of null infinity have a very simple form. Adopting a standard coordinate system $x^\mu = (x, y, z, t)$ for Minkowski space and Bondi coordinates $x^a = (u, \zeta, \bar\zeta)$ for $\mathcal{I}$, with $\zeta, \bar\zeta$, stereographic coordinates on the sphere, the cuts are described or given by

$$u = Z(x^\mu, \zeta, \bar\zeta) = x^\mu l_\mu(\zeta\bar\zeta)$$

with

$$l_\mu(\zeta, \bar\zeta) = \frac{1}{2\sqrt{2}P}(1 + \zeta\bar\zeta, \zeta + \bar\zeta, i(\zeta - \bar\zeta), -1 + \zeta\bar\zeta); \quad P = (1 + \zeta\bar\zeta),$$

i.e. each cut is described by a function on the sphere which is a linear combination of the $l = 0$ and $l = 1$ spherical harmonics. The Minkowski space cuts are thus global sections of $\mathcal{I}$.

Unfortunately, for a generic space-time, the cuts will not define global sections of $\mathcal{I}$ above S^2. It is clear from linearized theory or e.g. the Schwarzschild solution (with an interior solution) that, due to the presence of Weyl curvature, null cones will, generically, develop caustics and self-intersections. Thus, the cuts associated with these cones will also self-intersect and certain portions of each cut will lie to the 'future' of other portions.

It is therefore useful to regard the cuts as embedded submanifolds in a larger space, namely $T^*(\mathcal{I})$, the cotangent bundle of $\mathcal{I}$ restricted to forms w_a normalized so that $w_a n^a = 1$. More accurately this larger space is the five dimensional contact bundle, $C(\mathcal{I})$ with local coordinates (u, x^i, w_i) and contact form, $w = du + w_i dx^i$. (See (6)). Now construct the natural bundle $C(\mathcal{I}) \to \mathcal{I} \to S^2$. Then we can conclude that cuts are smooth, embedded submanifolds of $C(\mathcal{I})$, such that the projection of each cut into the base space is all of S^2.

In the following we discuss the relationship between a *four-parameter* family of global cuts of $C(\mathcal{I})$ and the conformal geometry of an asymptotically simple space-time.

This relationship is given by the following: *Given two cuts constructed from two neighboring interior points x^μ and x'^μ of an asymptotically simple space-time, the two points are null separated if and only if the cuts are tangent to each other along some generator of $\mathcal{I}$ or equivalently they have a point in common in $C(\mathcal{I})$.*

This is clear, since if the two points are null separated, they lie along a null geodesic, and at the intersection of the geodesic with $\mathcal{I}$ the cuts for the two points are both orthogonal to the incoming null direction, and hence are tangent to each other. This point of $\mathcal{I}$, say (u, x^i), and the common tangency given by the gradient of the mapping defining either cut, namely $\partial_i Z$, determine a point of $C(\mathcal{I})$ with coordinates $(u, x^i, w_i = \partial_i Z)$. The converse follows by essentially reversing the above argument.

We thus have the major result that knowledge of the light cone cuts of $\mathcal{I}$ is equivalent to knowledge of the conformal geometry of the space-time.

Summarizing our results: An appropriate four parameter family of local cuts of $\mathcal{I}$ or four parameter family of embedded submanifolds of $C(\mathcal{I})$, defines a space-time; the class of cuts that are tangent to each other at $\mathcal{I}$ (or have a point in common on $C(\mathcal{I})$) define null geodesics as well as points that are null separated and thus yield a conformal structure on the space-time.

We have seen the means of picking out null geodesics directly from the geometry of the cuts or sections of $C(\mathcal{I})$. We now show that it is also possible to choose *individual points* on the geodesics by introducing a smooth, monotonic parameter to label each point on the null geodesic. This parameter is found from the following considerations: Choose two points x^μ and x'^μ on a common null geodesic that intersects $\mathcal{I}$ at (u, x^i). Consider now the behaviour of the light cone cuts $Z(x^\mu)$ and $Z'(x'^\mu)$ in the neighborhood of (u, x^i). At (u, x^i), Z and Z' have the same values of u and $w_i = \partial_i Z$. However, the mean curvatures R and R' of the cuts are different (see §2.1). Moreover, in the next section, we will show that R is a monotonic function of the affine length defined along the null geodesic. Thus, it can be used to label each point on that geodesic. Although R is not gauge invariant, the difference $R - R'$ is gauge invariant.

Notice that, since in the six-dimensional bundle $C(\mathcal{I}) \times R$ over S^2, the cuts never intersect each other, one can label all the cuts by simply choosing any one particular fiber $F_x = (u, w_i, R)$ above any point x^i on S^2. *Each fiber yields a coordinate system for some region of the space-time.* The coordinates (u, w_i, R), *for fixed x^i*, have the following meaning: for a fixed u, the values w_i and R give the past light cone from the point (u, x^i) at $\mathcal{I}$. The value w_i labels a null geodesic on that cone and R a particular point on that null geodesic. Note that the coordinates (u, w_i, R) are the generalization to curved space-time of the null-plane coordinates of flat space associated with the generator x^i of $\mathcal{I}$.

In the next section we well show that the S^2 family of coordinates $\theta^\mu(x^i) =:$ (u, w_i, R) can be used as 'global' coordinates for the space-time in the following sense. Given a point x^μ in the interior of the space-time, one can always find a neighborhood U of that point, and find a value of x^i (i.e. a fiber over S^2) for which $\theta^\mu(x^i)$ is a good coordinate system on U.

3 An Atlas for the Space-Time

We now show that the family of coordinate systems $\theta^\mu(x^i)$ forms an atlas for an asymptotically simple and empty space-time. (See [2] for the definition of asymptotically simple and empty. We also fully expect the results of this paper to hold under less restrictive matter fall-off conditions, such as those satisfied by smooth Maxwell fields at $\mathcal{I}$.) The idea is to show that given any point x^μ on the space-time, one can always find some generator x^i, of $\mathcal{I}$, such that $\theta^\mu(x^i)$ is a (good) coordinate system in a neighborhood of x^μ.

The main result of this section can thus be stated as

Theorem 3.1 *The coordinate systems $\theta^\mu(x^i)$ parametrized by the S^2 set of generators of $\mathcal{I}$, i.e. x^i, form an atlas for an asymptotically simple and empty space-time.*

We outline the basic idea in the proof of the above contention, before giving the details. We will first show that for any point x^μ, in an asymptotically flat space-time, there will be a null geodesic from x^μ that gets to $\mathcal{I}$ (at some point, say p) before reaching a conjugate point to x^μ. It will follow that going backwards along the same geodesic from that point p on $\mathcal{I}$, there will also be no conjugate points between p and x^μ. The local coordinates $\theta^\mu(x^i)$, of the point x^μ, are then the value u at p, the w_i labelling the geodesic through x^μ and the value of R labelling points on the geodesic. $\theta^\mu(x^i)$ is then extended to a neighborhood of x^μ.

We first state some results concerning the behavior of light cones from arbitrary points of an aymptotically simple and empty space-time.

Lemma 3.2 *Let (M, g_{ab}) be an asymptotically simple and empty space-time and let x^μ be a point in M. Then there exists a future-directed null geodesic λ from x^μ such that λ has no point conjugate to x^μ, up to and including $\mathcal{I}^+$.*

Proof We will assume the contrary and arrive at a contradiction. Thus we assume that every future-directed null geodesic from x^μ contains a point conjugate to x^μ before or at $\mathcal{I}^+$. First consider $\dot{J}^+(x^\mu, M)$, the boundary (in M) of the set of points in M that can be reached from x^μ by future-directed causal curves in M. Since (M, g_{ab}) is asymptotically simple and empty, $\dot{J}^+(x^\mu, M)$ is generated by null geodesics that have x^μ as their past endpoint [3, Theorem 6.9.1]. Now define $\overline{M} = M \cup \mathcal{I}^+$ and consider $\dot{J}^+(x^\mu, \overline{M})$, the boundary (in $\overline{M}$) of the causal future of x^μ in $\overline{M}$. One easily shows that any point in $\dot{J}^+(x^\mu, \overline{M})$ is the limit of a sequence of points in $\dot{J}^+(x^\mu, M)$, and since the space of null directions from x^μ is compact, this implies that $\dot{J}^+(x^\mu, \overline{M})$ is also generated by null geodesics that have x^μ as their past endpoint. Also, these null generators must exit $\dot{J}^+(x^\mu, \overline{M})$ before reaching, or at, a point conjugate to x^μ, since if a conjugate point exists between two points of a null geodesic, that null geodesic can be deformed to a timelike curve connecting the points [2, Proposition (4.5.12)] and the second point thus cannot be on the boundary. Since our assumption is that all geodesics from x^μ contain points conjugate to x^μ at or before reaching $\mathcal{I}^+$, this implies that $\dot{J}^+(x^\mu, \overline{M}) \cap \mathcal{I}^+$ is composed of points y^μ in $\mathcal{I}^+$ such that x^μ and y^μ are connected by a null geodesic and x^μ and y^μ are conjugate.

However, [2, Theorem (6.9.3)] asserts that $\dot{J}^+(x^\mu, \overline{M})$ intersects every generator of $\mathcal{I}^+$ once. We now show that this contradicts the properties of the set $\dot{J}^+(x^\mu, \overline{M})$ that we have shown above. Consider the map, $\mathcal{S}: S_x^2 \to S^2$ that relates the future-directed null directions, S_x^2, at x^μ to the null generators of $\mathcal{I}$, (points on the S^2) via the intersection of the associated null geodesics (from x^μ) with those generators. Since $\mathcal{S}$ is a smooth map, those null directions at x^μ whose corresponding geodesic curves intersect $\mathcal{I}^+$ at a conjugate point are singular points of the map, by the meaning of 'conjugate point'. Now if $\dot{J}^+(x^\mu, \overline{M})$ were to intersect every generator of $\mathcal{I}^+$, then the union of all the images of the singular points of $\mathcal{S}$ would be all of S^2. But by Sard's Theorem, the union of the images of the singular points of $\mathcal{S}$ must be a measure zero on S^2, and thus we have our contradiction. $\qquad\square$

Corollary 3.3 *Let λ be a null geodesic from x^μ, with tangent l^μ at x^μ, satisfying the property of Lemma 3.2. Then there exists some neighborhood of (x^μ, l^μ) in the bundle of null vectors over (M, g_{ab}) satisfying the properties of Lemma 3.2.*

Proof Solutions of the Jacobi equation depend smoothly on their initial data and on the curve along which they are integrated. Since along λ solutions of the Jacobi equation, that vanish at x^μ, do not go to zero before reaching, or at $\mathcal{I}^+$, there is some neighborhood of the curve and of the initial data for which solutions of the Jacobi equation have the same property. $\qquad\square$

Lemma 3.4 *Let $\rho(s_1, s)$ be the (unphysical) divergence (more accurately, $-\frac{1}{2}\nabla_\mu l^\mu$) of the light cone with apex at s_1, evaluated at a point s along a null geodesic λ with s_1 and s affine parameters along λ. If the point s_1 has no conjugate point up to s (inclusive), then*

$$d\rho(s_1, s)/ds_1 < 0.$$

The proof is given in Appendix B.

Lemma 3.5 *Given a null geodesic λ, and a point x^μ such that x^μ has no future conjugate point along λ, then points y^μ, to the future of x^μ on λ, also have no future conjugate points along λ.*

The proof follows immediately from Lemma 3.4.

Lemma 3.6 *Given two points, x^μ and y^μ, in the space-time joined by a null geodesic λ, if no point conjugate to x^μ lies between x^μ and y^μ, then no point conjugate to y^μ lies between x^μ and y^μ. This applies, in particular, when y^μ becomes a point of $\mathcal{I}$.*

The proof follows from the Sachs-Penrose theorem (which relates the properties of a future light-cone congruence to a past one), see Appendix A, and Lemma 3.5.

Finally, we show that the function R defined at $\mathcal{I}$, as the mean curvature of the cut, is well behaved as a function of s, the value of the affine parameter of the apex of the cone associated with the cut. More specifically we have

Lemma 3.7 *On a null geodesic from $\mathcal{I}$, until a conjugate point is reached, dR/ds is strictly less than zero, where s is the affine parameter.*

Proof We show (see Appendix C) that $R(s) = \rho(s, \mathcal{I})$ where $\rho(s, \mathcal{I})$ is the (unphysical) divergence of the light cone congruence, with apex at s, evaluated at $\mathcal{I}$. Therefore, it follows from the corollary to Lemma 3.4 that R is a strictly decreasing function of s and can thus be used to label points on the geodesic λ. $\square$

Proof of Theorem 3.1 Lemmas 3.2, 3.4, 3.5 and 3.6 guarantee the existance of a congruence of null geodesics from *some* generator x^i of $\mathcal{I}$, to a neighborhood of any point x^μ in the space-time such that no points conjugate to points on the generator exist between the generator and the neighborhood. Therefore, u and w_i, from their definitions, uniquely label the null geodesics through the neighborhood of x^μ. Finally, Lemmas 3.4 and 3.7 show that R is a good coordinate to label points along these null geodesics.

3.1 An Application

As an application of this formalism we will display an important class of solutions of the (null) geodesic deviation equation in terms of the $\theta^\mu(x^i)$ coordinates. The associated coordinate forms and vector basis are $d\theta^\mu$ and $\partial/\partial\theta^\mu$.

The geodesic deviation equation [8, 3] can be written as

$$D^2 M^{(i)} = \mathcal{R}(l, M^{(i)}, l)$$

where l^μ is the null vector tangent to λ, $\mathcal{R}$ is the Riemann tensor, and $M^{(i)}$, for $i = 1, 2$, are two linearly independent vectors chosen so that they are proportional to E_i at $\mathcal{I}$.

In terms of our coordinate basis, the solutions $M^{(i)}(s_1, s)$, i.e. the deviation at s, corresponding to a light cone congruence with apex at s_1, can be written (up to terms proportional to l^μ) as [3]

$$M^{(i)}(s_1, s) = [R(s_1) - R(s)]\partial/\partial w_i + g^{ik}[\sigma_{kj}(s_1) - \sigma_{kj}(s)]\partial/\partial w_j \tag{10}$$

where $R(s_1)$ and $\sigma_{kj}(s_1)$ $[R(s)$ and $\sigma_{kj}(s)]$ are the curvatures of the light cone cuts (see §2.1) coming from an apex at $s_1[s]$.

As we can see from (10) the non-trivial part of $M^{(i)}$ is given by $\sigma_{kj}(s)$ and represents a measure of the gravitational field along the null geodesic λ.

In Minkowski space, $\sigma_{kj}(s) = 0$ and the two deviation vectors read

$$M^{(i)} = (R - R')\partial/\partial w_i = (s_1 - s)\partial/\partial w_1.$$

4 Conclusion

The main point we have been trying to emphasize is that given a space-time, then the points of that space-time can be reinterpreted as a four-parameter set of (light-cone) cuts of $\mathcal{I}$ and the conformal geometry at these points can be reconstructed by analyzing the geometry of the cuts. The analysis is done *completely* at $\mathcal{I}$, a point to which we will shortly return.

These observations suggest another point of view or approach to general relativity where instead of beginning with a space-time manifold and its usual structures, we consider $\mathcal{I}$ and some particular privileged four-parameter family of cuts to be the primary objects, from which the space-time manifold and metric will appear as derived concepts. We now briefly outline this construction.

We begin with our basic structure, a two dimensional sphere S^2 (the sphere of null directions) with coordinates x^i and a line bundle, which is to represent $\mathcal{I}$, above it, with fiber coordinate u. On $\mathcal{I}$, cuts or privileged sections $u = Z(x^i)$ are to be obtained as a solutions of the differential equation [1]

$$\nabla_i \nabla_j Z - \tfrac{1}{2} g_{ij} \nabla^2 Z = \Lambda_{ij}(Z, \nabla_i Z, \nabla^2 Z, x^i) \tag{11}$$

where ∇_i is the covariant derivative on the sphere, and Λ_{ij} are two functions of the six variables, $(Z, \nabla_i Z, \nabla^2 Z, x^i)$ assumed to be given (but which must in the future be determined) and which satisfy the integrability conditions of (11). Note that from the Sachs theorem on the transformation of asymptotic shear, it can be shown that Λ_{ij} has the simple geometric meaning of being the difference between the Bondi shear Σ_{ij} and the shear of the cut, $u = Z$, namely σ_{ij}, i.e. from (4) and (9),

$$\Lambda_{ij} = \Sigma_{ij} - \sigma_{ij}. \tag{12}$$

If one is given arbitrarily the values $Z, \nabla_i Z, \nabla^2 Z$, at a given point x_0^i, one can obtain (using (11)) all the coefficients in the Taylor series expansion of the solution $Z(x^i)$. Thus, the solution space of (11) is four dimensional and one can thus write the solution as $Z(x^\mu, x^i)$ where x^μ are the four parameters identifying solutions and x^i are points on the sphere. The space-time manifold is to be identified with this solution space.

The next step is to construct, on the solution space x^μ, a (conformal) metric $g_{\mu\nu}$ as a functional of Z and its derivatives, i.e., to obtain $g_{\mu\nu}\{Z\}$. This construction, which is based on the assumption that $\partial Z/\partial x^\mu$ is a null vector of the solution space, has been carried out in detail [1]. Roughly, the basic idea for the construction is as follows: if $u = Z(x^\mu, \zeta, \bar\zeta)$ is the solution of (11), representing the intersection of the light-cone from x^μ with $\mathcal{I}$, then it is easy to see that holding $u, \zeta, \bar\zeta$ fixed and allowing x^μ to vary, defines a characteristic surface (the past light-cone of $u, \zeta, \bar\zeta$). Taking the gradient of that surface at x^μ yields a null covector. If $\zeta, \bar\zeta$ range over the sphere, the gradient ranges over the light-cone of x^μ, thus defining the conformal metric.

It must be pointed out that an arbitray choice of

$$\Lambda_{ij}(Z, \nabla_i Z, \nabla^2 Z, x^i)$$

will in general not yield a Lorentzian metric. To obtain such a metric one has to impose certain restrictions or 'compatibility conditions' on the allowed Λ_{ij} [1, 3].

Finally, the idea is to further restrict the class of allowable Λ_{ij} so that the metric of the solution space satisfies the conformal Einstein equations. Aside from one important special case, (briefly discussed below), which has been solved, it is the formulation of this restriction which has not yet been satisfactorily accomplished. However, there appear to be several promising approaches to the problem [3, 11] which also will be briefly discussed below.

1. The problem of vacuum anti-self-dual (self-dual) space-times has been solved [9, 10] in this context. If we consider equations (11) and (12) and select one of the two components (say the 11 component) and take $\sigma_{11} = 0$ (the other component σ_{22} is determined by the integrability condition), we obtain what has been referred to as the 'good cut equation'.

In 'standard' notation it becomes

$$\eth^2 Z = \sigma^0(Z, \zeta, \bar{\zeta}) \qquad (13)$$

where $\eth$ and σ^0 are, respectively the edth operator and the Bondi shear. Solutions of (13), which are regular functions on the $(\zeta, \bar{\zeta})$ sphere, form a four-parameter set, $Z(x^\mu, \zeta, \bar{\zeta})$, which determine a vacuum anti-self-dual metric.

2. One approach [11] to finding equations for the Λ_{ij} which are (conformally) equivalent to the Einstein vacuum equations is to consider two conformally related metrics $g_{\mu\nu}$ and $\tilde{g}_{\mu\nu} = \Omega^2 g_{\mu\nu}$. If the Ricci tensor of $g_{\mu\nu}$ vanishes, then the Ricci tensor of $\tilde{g}_{\mu\nu}$ is expressed as a function of Ω and its derivatives. If these equations are *considered* as equations for the determination of an Ω to make the Ricci tensor of $g_{\mu\nu}$ vanish, they *must* satisfy integrability conditions. These conditions, which are expressed completely in terms of Λ_{ij} and are quite involved and difficult to analyse, are our conformal Einstein equations for the Λ_{ij}.

3. Although the above approach is conceptually simpler, an alternative approach [11] based on the use of the (infinitesimal) holonomy operator, appears to be in practice considerably simpler. This second approach uses as the basic variables the cut function Z and the Λ_{ij} as well as the pair of holonomy operators $H^\pm$. The $H^\pm$ are constructed by parallel propagation around two closed null curves. Each closed curve is obtained by choosing a field point x^μ as the 'starting point' and taking two neighboring null geodesics from x^μ to $\mathcal{I}$ and finally connecting them there to form the closed curve. The difference between the two curves is that near $\mathcal{I}$, the two-flats formed by the pairs of geodesics and their deviation vectors are respectively self-dual and anti-self-dual. (A major technical complication is that, due to Weyl curvature, these two-flats do not remain self-dual and anti-self-dual as the geodesics 'return' to the field point x^μ.) One can construct equations [11] that relate the different variables Z, Λ, and $H^\pm$, which though complicated, appear to be quite amenable to a perturbation calculation. A model for these ideas to be tested on, is the reformulation of vacuum Maxwell theory on an asymptotically simple space-time, in terms of Z and the Maxwell holonomy $H^\pm$, where many similar problems arise. This has a clear and pretty solution [12].

A   Appendix

Geodesic deviation applied to null geodesics, is a particularly useful tool to relate properties of null infinity, $\mathcal{I}$, to the interior local geometry of a space-time and is basic in our proof of Lemmas 3.4, 3.5 and 3.6. The geodesic deviation equation (for null geodesics) can be written as [3, 8]

$$D^2 X = -QX \tag{14}$$

where D is the derivative with respect to the affine length defined along a null geodesic λ, i.e. $D = l^\mu \nabla_\mu$, X is a two by two matrix

$$X = \begin{pmatrix} \xi & \bar{\eta} \\ \eta & \bar{\xi} \end{pmatrix}$$

describing a complex deviation vector M, with ξ and η the components of the deviation vector along a pair of normalized parallel propagated basis vectors m_μ and $\bar{m}_\mu$ (see e.g. [8]) and Q represents the Weyl and Ricci tensor components given by

$$Q = \begin{pmatrix} \Phi & \Psi \\ \bar{\Psi} & \Phi \end{pmatrix}$$

with $\Phi = R_{\mu\nu} l^\mu l^\nu$, $\Psi = C_{\mu\nu\alpha\beta} l^\mu m^\nu l^\alpha m^\beta$.

Often, one wants to relate two different deviation vectors (or congruences) defined along the same null geodesic λ. Let X_1 and X_2 be two null congruences defined along λ. A very useful result [7], that is easily checked via (14), is that

Lemma A.1 *The function* C *defined by*

$$C = X_1^* D X_2 - D X_1^* X_2 \tag{15}$$

is constant along λ. *(* * *means adjoint).*

An important application of this lemma is the Sachs-Penrose reciprocity theorem. Assume that X_1 is a deviation vector associated with a null cone congruence with apex at a point p_1 and X_2 another deviation vector associated with the null cone congruence with apex at p_2. Then using the constancy of C, the fact that X_1 and X_2 vanish at p_1 and p_2 respectively, and that one can (right) rescale any X so that DX_1 and DX_2 equal I at p_1 and p_2 respectively, we have that [7]

Theorem A.2 (Sachs-Penrose) *The quantity* X_1 *at the point* p_2 *is related to* X_2 *at* p_1 *according to*

$$X_1(\text{at } p_2) = -X_2(\text{at } p_1).$$

Corollary A.3 *From this it follows that if* X_1 *is a non-singular matrix (caustic free) at* p_2, *so is* X_2 *at* p_1.

This is used in proving Lemma 3.6.

B   Appendix

We prove Lemma 3.4.

One can easily show, by integrating (15) that if $X = X_1$ is a solution of (14), then, (with $Y = X_2$)

$$Y = X(s)[A + \int^s (X^*X)^{-1}ds']B$$

is the general solution, with A and B constant matrices.

In particular, given some deviation vector X, a light cone congruence Y with apex at $s = s_1$ can be written as

$$Y(s,s_1) = X(s) \int_{s_1}^s (X^*X)^{-1}ds' = X(s)\mathcal{B}(s,s_1). \tag{16}$$

We have put $B = I$ using the right rescaling freedom.

We now use (16) to prove Lemma 3.4. The optical parameters $\rho(s,s_1)$ and $\sigma(s,s_1)$ are related to Y via

$$P = \begin{pmatrix} \rho & \sigma \\ \bar{\sigma} & \rho \end{pmatrix} = -DY \cdot Y^{-1}. \tag{17}$$

We use (16) to analyze the behaviour of $P(s,s_1)$ at a definite point s (in particular at $\mathcal{I}$) when the apex s_1 is varied. Substituting (16) into (17), we obtain

$$P(s,s_1) = P_0(s) - (X\mathcal{B}X^*)^{-1} \tag{18}$$

with $P_0 = -DX \cdot X^{-1}$. Taking the derivative of (18) with respect to the variable s_1, we obtain after a simple calculation an expression of the form

$$D_1 P = -SS^*.$$

This implies that the diagonal terms on the right side are positive and hence we have

$$D_1\rho(s,s_1) < 0,$$

which proves Lemma 3.4.

C   Appendix

We describe the relationship of the (non-physical) divergence, ρ, of the cone N_x, to the trace of the second fundamental form of the associated light-cone cut $C_x = N_x \cap \mathcal{I}$.

Let m^μ be a parallel propagated null vector tangent to the null cone and tangent to the cut at $\mathcal{I}$. The divergence of the generating vectors of the cone, ρ, is defined by

$$\rho = m^\mu \bar{m}^\nu \nabla_\mu F_\nu, \tag{19}$$

where the null cone is described by $F(x^\mu) = 0$ and $F_\mu = \nabla_\mu F$. In the neighborhood of $\mathcal{I}$, $F(\Omega, u, x^i)$ can be expressed as

$$F = F^0(u, x^i) + \Omega F^1(u, x^i) + O(\Omega^2) \tag{20}$$

where $F^0 = 0$ describes the null cone cut. From (20) follows

$$\nabla_\mu F \equiv F_\mu = F^0_\mu + n_\mu F^1 + \Omega F^1_\mu,$$

and

$$\nabla_\mu F_\nu = \nabla_\mu F^0_\nu + \nabla_\mu n_\nu F^1 + 2n_{(\nu} F^1_{\mu)} + \Omega \nabla_\mu F^1_\nu \tag{21}$$

with $n_\mu = \nabla_\mu \Omega$. It follows from (21) that at $\mathcal{I}$

$$\nabla_\mu F_\nu = \nabla_\mu F^0_\nu + 2n_{(\mu} F^1_{\nu)},$$

where we have used a Bondi conformal factor Ω such that $\nabla_\mu n_\nu = 0$ at $\mathcal{I}$. Finally, evaluating (19) at $\mathcal{I}$ one obtains (see (7) and (8))

$$\rho = m^\mu \bar{m}^\nu \nabla_\mu F^0_\nu = -R.$$

This result shows that one can always choose a conformal frame, i.e. a choice of conformal factor, such that the divergence ρ, in that frame, equals the trace of the second fundamental form of the cut. Since ρ is an a monotonic function of the affine length s, (Lemma 3.4) so is R and thus we have Lemma 3.7.

Acknowledgements

We are pleased to acknowledge enlightening conversations with L. Mason, J. Frauendiener and O. Ruella. We also thank the anonymous referee for several valuable comments.

In addition we acknowledge and thank the NSF for support under grant #PHY-8803073.

References

[1] Kozameh, C.N. & Newman, E.T., J. Math. Phys **24** (1983), 2481.

[2] Hawking, S.W. & Ellis, G.F.R., *The Large Scale Structure of Space-Time*, Cambridge University Press (1973).

[3] Kozameh, C.N., & Newman, E.T., In *Topological Properties and Global Structure of Space-Time* (eds Bergmann, P. & de Sabbatta, V.), Plenum Press (1986).

[4] Geroch, R., In *Asymptotic Structure of Space-Time* (eds Esposite, P. & Witten, L.), Plenum Press (1977).

[5] Ashtekar, A., J. Math. Phys. **22** (1981), 2885.

[6] O'Neil, B. *Elementary Differential Geometry*, Academic Press (1966).

[7] Penrose, R., In *Perspective in Geometry and Relativity* (ed. Hoffman, B.), Indiana University Press, Bloomington (1966).

[8] Penrose, R. & Rindler, W., *Spinors and Space-Time*, Cambridge University Press, Cambridge (1984).

[9] Ko, M., Newman, E.T. & Tod, K.P., *The theory of H-Space*, Phys. Rep. **71** (1981), 53.

[10] Hansen, R.O., Newman, E.T., Penrose, R. & Tod, K.P., Proc. R. Soc. Lond. **A363** (1978), 445.

[11] Kozameh, C.N. & Newman, E.T., In *Asymptotic Behaviour of Mass and Spacetime Geometry* (ed. Flaherty, F.J.), Lecture Notes in Physics **202**, Springer-Verlag, New York, (1984).
Kozameh, C.N. & Newman, E.T., J. Math. Phys. **31** (1985), 801.

[12] Kozameh, C.N. & Newman, E.T., *Maxwell fields on Asymptotically Simple Space-Times*, in preparation.

Twistors, Massless Fields and the Penrose Transform

T.N. Bailey   M.A. Singer

1 Introduction and Notation

The aim of this review is twofold. Firstly we wish to discuss some of the advances in the twistor theory of massless fields that have happened since the original cohomological treatment by Eastwood, Penrose and Wells [20], particularly in the areas of non-analytic fields and their relation to quantum field theory, and fields with sources. Another advance is that representation theory methods have been introduced which give a particularly effective means of computing isomorphisms between cohomology and solutions of field equations. These methods have been studied in considerable generality [9] and the second aim of this review is to present the particular cases of interest to twistor physics in sufficient detail to enable non-representation theorists to perform the computations.

We will give only the barest outline of the background here, as it is our intention to take the original cohomological treatment of twistor theory by Eastwood, Penrose and Wells [20] as our starting point. Twistor theory originated as a reformulation of relativistic physics which, it was hoped, would make directions for progress in physics more apparent. From a mathematical point of view, the fundamental observation is that if one takes Minkowski space and complexifies and compactifies it, then it can be identified as the Grassmannian $M = \mathrm{Gr}_2(T)$, i.e. the space of (complex) two dimensional subspaces of T, a four (complex) dimensional vector space. If a volume form is chosen on T, then the group $SL(4, \mathbb{C})$ acts as the group of conformal symmetries on M—as a result, the twistor formalism is particularly well suited to conformally invariant problems.

One of the principal attractions of this viewpoint was the fact that solutions of various conformally invariant equations such as the conformally invariant wave equation and Maxwell's equations could be generated by means of contour integrals of homogeneous holomorphic functions on certain regions in T—see e.g. [31]. Questions which naturally arise are firstly what does this mean mathematically, and secondly does one get all solutions this way.

The mathematical picture presented by Eastwood, Penrose and Wells [20] is as follows. Let P be the projective space of T and let F denote the space of $(1,2)$-flags in T—that is the parameter space of '1-dimensional subspaces inside 2-dimensional subspaces of T'. It turns out that F can also be identified as the total space of the projectivised bundle of (primed) spinors on M. Then one has a natural double fibration

$$(1)$$

where the maps are the obvious 'forgetful' ones.

The interpretation of the contour integral formulae is that there are certain isomorphisms one can prove between sheaf cohomology groups on certain regions on P and the kernels or cokernels of certain conformally invariant differential operators on corresponding regions of M.

In §2, we present an up-to-date version of this transform which is more general and easier to compute with than the original. The rest of the review concerns work that attempts to put into this frame-work certain problems that the original version did not consider. In §3, we consider the problem of describing fields on real (rather than complex) Minkowski space. The appropriate notion here is that of hyperfunctions, which are a class of generalised functions on a real manifold given as boundary values of holomorphic functions from its complexification. The corresponding picture in twistor space is that the choice of a preferred 'real (compactified) Minkowski space' $M \subset \mathsf{M}$ is equivalent to the choice of a pseudo-Hermitian form Φ of signature $++--$ on T, the 'null cone' of which defines a (real) codimension 1 submanifold $P \subset \mathsf{P}$. Hyperfunction fields on M then correspond to relative cohomology classes of the pair $(\mathsf{P}, \mathsf{P} \setminus P)$—or roughly speaking, boundary values of cohomology classes on P. Also in §2, we consider how the various structures one needs for quantum field theory appear in this representation of fields as cohomology classes.

In §3, we consider the problem of describing fields with sources on a world-line. It turns out that such fields are actually defined on a certain double cover of a region in M and that the appropriate twistor space is a non-Hausdorff manifold. Again however a cohomological interpretation of these fields is possible. This is an example of the general problem of using double fibrations such as (1) in cases where the fibres of the map μ restricted to the relevant region have non-trivial topology, and we discuss this general problem here also.

Our conventions generally follow [38, 39, 20], in particular in the use of abstract indices. We use the same notation for a vector bundle and its sheaf of sections. We use M, P, etc. for complex, compactified Minkowski space and projective twistor space. When we consider real Minkowski space and the corresponding (5 real dimensional) null twistors we denote them by M, P, and so on.

We have decided for simplicity throughout to assume that 'primed' and 'unprimed' conformal weights [20] have been identified. Equivalently, we have a preferred volume form on non-projective twistor space T which identifies $\mathcal{O}(-4)$ with the holomorphic 3-forms Ω^3 on P.

The authors would like to express their gratitude to M.G. Eastwood.

2 The Penrose Transform

2.1 *Introduction*

In this section we will discuss the improvements and extensions to the Penrose transform that have occurred since the original treatment in [20], in so far as they apply to the case of 4-dimensional space-time. For simplicity, we consider only fields on regions of complex Minkowski space.

The Penrose transform is a machine for proving isomorphisms between the cohomology of the natural vector bundles on P and kernels or cokernels of conformally invariant differential operators on M. We start with the double fibration (1) and work locally with some (suitably convex) region $U \subset \mathsf{M}$, $\nu^{-1}(U) = U' \subset \mathsf{F}$ and $\mu(U') = U'' \subset \mathsf{P}$.

We now outline the method of the transform. Let $V \to \mathsf{P}$ be some homogeneous vector bundle (i.e. some vector bundle of tensors on P, tensored with one of the line bundles $\mathcal{O}(n)$ of homogeneous functions). We assume until we explicitly state the contrary that the region U is 'sufficiently convex' for there to be an isomorphism

$$H^*(U'', V) = H^*(U', \mu^{-1}(V)) \tag{2}$$

where $\mu^{-1}V$ denotes the inverse image sheaf [12]. This part of the transform is called the *pull-back mechanism* and will be discussed further in §4.2. We are, as we remarked in the introduction, using the same notation for a vector bundle and its sheaf of sections.

We construct a resolution

$$0 \longrightarrow \mu^{-1}(V) \longrightarrow \mathcal{R}^\bullet, \tag{3}$$

of $\mu^{-1}(V)$ by vector bundles, where the maps $\mathcal{R}^i \to \mathcal{R}^{i+1}$ are differential operators. The cohomology groups $H^*(U', \mu^{-1}(V))$ can then be computed in terms of the $H^*(U', \mathcal{R}^\bullet)$ by means of a spectral sequence or (what is essentially the same thing) breaking up (3) into a number of short exact sequences and using the resulting long exact cohomology sequences (see the discussion in [20]). The cohomology groups $H^*(U', \mathcal{R}^\bullet)$ can be interpreted as sections of conformally weighted spinor sheaves over $U \subset \mathsf{M}$ and the differential operators connecting them are conformally invariant parts of powers of the Levi-Civita connection. Thus we arrive at an interpretation of $H^*(U'', V)$ in terms of spinor fields on U.

The main improvements in the methods we will discuss over those in [20] is the use of the 'relative Bernstein–Gelfand–Gelfand (BGG) resolution' [11, 35, 9] for $\mathcal{R}^\bullet$, rather than the relative de Rham resolution. One would have the latter in any double fibration, but the special structure of the fibres for the ones we consider (see §2.4.2) means that one has the former also, and since the BGG resolutions turn out to be 'reduced' versions of the de Rham resolutions, it is more efficient to use them. These methods also work immediately for the non-linear graviton and curved ambitwistor spaces—and so we shall present them in this general form.

2.2   *The Penrose Transform for Curved Ambitwistor Space*

We will consider the Penrose transform first for the simpler case of the space of null geodesics (curved ambitwistor space) $\mathcal{A}$, following Eastwood [18]. The total space of the bundle of null directions $\mathcal{G}$ over a complex conformal spin 4-manifold $\mathcal{M}$ has fibre the Cartesian product of the fibres of the projective spin bundles $\mathsf{P}(\mathcal{O}_A) \times \mathsf{P}(\mathcal{O}_{A'})$ and there is a double fibration

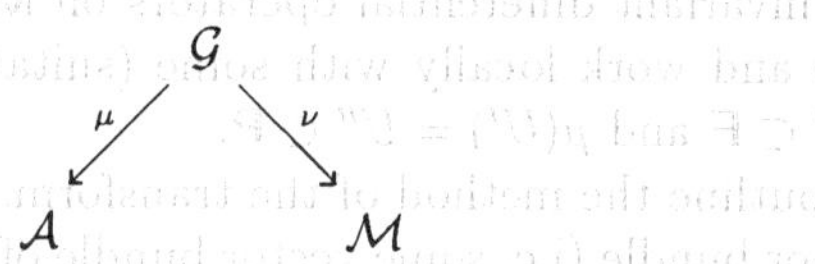

We will work with some local 'sufficiently convex' region $U \subset \mathcal{M}$ and corresponding regions $\nu^{-1}(U) = U' \subset \mathcal{G}$ and $\mu(U') = U'' \subset \mathcal{A}$.

There are natural line bundles $\mathcal{O}(p,r)[q]$ on $\mathcal{G}$, whose sections have homogeneity p, r in homogeneous co-ordinates $[\pi_{A'}], [\eta_A]$ on the fibre of $\mathcal{G} \to \mathcal{M}$ respectively and have conformal weight q (thought of as pulled back from $\mathcal{M}$). We have natural tautological objects $\pi^{A'} \in \Gamma(\mathcal{G}, \nu^*\mathcal{O}^{A'} \otimes \mathcal{O}(1,0)[-1])$ and $\eta^A \in \Gamma(\mathcal{G}, \nu^*\mathcal{O}^A \otimes \mathcal{O}(0,1)[-1])$.

We now choose a conformal scale on $U \subset \mathcal{M}$. The lift of the resulting Levi-Civita connection acts in the horizontal distribution on $\mathcal{G}$ and in particular, the differential operator $\nabla = \eta^A \pi^{A'} \nabla_{AA'}$ acts along the (1-dimensional) fibres of $\mathcal{G} \to \mathcal{A}$ and has weight $(1,1)[-2]$ (in the sense that it maps $\mathcal{O}(p,r)[q]$ to $\mathcal{O}(p+1, r+1)[q-2]$).

Under the conformal rescaling $g_{ab} \mapsto \Omega^2 g_{ab}$ the operator ∇ changes according to $\tilde{\nabla} = \nabla + q\Upsilon$ when acting on $\mathcal{O}(p,r)[q]$, where $\Upsilon = \Omega^{-1}\eta^A\pi^{A'}\nabla_{AA'}\Omega$. Also, corresponding to each choice of conformal scale we have a section Φ of $\mathcal{O}(2,2)[-4]$ which encodes the trace-free part of the Ricci tensor according to $\Phi = -\frac{1}{2}\eta^A\eta^B\pi^{A'}\pi^{B'}R_{AA'BB'}$. This changes under rescaling according to $\tilde{\Phi} = \Phi - \nabla\Upsilon + \Upsilon^2$. It is worth remarking that these formulae for conformal rescalings are much simpler than the corresponding ones on $\mathcal{M}$ [38] because the indices are eliminated—this is why the space $\mathcal{G}$ is so useful for studying conformal invariance.

Table 1: Information for the Penrose transform from ambitwistor space

<table>
<tr><td colspan="2" align="center">Resolution</td></tr>
<tr><td colspan="2">

$$0 \longrightarrow \mu^{-1}\mathcal{O}(p,q,r) \longrightarrow \mathcal{O}(p,r)[q]$$
$$\xrightarrow{D^{q+1}} \mathcal{O}(p+q+1,r+q+1)[-q-2] \longrightarrow 0$$

</td></tr>
<tr><td colspan="2" align="center">Differential operators</td></tr>
<tr><td colspan="2">

$$D^1 = \nabla$$
$$D^2 = \nabla^2 + \Phi$$
$$D^3 = \nabla^3 + 4\Phi\nabla + 2(\nabla\Phi)$$
$$D^4 = \nabla^4 + 10\Phi\nabla^2 + 10(\nabla\Phi)\nabla + 3(\nabla^2\Phi + 3\Phi^2)$$

</td></tr>
<tr><td colspan="2" align="center">Direct images</td></tr>
<tr><td>

$p \geq 0, r \geq 0;$

</td><td>

$\nu_*^0 \mathcal{O}(p,r)[q] = \mathcal{O}^{\overbrace{(A'\cdots L')}^{p}\overbrace{(A\cdots L)}^{r}}[q]$

</td></tr>
<tr><td>

$p \leq -2, r \geq 0;$

</td><td>

$\nu_*^1 \mathcal{O}(p,r)[q] = \mathcal{O}^{\overbrace{(A'\cdots L')}^{-p-2}\overbrace{(A\cdots L)}^{r}}[p+q+1]$

</td></tr>
<tr><td>

$p \geq 0, r \leq -2;$

</td><td>

$\nu_*^1 \mathcal{O}(p,r)[q] = \mathcal{O}^{\overbrace{(A'\cdots L')}^{p}\overbrace{(A\cdots L)}^{-r-2}}[r+q+1]$

</td></tr>
<tr><td>

$p \leq -2, r \leq -2;$

</td><td>

$\nu_*^2 \mathcal{O}(p,r)[q] = \mathcal{O}^{\overbrace{(A'\cdots L')}^{-p-2}\overbrace{(A\cdots L)}^{-r-2}}[p+q+r+2]$

</td></tr>
<tr><td colspan="2" align="center">All others vanish</td></tr>
</table>

For each $q > 0$ there is a natural conformally invariant order $(q+1)$
differential operator acting on the bundle $\mathcal{O}(p,r)[q]$ on $\mathcal{G}$ obtained by adding
lower order 'correction terms' involving Φ and its derivatives to ∇^{q+1} (we will
explain some of the geometry behind this in §2.4.2). It is easy to verify that
$D^1 = \nabla$ and $D^2 = \nabla^2 + \Phi$ and a proof of existence and algorithm for their
construction can be found in [25]. Several more are given in table 1 where all
the essential information for computing the transform has been gathered.

We use these operators to define conformally invariant vector bundles on
$\mathcal{A}$. To be precise, the rank $(q+1)$ vector bundle $\mathcal{O}(p,q,r) \twoheadrightarrow \mathcal{A}$ is defined by

$$\mu^{-1}\mathcal{O}(p,q,r) = \mathrm{Ker}\left\{ \mathcal{O}(p,r)[q] \xrightarrow{D^{q+1}} \mathcal{O}(p+q+1,r+q+1)[-q-2] \right\}$$

so that on $\mathcal{G}$ we have as the resolution $\mathcal{R}^\bullet$ of $\mu^{-1}\mathcal{O}(p,q,r)$ the short exact
sequence

$$0 \longrightarrow \mu^{-1}\mathcal{O}(p,q,r) \longrightarrow \mathcal{O}(p,r)[q]$$
$$\xrightarrow{D^{q+1}} \mathcal{O}(p+q+1,r+q+1)[-q-2] \longrightarrow 0 \tag{4}$$

which is repeated in table 1 for reference.

Given the isomorphism (recall (2))

$$H^*(U'',\mathcal{O}(p,q,r)) = H^*(U',\mu^{-1}\mathcal{O}(p,q,r))$$

we can compute $H^*(U'', \mathcal{O}(p,q,r))$ by using the long exact cohomology sequence of (4). To make the passage to $\mathcal{M}$, we need to interpret the cohomology groups $H^*(U', \mathcal{O}(p,r)[q])$ on U. The *n-th direct image sheaves,* $\nu_*^n \mathcal{O}(p,r)[q]$ over $\mathcal{M}$ are defined by

$$\Gamma(U, \nu_*^n \mathcal{O}(p,r)[q]) = H^n(U', \mathcal{O}(p,r)[q])$$

and it is easy to compute them (using Serre duality on $\mathbf{CP}^1$)—the non-vanishing ones are given in table 1. The induced operators between the direct images on $\mathcal{M}$ can be computed, but for small values of q can be guessed. The question of how the bundles $\mathcal{O}(p,q,r)$ are related to the structure of $\mathcal{A}$ will be addressed in §2.2.2.

2.2.1   Examples

We will now calculate two typical examples—several more can be found in [18]. The first is to compute $H^1(U'', \mathcal{O}(-4,0,0))$, where the resolution on $\mathcal{G}$ is

$$0 \longrightarrow \mu^{-1}\mathcal{O}(-4,0,0) \longrightarrow \mathcal{O}(-4,0)[0] \xrightarrow{D'} \mathcal{O}(-3,1)[-2] \longrightarrow 0.$$

and from table 1 the sheaves in the resolution have only first direct images. The only non-trivial part of the long exact sequence is thus

$$0 \longrightarrow H^1(U', \mu^{-1}\mathcal{O}(-4,0,0)) \longrightarrow \Gamma(U, \mathcal{O}^{(A'B')}[-3])$$

$$\xrightarrow{D'} \Gamma(U, \mathcal{O}^{AA'}[-4]) \longrightarrow H^2(U', \mu^{-1}\mathcal{O}(-4,0,0)) \longrightarrow 0$$

and so, rearranging the indices, $H^1(U'', \mathcal{O}(-4,0,0))$ is the kernel of a first order differential operator from $\mathcal{O}_{(A'B')}[-1]$ to $\mathcal{O}_{AA'}[-2]$. We guess this to be the massless field operator since there is nothing else it could be—but one could check explicitly.

The second example is to compute $H^0(U'', \mathcal{O}(0,1,0))$. The resolution in this case is

$$0 \longrightarrow \mu^{-1}\mathcal{O}(0,1,0) \longrightarrow \mathcal{O}(0,0)[1] \xrightarrow{D^2} \mathcal{O}(2,2)[-4] \longrightarrow 0$$

and thus we immediately see that $H^0(U'', \mathcal{O}(0,1,0))$ is isomorphic to the kernel of a second order conformally invariant differential operator from $\mathcal{O}[1]$ to $\mathcal{O}^{(AB)}_{(A'B')}[-2]$ on U. Given the form of D^2, we can guess this operator and it gives us the isomorphism

$$H^0(U'', \mathcal{O}(0,1,0)) = \left\{ \sigma \in \Gamma(U, \mathcal{O}[1]) : (\nabla^{(A}_{(A'} \nabla^{B)}_{B')} + \Phi^{AB}_{A'B'})\sigma = 0 \right\}.$$

If we conformally rescale so that $\sigma = 1$ we see that in that metric $\Phi^{AB}_{A'B'} = 0$, and so nowhere vanishing sections of $\mathcal{O}(0,1,0)$ correspond to Einstein metrics in the conformal class. In fact, $\mathcal{O}(0,1,0)$ is LeBrun's *Einstein bundle* ([33], discussed from this point of view in [18, 8].)

2.2.2   Identifying the Bundles $\mathcal{O}(p,q,r) \to \mathcal{A}$

We now turn to the problem of relating these bundles $\mathcal{O}(p,q,r)$, which we have essentially defined in terms of the transform, to the structure of $\mathcal{A}$. The reader who is more interested in the Penrose transform from $\mathcal{A}$ as a prototype for the transform from the non-linear graviton may want to skip this section.

The intrinsic local geometry of curved ambitwistor space is that of a 5-dimensional contact manifold—which means that it comes equipped with a rank 4 sub-bundle D of the tangent bundle that is in a well defined sense maximally non-integrable (see e. g. [1]). There is a short exact sequence of vector bundles

$$0 \longrightarrow D \longrightarrow T(\mathcal{A}) \xrightarrow{\ \alpha\ } \mathcal{O}(1,0,1) \longrightarrow 0,$$

where the map α takes a tangent to $\mathcal{A}$ at a given null geodesic, specified in the form of a Lie dragged connecting vector $q^{AA'}$ to a nearby null geodesic, to $\pi_{A'}\eta_A q^{AA'}$, which is constant along the geodesic and hence defines a point in the fibre of $\mathcal{O}(1,0,1)$. In the flat case, where ambitwistor space A is embedded as $Z^\alpha W_\alpha = 0$ in $\mathbf{P} \times \mathbf{P}^*$, the bundles $\mathcal{O}(p,0,r)$ are just the restrictions of the bundles whose sections are functions homogeneous of degree p,r in twistors and dual twistors respectively.

In general, let γ be a null geodesic in $\mathcal{M}$. The fibre of D over the point in $\mathcal{A}$ corresponding to γ consists of solutions to the Jacobi equation for Lie-dragged connecting vectors to nearby abreast null geodesics (see [39, §7.2] for this and what follows). This equation reduces to the matrix equation

$$\nabla^2 \begin{pmatrix} \zeta \\ \tilde{\zeta} \end{pmatrix} = - \begin{pmatrix} \Phi & \Psi \\ \tilde{\Psi} & \Phi \end{pmatrix} \begin{pmatrix} \zeta \\ \tilde{\zeta} \end{pmatrix}$$

where ζ and $\tilde{\zeta}$ are sections of $\mathcal{O}(1,-1)[1]$ and $\mathcal{O}(-1,1)[1]$ respectively and the Weyl curvature enters in the form of the quantities $\Psi = \eta^A \eta^B \eta^C \eta^D \Psi_{ABCD}$ and $\tilde{\Psi} = \pi^{A'}\pi^{B'}\pi^{C'}\pi^{D'}\tilde{\Psi}_{A'B'C'D'}$. We note that for the purposes of the Jacobi equation we are thinking of γ embedded in $\mathcal{M}$, but for the purposes of the bundles we think of it embedded in $\mathcal{G}$—this confusion is often quite helpful.

In the flat case where $\Psi = 0 = \tilde{\Psi}$, we see this system decouples to give us

$$D = \mathcal{O}(1,1,-1) \oplus \mathcal{O}(-1,1,1)$$

but in the general case, the bundles on the right-hand side are not directly related to D and so the bundles we have defined, and of which we can therefore take the Penrose transform, are not tied in an elementary way to the contact structure of $\mathcal{A}$.

A slightly different perspective on this problem is obtained by considering the bundle $\mathcal{O}(0,1,0) \to \mathcal{A}$, which as we observed in §2.2.1 is the Einstein bundle. An ambitwistor space together with a nowhere vanishing section of this bundle thus give a solution of the (complex) Einstein equations with cosmological constant, in much the same way as a non-linear graviton together with a non-vanishing section of $\Omega^1(2)$ give a half-flat solution. This does not however constitute a solution of the full Einstein equations in the sense that the non-linear graviton solves the half-flat equations precisely because it is not known how to characterise $\mathcal{O}(0,1,0)$ in terms of the local geometry of $\mathcal{A}$ (i.e. without first constructing the space-time). Some progress on this problem can be found in [7, 34].

Given the 'homogeneous function' bundles $\mathcal{O}(p,0,r)$ and the Einstein bundle $\mathcal{O}(0,1,0)$, the general bundle $\mathcal{O}(p,q,r)$ can be built up from the formulae

$$\begin{aligned}
\mathcal{O}(p,q,r) \otimes \mathcal{O}(s,0,t) &= \mathcal{O}(p+s,q,r+t) \\
\odot^n \mathcal{O}(0,1,0) &= \mathcal{O}(0,n,0) \\
\Lambda^2 \mathcal{O}(0,1,0) &= \mathcal{O}(1,0,1).
\end{aligned}$$

The first of these observations is trivial, the second follows from the fact (which can be explicitly verified for small n) that the product of n solutions of $D^2\sigma = 0$ is a solution of $D^{n+1}\sigma = 0$, and the third follows from the easily checked fact that if $D^2\sigma = 0 = D^2\tau$ then the quantity $\mu = \sigma\nabla\tau - \tau\nabla\sigma$ is a section of $\mathcal{O}(1,1)[0]$ and obeys $\nabla\mu = 0$ and hence defines a point in the fibre of $\mathcal{O}(1,0,1)$.

2.3 The Penrose Transform for the Non-linear Graviton

The situation for the non-linear graviton is very similar to the above, but somewhat more complicated, the main difference being that the resolutions are one sheaf longer thus making the use of a spectral sequence convenient. Our treatment essentially follows [4].

We start with a half conformally flat complex spin 4-manifold $\mathcal{M}$. The total space of the projective primed spin bundle $\mathcal{F}$ then fibres over $\mathcal{M}$ and over the twistor space $\mathcal{P}$ (a complex 3-manifold) which parametrises the α-surfaces:

$$\begin{array}{ccc}
 & \mathcal{F} & \\
{}^{\mu}\swarrow & & \searrow^{\nu} \\
\mathcal{P} & & \mathcal{M}
\end{array}$$

We will work locally with sets $U \Subset \mathcal{M}$, $U' \Subset \mathcal{F}$, $U'' \Subset \mathcal{P}$ as usual.

On $\mathcal{F}$ we have natural vector bundles $\mathcal{O}^{(A\cdots L)}(p)[q]$ of conformally weighted spinor fields (pulled back from $\mathcal{M}$), homogeneous of degree p in homogeneous fibre co-ordinates $[\pi_{A'}]$. We also have the tautological object $\pi^{A'}$ which is a section of $\nu^* \mathcal{O}^{A'} \otimes \mathcal{O}(1)[-1]$.

After a choice of conformal scale on U, we have the (lifted) operator $\nabla_A = \pi^{A'}\nabla_{AA'}$ on $U' \Subset \mathcal{F}$ with weight $(1)[-1]$, acting tangentially to the fibres of $\mathcal{F} \to \mathcal{P}$. The trace-free Ricci curvature is encoded as a section $\Phi_{AB} = -\frac{1}{2}\pi^{A'}\pi^{B'}R_{AA'BB'}$ of $\mathcal{O}_{(AB)}(2)[-2]$ on U'.

For $q, r \geq 0$, the operator

$$\mathcal{O}^{\overbrace{(A\cdots D)}^{r}}(p)[q] \longrightarrow \mathcal{O}^{\overbrace{(A\cdots DE\cdots L)}^{r+q+1}}(p+q+1)[-q-2]$$
$$\phi^{A\cdots D} \longmapsto \nabla^{(E}\cdots\nabla^{L}\phi^{A\cdots D)}$$

has a modification by lower order terms involving Φ_{AB} and its derivatives which is conformally invariant. It is the zeroth direct image of the operator D^{q+1} on $\mathcal{G}$ of §2.2 under the natural fibration $\mathcal{G} \to \mathcal{F}$. We will risk confusion by writing D^{q+1} for this operator on $\mathcal{F}$ also.

Just as for the ambitwistor case we may define vector bundles $\mathcal{O}(p,q,r) \mapsto \mathcal{P}$ for $q, r \geq 0$ by

$$\mu^{-1}\mathcal{O}(p,q,r) = \mathrm{Ker}\left\{\mathcal{O}^{\overbrace{(A\cdots D)}^{r}}(p)[q] \xrightarrow{D^{q+1}} \mathcal{O}^{\overbrace{(A\cdots DE\cdots L)}^{r+q+1}}(p+q+1)[-q-2]\right\}$$

and the 'relative BGG' resolution $\mathcal{R}^{\bullet}$ of $\mu^{-1}\mathcal{O}(p,q,r)$ is

$$0 \to \mu^{-1}\mathcal{O}(p,q,r) \to \mathcal{O}^{\overbrace{(A\cdots D)}^{r}}(p)[q] \xrightarrow{D^{q+1}} \mathcal{O}^{\overbrace{(A\cdots DE\cdots L)}^{r+q+1}}(p+q+1)[-q-2]$$
$$\xrightarrow{E^{r+1}} \mathcal{O}^{\overbrace{(F\cdots L)}^{q}}(p+q+r+2)[-q-r-3] \longrightarrow 0 \tag{5}$$

where the order $r+1$ differential operator E^{r+1} is a conformally invariant modification by lower order terms of

$$\phi^{A\cdots L} \longmapsto \nabla_{A}\cdots\nabla_{E}\phi^{A\cdots L},$$

which can be constructed as the first direct image of D^{r+1} under $\mathcal{G} \to \mathcal{F}$. The sequence (5) is repeated in table 2 where we have gathered all the useful calculational information from this section.

The operators D^{n}, E^{n} are essentially the same in the sense that they are given by the same formulae but for D^{n} the indices must be raised and symmetrised into any existing indices whereas for E^{n} the indices are contracted into those of the operand. The first few are given in table 2—but they can immediately be deduced from the expressions for D^{n} in table 1. The direct images of the bundles $\mathcal{O}^{(A\cdots D)}(p)[q]$ on $\mathcal{M}$ are easy to calculate and are also listed in table 2.

We now have all the data needed in order to use a spectral sequence to compute $H^{*}(U'', \mathcal{O}(p,q,r)) = H^{*}(U', \mu^{-1}\mathcal{O}(p,q,r))$ in terms of spinor fields on U. We will illustrate this by example in the next section.

Table 2: Information for the Penrose transform from a non-linear graviton.

Resolution (initial zero omitted)		
$\mu^{-1}\mathcal{O}(p,q,r) \to \mathcal{O}^{\overbrace{(A\cdots D)}^{r}}(p)[q] \overset{D^{q+1}}{\to} \mathcal{O}^{\overbrace{(A\cdots DE\cdots L)}^{r+q+1}}(p+q+1)[-q-2]$		
$\overset{E^{r+1}}{\longrightarrow} \mathcal{O}^{\overbrace{F\cdots L}^{q}}(p+q+r+2)[-q-r-3] \longrightarrow 0.$		
Differential operators		
$\begin{aligned} D^1 &= E^1 = \nabla_A \\ D^2 &= E^2 = \nabla_{(A}\nabla_{B)} + \Phi_{AB} \\ D^3 &= E^3 = \nabla_{(A}\nabla_B\nabla_{C)} + 4\Phi_{(AB}\nabla_{C)} + 2(\nabla_{(A}\Phi_{BC)}) \\ D^4 &= E^4 = \nabla_{(A}\nabla_B\nabla_C\nabla_{D)} + 10\Phi_{(AB}\nabla_C\nabla_{D)} + 10(\nabla_{(A}\Phi_{BC)})\nabla_{D)} \\ & \qquad\qquad +3(\nabla_{(A}\nabla_B\Phi_{CD)} + 3\Phi_{(AB}\Phi_{CD)}) \end{aligned}$		
Direct images		
$p \geq 0 \quad \nu_*^0\mathcal{O}^{(A\cdots D)}(p)[q] = \mathcal{O}^{\overbrace{(A'\cdots L')}^{p}(A\cdots D)}[q]$		
$p \leq -2 \quad \nu_*^1\mathcal{O}^{(A\cdots D)}(p)[q] = \mathcal{O}^{\overbrace{(A'\cdots L')}^{p-2}(A\cdots D)}[p+q+1]$		
All others vanish		

2.3.1   Examples

We can compute the cohomology of $\mu^{-1}\mathcal{O}(p,q,r)$ using the spectral sequence

$$E_1^{p,q} = \Gamma(U, \nu_*^q(\mathcal{R}^p))) \Rightarrow H^{p+q}(U', \mu^{-1}\mathcal{O}(p,q,r)).$$

To deal with difficult cases, some understanding of the mechanism of spectral sequences is required (see e.g. [27]), but in simple cases one can get by without, as we hope the examples will demonstrate.

We will first do an example which shows all the features of the general case, that of computing $H^1(U'', \mathcal{O}(-2,1,0))$—we shall show in §2.3.2 that $\mathcal{O}(-2,1,0)$ is actually the bundle of holomorphic 1-forms on $\mathcal{P}$. This calculation was originally done in [17]. The resolution (omitting the resolved sheaf and the zeros) in this case is

$$\mathcal{O}(-2)[1] \overset{D^2}{\longrightarrow} \mathcal{O}^{(AB)}(0)[-3] \overset{E^1}{\longrightarrow} \mathcal{O}^A(1)[-4].$$

One now writes the $\nu_*^q\mathcal{R}^p$ in an array with p running horizontally. The maps induced from the resolution also run horizontally:

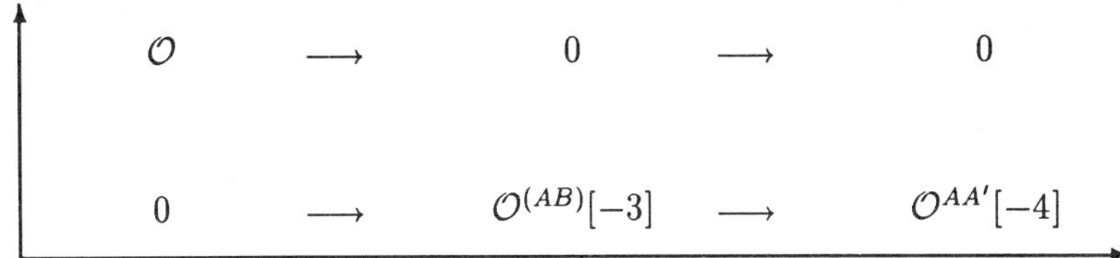

The next step is to take the cohomology at each stage (assuming zeros at the points off the diagram). This leaves the '$\mathcal{O}$' at the $(1,2)$ position unchanged. The map from $(2,1)$ to $(3,1)$ is induced from E^1 and hence a first order conformally invariant operator. Adjusting the indices, it must be the zero rest-mass operator $\mathcal{O}_{(AB)}[-1] \to \mathcal{O}_{AA'}[-2]$, or to put it another way, the part of the exterior derivative that takes the anti-self-dual 2-forms Ω^2_- to Ω^3. Given that U is sufficiently 'nice', this map is onto the closed 3-forms and therefore its image can be identified with $\Omega^4 = \mathcal{O}[-4]$. Thus on taking cohomology, the $(2,1)$ position becomes the anti-self-dual Maxwell fields and the $(3,1)$ position becomes $\mathcal{O}[-4]$. One gets induced maps at this stage running from (p,q) to $(p+2, q-1)$, the only one of which that can be non-trivial in any example is shown

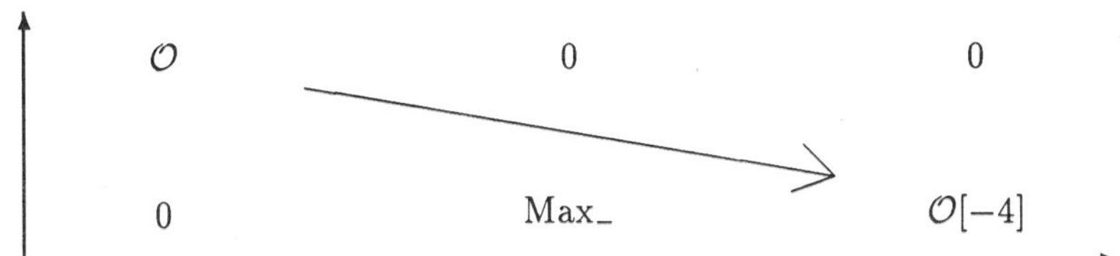

One now takes cohomology again, and gets induced maps from (p,q) to $(p+3, q-2)$ which *must* be trivial—the spectral sequence is said to have converged. What one finds now on the diagonal $p+q = n$ is a composition series for the n-th cohomology of the resolved sheaf—in this case $H^n(U'', \mathcal{O}(-2,1,0))$.

In our example, the $(2,1)$ entry is unchanged since no non-trivial map passes through it. The arrow from $\mathcal{O}$ to $\mathcal{O}[-4]$ turns out after some work to be a conformally invariant modification (using lower order correction terms involving the Ricci curvature) of the composition of the wave operator with itself which we shall write as $\square^2$ [22] (see also §2.4.1). Thus the kernel of this operator and Max_ give a composition series for $H^1(U'', \mathcal{O}(-2,1,0))$. Equivalently, there is an exact sequence

$$0 \longrightarrow \mathrm{Max}_-(U) \longrightarrow H^1(U'', \mathcal{O}(-2,1,0)) \longrightarrow \Gamma(U, \mathrm{Ker}(\square^2)) \longrightarrow 0. \qquad (6)$$

We note that in general we expect the induced operator from $(1,2)$ to $(3,1)$ to have order $q+r+2$. In this case however the order is increased by one because of our identification of $\mathrm{Im}(\Omega^2_- \to \Omega^3)$ with Ω^4 which involves a differentiation.

A similar calculation for $\mathcal{O}(-3,0,1)$ which is, as we shall show below, the bundle of holomorphic 2-forms, shows that its first cohomology is given by fields $\phi_{AA'}$ satisfying $\nabla^{A'}_{(A}\phi_{B)B'} = 0$ and a further condition which is a modification by lower order terms of $\Box\nabla^{AA'}\phi_{AA'} = 0$. This condition is a conformally invariant partial gauge fixing for the potential $\phi_{AA'}$ of a right-handed Maxwell field (see [22] where this is described with no reference to its twistorial origins). The gauge freedom in $\phi_{AA'}$ is the addition of $\nabla_{AA'}f$ where f is a solution of (conformally invariant) $\Box^2 f = 0$. We will comment further on this in §4.3.

Things are usually much easier than this. For example in the case of $\mathcal{O}(-4,0,0)$, the resolution omitting the zero and the resolved sheaf, becomes

$$\mathcal{O}(-4)[0] \xrightarrow{D^1} \mathcal{O}^A(-3)[-2] \xrightarrow{E^1} \mathcal{O}(-2)[-3].$$

The first stage of the spectral sequence thus gives

$$\begin{array}{ccccc} \mathcal{O}^{(A'B')}[-3] & \longrightarrow & \mathcal{O}^{AA'}[-4] & \longrightarrow & \mathcal{O}[-4] \\[2em] 0 & \longrightarrow & 0 & \longrightarrow & 0 \end{array}$$

After taking cohomology, the sequence will have converged, so we see that $H^1(U'', \mathcal{O}(-4,0,0))$ is the kernel of a first order operator from $\mathcal{O}_{(A'B')}[-1]$ to $\mathcal{O}_{AA'}[-2]$—clearly the zero rest-mass operator as expected.

One now takes cohomology again, and gets induced maps from (p,q) to $(p+2,q-2)$ which must be $\mathcal{P}$ ... [illegible]

2.3.2 Identifying the Bundles $\mathcal{O}(p,q,r) \to \mathcal{P}$

In the case of the non-linear graviton, the bundles $\mathcal{O}(p,q,r)$ turn out to precisely all the irreducible parts of tensor powers of the tangent and cotangent bundles as we will now explain. The information is summarised in two tables for reference.

We begin by observing that $\mathcal{O}(p,0,0)$ is just the bundle of 'homogeneous functions' usually denoted by $\mathcal{O}(p)$. The tangent bundle $T(\mathcal{P})$ and the bundles of n-forms $\Omega^n(\mathcal{P})$ are given by

Forms and tangent bundle		
$T(\mathcal{P})$	$=$	$\mathcal{O}(1,0,1)$
$\Omega^1(\mathcal{P})$	$=$	$\mathcal{O}(-2,1,0)$
$\Omega^2(\mathcal{P})$	$=$	$\mathcal{O}(-3,0,1)$
$\Omega^3(\mathcal{P})$	$=$	$\mathcal{O}(-4,0,0)$
$\mathcal{O}(n)$	$=$	$\mathcal{O}(n,0,0)$

To see the first of these, note that a point in the fibre of $\mathcal{O}(1,0,1)$ corresponds to a solution of $\nabla^{(A}\omega^{B)} = 0$ on (the appropriate leaf of) $\mathcal{F} \to \mathcal{P}$, and hence to a solution of $\pi^{A'}\nabla_{A'}^{(A}\omega^{B)} = 0$ on the corresponding α-surface in $\mathcal{M}$. If we think of $\pi_{A'}$ as a parallely propagated spinor defining the α-surface, and choose a spinor $\mu^{A'}$ satisfying $\pi_{A'}\mu^{A'} = 1$, then this equation is exactly the condition for the vector $\omega^A\mu^{A'}$ (defined modulo tangents to the α-surface) to be a connecting vector to a neighbouring α-surface, and hence to define a tangent vector to $\mathcal{P}$.

We now identify the dual of $\mathcal{O}(1,0,1)$ as $\mathcal{O}(-2,1,0)$ by noting that an element of the fibre of the latter is a solution on (the appropriate leaf of) $\mathcal{G}$ of $(\nabla_{(A}\nabla_{B)} + \Phi_{AB})\sigma = 0$. The quantity $2\omega^A\nabla_A\sigma - \sigma\nabla_A\omega^A$ is a section of $\mathcal{O}(0,0,0)$ which some manipulation shows to be constant. This pairing gives the required duality.

Similarly, if σ, τ both represent points in the fibre of $\mathcal{O}(-2,1,0)$, we can form $\beta^A = \sigma\nabla^A\tau - \tau\nabla^A\sigma$, which obeys the correct equation to represent a point in the fibre of $\mathcal{O}(-3,0,1)$, and thus identifies this bundle with Ω^2.

Ad hoc arguments of this sort usually suffice to find one's way around the more elementary bundles. More generally, we have the relationships

Tensor products of basic bundles		
$\mathcal{O}(p,q,r) \otimes \mathcal{O}(s,0,0)$	$=$	$\mathcal{O}(p+s,q,r)$
$\odot^n\mathcal{O}(0,1,0)$	$=$	$\mathcal{O}(0,n,0)$
$\odot^n\mathcal{O}(0,0,1)$	$=$	$\mathcal{O}(0,0,n)$

We have thus identified $\mathcal{O}(n,0,n)$ as $\odot^n T(\mathcal{P})$ and $\mathcal{O}(-2m,m,0)$ as $\odot^m\Omega^1(\mathcal{P})$. The most general irreducible tensor bundle on $\mathcal{P}$ is

$$\mathcal{O}(p,q,r) = \text{Trace-free part of} \left\{\odot^r T(\mathcal{P}) \otimes \odot^q\Omega^1(\mathcal{P}) \otimes \mathcal{O}(p-r+2q)\right\}.$$

In the flat case, it is sometimes useful to be able to express the bundles $\mathcal{O}(p,q,r)$ and hence the tensor powers of the tangent and cotangent bundles of $\mathbf{P}$ in terms of 'homogeneous' twistor indices.

To be more precise, let $\mathcal{O}^\alpha \to \mathbf{P}$ denote the product bundle $\mathbf{T}^\alpha \times \mathbf{P} \to \mathbf{P}$ where $\mathbf{T}^\alpha$ denotes non-projective twistor space. There are two (dual) short exact sequences that relate these bundles to the tangent and cotangent bundles:

$$0 \longrightarrow \mathcal{O} \xrightarrow{\times Z^\alpha} \mathcal{O}^\alpha(1) \longrightarrow T \longrightarrow 0$$
$$0 \longrightarrow \Omega^1 \longrightarrow \mathcal{O}_\alpha(-1) \xrightarrow{\times Z^\alpha} \mathcal{O} \longrightarrow 0. \tag{7}$$

One can think of $\times Z^\alpha$ as multiplication by the homogeneous co-ordinates or, slightly more geometrically, one can regard Z^α as the tautological section of $\mathcal{O}^\alpha(1)$.

More generally, define the bundles $\hat{\mathcal{O}}_{\epsilon\cdots\kappa}^{\alpha\cdots\delta}$ to have sections symmetric in both sets of indices and trace-free. Because of the maps given by Z^α, these bundles are not irreducible. The irreducible parts are given by sections of them satisfying $Z^\epsilon f^{\alpha\cdots\delta}_{\epsilon\cdots\kappa} = 0$, modulo those of the form $Z^{(\alpha}g^{\beta\cdots\delta)}_{\epsilon\cdots\kappa}$. If we denote this reduced bundle by $\mathcal{D}^{\alpha\cdots\delta}_{\epsilon\cdots\kappa}$ then we have the isomorphism

$$\mathcal{O}(p,q,r) = \mathcal{D}^{\overbrace{\alpha\cdots\delta}^{r}}_{\underbrace{\epsilon\cdots\kappa}_{q}}(p+q). \tag{8}$$

This analysis also works for the general non-linear graviton where $\mathcal{O}^\alpha$ is no longer a product bundle but is instead the Ward transform of the local twistor bundle.

2.4   *Geometric Aspects of the Penrose Transform*

We will make some observations here on some geometric matters related to the Penrose transform.

2.4.1   Conformally Invariant Operators

One of the reasons one is interested in the Penrose transform is that it generates conformally invariant differential operators, the existence and classification of which are of some interest both mathematically and physically. They are the subject of other articles in this volume and so we will be brief here. We have encountered several such operators already, the massless field operators, the conformally invariant wave operator,

$$\Box + R/6 : \mathcal{O}[-1] \longrightarrow \mathcal{O}[-3],$$

the 'Einstein' operator

$$\nabla^{(A}_{(A'}\nabla^{B)}_{B')} + \Phi^{AB}_{A'B'} : \mathcal{O}[1] \longrightarrow \mathcal{O}^{(AB)}_{(A'B')}[-1],$$

and the conformally invariant modification of the square of the wave operator $\mathcal{O} \to \mathcal{O}[-4]$ which arose in §2.3.1. Written out in full it is

$$\Box^2 : \phi \longmapsto \nabla_b[(\nabla^b\nabla^a - 2R^{ab} + \tfrac{2}{3}Rg^{ab})\nabla_a\phi].$$

Note that in flat space, in a flat metric, the operator is given by just its highest order part. If one conformally rescales, to a non-flat metric in the conformal class, the lower order terms involving curvature appear as above. One might hope that curved analogues of all flat operators could be generated in this way, but this is not so.

In the flat case the classification of invariant differential operators is a problem in representation theory—one is looking for $SL(4,\mathbb{C})$ equivariant operators between homogeneous vector bundles on M regarded as an $SL(4,\mathbb{C})$

homogeneous space. The problem is solved using the machinery of representation theory and an account is given in [21] and in more generality in [9, 8]. In [21, 8], the local twistor connection [16, 39] is used to investigate the question as to whether these operators have *curved analogues*—i.e. conformally invariant differential operators in the curved case (with co-efficients that will generally involve the Ricci curvature) which have the same symbol (the models being the examples above). The conclusion reached in these references, that *all* of them do have curved analogues, is however erroneous.

The flat space invariant operators fall into two classes—standard and non-standard. The standard operators are those which are obtained as direct images of the operators D^n from $\mathsf{G} = \mathsf{F}_{1,2,3}(\mathbf{C}^4)$, the flat space version of $\mathcal{G}$, or as compositions of the same. The non-standard operators are basically those involving the wave operator in some way. They are also distinguished by the fact that they appear when the spectral sequence for the transform from P fails to converge at the first step.

The standard operators all have curved analogues which can be generated as direct images of the D^n from $\mathcal{G}$ as in §2.2 [25, 26] and it is noteworthy that the correction terms involve only the trace-free part of the Ricci curvature. It would seem at present that some non-standard operators fail to have curved analogues—see the article by Baston and Eastwood in this volume.

Of relevance to the arguments in [21], but also of more general interest is the BGG resolution on M. These diagrams of bundles and differential operators on M are given by

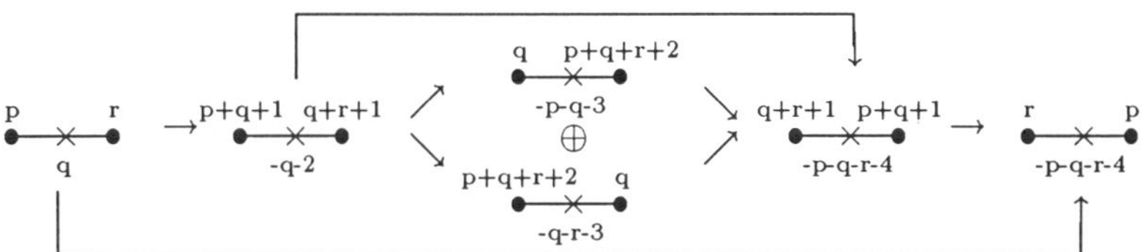

We are using the notation (described more fully in the appendix)

$$\underset{q}{\overset{p\quad r}{\bullet\!\!\!\times\!\!\!\bullet}} = \mathcal{O}^{\overbrace{A'\cdots L'}^{p}\overbrace{A\cdots L}^{r}}[q]$$

with some of the numbers below the diagram for reasons of space. We note that this only makes sense provided $p, r \geq 0$, and when this convention is violated, the entry should be replaced by the zero sheaf. The case $p = q = r = 0$ is (ignoring the two longer arrows) just the de Rham resolution and the 'generalised de Rham resolutions' of Buchdahl [13] are also a special case— in fact, if $p, q, r \geq 0$ then these sequences provide resolutions of the locally constant sheaf of $(0, p, p + q, p + q + r)\mathsf{T}$ valued functions in the notation of the appendix.

The maps in the BGG resolutions give a complete list of all the conformally invariant differential operators on M. The medium length arrows are just the composition of the short arrows via a direct summand if these exist. If the summands do not exist however they give something new. The non-standard operators are given by the 'long hops' from the first to the last bundle, together with these interesting cases of the medium length arrows.

In curved space, the longer arrows may not exist, but as we remarked above, the short arrows, which are all standard operators, do have curved analogues. The sequences are no longer exact or even complexes. They do however provide a useful 'index' of possible conformally invariant operators. There are also conformally invariant operators in the curved case which have no 'flat analogue' because, for example, their symbol may involve the Weyl curvature.

A case of some interest is $p = r = 1, q = 0$, where the various terms provide homes for conformal Killing vectors, linearised metrics, Weyl spinors and Bach tensors, and the differential operators are those one would consider in linearised gravity—see [10].

2.4.2   Projective Structure of α-surfaces and Null Geodesics

The BGG resolutions we have been using in the sections above are a feature of homogeneous spaces. It is perhaps no surprise then that they appear in the 'flat' versions of the Penrose transform, but it is somewhat surprising that they should feature in the curved versions we have been discussing. The reason that they do is that the fibres of the maps $\mathcal{G} \to \mathcal{A}, \mathcal{F} \to \mathcal{P}$ have exactly the same intrinsic structure in the curved case, namely that of a piece of the projective line and projective plane respectively, that they have in the flat case.

This is most easily seen for the non-linear graviton. If one fixes an α-surface Σ_p corresponding to $p \in \mathcal{P}$, the points on Σ_p correspond to curves passing through p. Each of these curves is determined by its projective tangent vector at p and so give an embedding $\Sigma_p \to \mathsf{P}(T_p(\mathcal{P}))$.

To turn now to the ambitwistor case, a null geodesic γ_p also comes with a natural embedding (at least locally) into $\mathbf{CP}^1$. This is most easily seen by considering the conformally invariant equation

$$(\nabla^2 + \Phi)\sigma = 0 \tag{9}$$

on sections over the geodesic of $\mathcal{O}[1]$. The solution space of this equation is just the fibre of $\mathcal{O}(0,1,0)$ over the point $p \in \mathcal{A}$ corresponding to γ_p. The null geodesic γ_p embeds in the projective fibre $\mathsf{P}(\mathcal{O}(0,1,0)|_p)$ of the rank 2 vector bundle $\mathcal{O}(0,1,0)$ by mapping each point to the 1-dimensional subspace of solutions which vanish at that point.

This *projective structure* of a null geodesic can also be seen by constructing a 3-dimensional family of preferred meromorphic functions on γ_p with the freedom of $\mathrm{SL}(2,\mathbf{C})$—these being the 'affine co-ordinates'. These functions can be specified in two equivalent ways, either they are the quotients of solutions of equation (9), or they are the affine parameters in metrics in the conformal class for which $\Phi = 0$. The operators D^{q+1} on $\mathcal{G}$ are just the familiar operators $\eth^{q+1} : \mathcal{O}(q) \to \mathcal{O}(-q-2)$ on the Riemann sphere (see [23]). They appear more complicated because the operator ∇ is not $\partial/\partial\zeta$ where ζ is an affine co-ordinate.

2.5 *Other Approaches to the Penrose Transform*

We mention in passing two other approaches to the Penrose transform. The first of these is the approach via Dolbeault cohomology—that is the representing of cohomology classes by differential forms rather than Čech cocycles. It is our view that there is rather more scope for these methods in twistor theory than the literature might suggest. There is already an excellent review article on this subject (Woodhouse [45]) which is concerned particularly with applications in the Euclidean (i.e. positive definite) version of twistor theory, and so we will not discuss the subject further here.

Another approach to taking the Penrose transform is via the local twistor bundle and was suggested by Mason [36]. The basic observation underlying the approach is that the elementary cases of the Penrose transform from homogeneous functions to massless fields can be generalised to give the transform of bundles such as $\mathcal{O}_\alpha(-1), \mathcal{O}^\alpha(1)$, etc. We will illustrate this method by calculating the Penrose transform of $H^1(U'', \Omega^1)$ which we considered also in §2.3.1. We work in a flat metric on $U'' \subset \mathbf{M}$.

The long exact cohomology sequence of (7) gives us

$$0 \to H^0(U'', \mathcal{O}) \to H^1(U'', \Omega^1) \to H^1(U'', \mathcal{O}_\alpha(-1)) \to H^1(U'', \mathcal{O}). \qquad (10)$$

It is easy to see that the Penrose transform of $H^1(U'', \mathcal{O}_\alpha(-1))$ is fields $\phi_{B\alpha}$, where α is a (dual) *local twistor* index (the dual local twistor bundle is just the product bundle $\mathbf{T}^* \times \mathbf{M}$ with the obvious flat connection in the flat case; see e.g.[39, 16] for this and what follows), of conformal weight -1, satisfying $\nabla^B_{B'}\phi_{B\alpha} = 0$ and ∇ denotes the tensor product of the local twistor and spin connections.

The definition of local twistors and their connection then give us that $\phi_{B\alpha}$ is equivalent to a pair of fields $\xi_{BA'}$, η_{BA}, sections of $\mathcal{O}_{BA'}$ and $\mathcal{O}_{BA}[-1]$ respectively, satisfying

$$\nabla^B_{B'}\xi_{BA'} - i\epsilon_{B'A'}\eta_B{}^B = 0, \quad \nabla^B_{B'}\eta_{BA} = 0. \qquad (11)$$

Under change of conformal scale, we have $\hat\xi_{BA'} = \xi_{BA'}$ and $\hat\eta_{BA} = \eta_{BA} - i\Upsilon_{AA'}\xi_B^{A'}$.

The symmetric part of the first of equations (11) tell us that $\xi_{BA'}$ is the potential for a left-handed Maxwell field, and indeed this gives us the rightmost map in (10) since we know that $H^1(U'', \mathcal{O})$ is isomorphic to potentials modulo gauge for such fields. Since we are calculating $H^1(U'', \Omega^1)$, we are only interested in the kernel of this map—i.e. the case where $\xi_{BA'}$ is in fact pure gauge, so that $\xi_{BA'} = \nabla_{BA'} f$ for some function f.

If we now write $\eta_{BA} = \rho_{BA} + \tau \epsilon_{BA}$ with $\rho_{BA} = \rho_{(BA)}$, and note that $H^0(U'', \mathcal{O}) = \mathbb{C}$, the system of equations representing $H^1(U'', \Omega^1)/\mathbb{C}$ is

$$\nabla^B_{B'} \rho_{BA} + \nabla_{B'A} \tau = 0, \quad \Box f + 4i\tau = 0.$$

We can differentiate the second of these equations to eliminate τ. Furthermore, note that f is only defined up to an additive constant. Whilst it is not clear from this analysis (and this is a slight weakness in the scheme), this constant can be identified with the image of $H^0(U'', \mathcal{O})$ and the end result is that the Penrose transform of $H^1(U'', \Omega^1)$ can be identified with solutions on U of the equation

$$\nabla_{AA'} \Box f - 4i\nabla^B_{A'} \rho_{BA} = 0.$$

Applying $\nabla^{AA'}$ we see that $\Box^2 f = 0$, and it is thus clear how this result fits in with our calculations in §2.3.1. An advantage of this method is that (apart from the question of the constants), one actually gets a precise description of the extension described using our previous methods only by a composition series. Bailey and Eastwood have recently observed that it is possible to obtain this result also using our previous methods applied to a different resolution on F.

In the curved case, $\mathcal{O}_\alpha$ etc. are no longer product bundles but can be defined to be the Ward transform of the local twistor bundles on space-time and the scheme still works.

3   Singular Massless Fields

One of the areas in the twistor description of massless fields that has advanced since the introduction of cohomology in [20] is the twistor description of fields which are not analytic. In the next section, we will discuss fields with sources on a given world-line and in this section we consider distributional massless fields on real Minkowski space.

3.1   *Relative Sheaf Cohomology*

We give a very brief sketch of relative sheaf cohomology which appears in several contexts in this review. The reader is referred to the standard texts for a fuller treatment (e.g. [24, 12] or [32] which is well adapted to the applications in twistor theory.)

We are given a closed (or relatively closed) subset F of a space X and a sheaf $\mathcal{S}$ on X. We choose an open cover $\mathcal{U}$ of X with a sub-cover $\mathcal{U}'$ of $X \setminus F$. A *relative Čech cochain* is a Čech cochain with respect to the cover $\mathcal{U}$ subject to the condition that it vanishes when restricted to the sub-cover $\mathcal{U}'$. There is thus an exact sequence

$$0 \longrightarrow C_F^p(X,\mathcal{S}) \longrightarrow C^p(X,\mathcal{S}) \longrightarrow C^p(X \setminus F,\mathcal{S}) \longrightarrow 0$$

where $C_F^p(X,\mathcal{S})$ is the group of relative cochains. The relative cochains inherit a coboundary operator from the ordinary cochains and the limit over finer and finer covers of the homology of $C_F^*(X,\mathcal{S})$ gives the *relative cohomology groups* $H_F^p(X,\mathcal{S})$. Just as for the familiar case, there is no need to take the limit since one has the *relative Leray theorem* which states that if $H^p(U,\mathcal{S}) = 0, p \geq 1$ for each set U in the cover $\mathcal{U}$, then that cover suffices to compute the relative cohomology.

The long exact cohomology sequence of the defining short exact sequence above gives the *relative cohomology exact sequence*

$$0 \to H_F^0(X,\mathcal{S}) \to H^0(X,\mathcal{S}) \to H^0(X \setminus F,\mathcal{S}) \to H_F^1(X,\mathcal{S}) \to \dots,$$

where the maps from cohomology on X to that on $X \setminus F$ are restriction.

Another important result is the *excision theorem* which states, roughly speaking, that the relative cohomology depends only on the immediate neighbourhood of the embedding of F in X. More precisely, given an open subset $X' \subset X$ such that $\overline{(X \setminus X')} \cap F = \emptyset$, there is a canonical isomorphism $H_F^n(X,\mathcal{S}) = H_F^n(X',\mathcal{S})$.

3.2 *Hyperfunctions*

Hyperfunctions are a class of generalised functions; we will only sketch the theory here referring the reader to [32] (see also [28] for a treatment avoiding cohomology) for the details. Hyperfunctions on a real manifold are boundary values of holomorphic functions defined on regions in its complexification and are, in an appropriate sense, the duals of the real-analytic functions and (as usual with generalised functions) given that the space of real-analytic functions is 'small' we expect the space of hyperfunctions to be 'large'. This is indeed the case, and the space of hyperfunctions strictly contains most other common spaces of distributions.

If $F \subset \mathbf{R}$ is a closed interval, the hyperfunctions on F are given by

$$\mathcal{B}(F) = \frac{\mathcal{O}(\mathbf{C} \setminus F)}{\mathcal{O}(\mathbf{C})}$$

and if f is an anaslytic (in the real sense) complex function on F, the pairing with a hyperfunction represented by the holomorphic function ϕ on $\mathbf{C} \setminus F$ is given by the contour integral

$$\oint \phi(z) f(z)\, \mathrm{d}z$$

where the function f has been extended holomorphically a little off F and the contour in $\mathbb{C} \setminus F$ circles once around F sufficiently close to lie in the domain of definition of f.

If F is the whole real line, then $\mathbb{C} \setminus \mathbb{R}$ splits into the upper and lower half planes $\mathbb{C}^{\pm}$. A hyperfunction ϕ is then given by holomorphic functions $\phi_{\pm}$ on $\mathbb{C}^{\pm}$ modulo entire functions on $\mathbb{C}$. One thinks of the hyperfunction as being the difference in the boundary values of $\phi_{\pm}$ on $\mathbb{R}$ and there are thus *boundary value maps* $b_{\pm} : \mathcal{O}(\mathbb{C}^{\pm}) \to \mathcal{B}(\mathbb{R})$. We write $\phi = b_+(\phi_+) - b_-(\phi_-)$. The well known distribution $1/(x + i0)$ on $\mathbb{R}$ is then given as the boundary value of a holomorphic function on $\mathbb{C}^+$ as $b_+(1/x)$.

In higher dimensions, holomorphic functions on $\mathbb{C}^n \setminus \mathbb{R}^n$ necessarily extend analytically accross $\mathbb{R}^n$ and so one has to take boundary values from 'smaller' regions than this. One takes regions which are the Cartesian product of some open set $U \subset \mathbb{R}^n$ with the interior of some cone in the imaginary. Examples are the future and past tubes $\mathsf{M}^{\pm}$ in Minkowski space. The equivalence of boundary values of different functions from different regions (which was just the freedom of entire functions for $n = 1$) is rather more complicated in general. It is most easily expressed by identifying the hyperfunctions on an open set $F \subset \mathbb{R}^n$ as the relative cohomology group $\mathcal{B}(F) = H^n_F(\mathbb{C}^n, \mathcal{O})$. It is easy to check from the relative cohomology exact sequence of §3.1 that this is consistent with our previous definition when $n = 1$ (it does not really matter in that definition whether F is open or closed).

All this goes through for a general real n-manifold X sitting in its complexification $\mathcal{X}$. Moreover, the hyperfunctions can take values in any analytic vector bundle V. They are given by boundary values of sections of the analytic continuation of the vector bundle from appropriate regions in the complex, the freedom being expressed by the cohomological definition,

$$\mathcal{B}_V(F) = H^n_F(\mathcal{X}, V),$$

where $F \subset X$ and the complex neighbourhood $\mathcal{X}$ of F must be chosen small enough so that the analytic continuation of V to a holomorphic vector bundle on $\mathcal{X}$ is possible.

Any analytic differential operator D can be extended to act on the hyperfunctions by analytically continuing it a little into the complex and insisting that its action commute with taking boundary values—symbolically, $D(b(\phi)) = b(D(\phi))$.

3.3 Hyperfunction Massless Fields

We can now consider hyperfunction solutions of the massless field equations. Let us begin on twistor space with the relative cohomology sequence for the null twistors $P \subset \mathbb{P}$ with sheaf $\mathcal{O}(-n - 2), n \geq 0$. One knows that $H^p(\mathbb{P}, \mathcal{O}(-n - 2)) = 0$ for $p = 1, 2$ and $\mathbb{P} \setminus P = \mathbb{P}^+ \cup \mathbb{P}^-$ and so the sequence gives us

$$H^1(\mathbf{P}^+, \mathcal{O}(-n-2)) \oplus H^1(\mathbf{P}^-, \mathcal{O}(-n-2)) = H^2_P(\mathbf{P}, \mathcal{O}(-n-2)). \qquad (12)$$

This sequence is in a well defined sense (see [40]) expressing the right-hand side as a difference in boundary values of the cohomology groups on the left-hand side.

Letting $\mathcal{Z}_{\mathcal{B},n}(M)$ denote hyperfunction massless fields of helicity $n/2$ on real compactified Minkowski space M, there is a 'Penrose transform'

$$\mathcal{P}_\mathcal{B} : H^2_P(\mathbf{P}, \mathcal{O}(-n-2)) \longrightarrow \mathcal{Z}_{\mathcal{B},n}(M)$$

obtained by applying the Penrose transform to the left-hand side of (12) to obtain massless fields $\phi_\pm$ on $\mathbf{M}^\pm$ and then taking the difference in the boundary values. The result is necessarily a hyperfunction solution of the equations since it is a difference in boundary values of holomorphic solutions. Since the future and past tubes stand in the same relationship to all affine co-ordinate patches, the boundary value can be taken on several affine co-ordinate patches on M and 'glued together' to give a solution defined on all of compactified real Minkowski space M.

The immediate question to ask of course is whether the map $\mathcal{P}_\mathcal{B}$ is an isomorphism. This was answered in the affirmative by Wells [44] by applying the transform machinery to relative cohomology classes. A more straightforward argument is given in [5].

One might also be interested in hyperfunction massless fields on real affine Minkowski space M^I. The relevant regions of twistor space are $\mathbf{P}^I$ and P^I given by removing the line I giving the infinity of the Minkowski space from $\mathbf{P}$ and P respectively. In [5], it is shown that there is a Penrose transform isomorphism

$$\mathcal{P}_\mathcal{B} : H^2_{P^I}(\mathbf{P}^I, \mathcal{O}(-n-2)) \longrightarrow \mathcal{Z}_{\mathcal{B},n}(M^I).$$

The existence of the map is easily deduced from the relative cohomology sequence for $P^I \subset \mathbf{P}^I$, which reduces to

$$0 \to H^1(\mathbf{P}^I, \mathcal{O}(-n-2)) \to \begin{matrix} H^1(\mathbf{P}^+, \mathcal{O}(-n-2)) \\ \oplus \\ H^1(\mathbf{P}^-, \mathcal{O}(-n-2)) \end{matrix} \to H^2_{P^I}(\mathbf{P}^I, \mathcal{O}(-n-2)) \to 0.$$

If $\phi \in H^2_{P^I}(\mathbf{P}^I, \mathcal{O}(-n-2))$, choose elements $\phi_\pm \in H^1(\mathbf{P}^\pm, \mathcal{O}(-n-2))$ so that $\phi_+ \oplus \phi_-$ maps to ϕ in the above sequence. Now take the Penrose transform to obtain holomorphic massless fields $\mathcal{P}\phi_\pm$ on $\mathbf{M}^\pm$ and the hyperfunction Penrose transform is obtained by taking the difference in boundary values:

$$\mathcal{P}_\mathcal{B}\phi = b(\mathcal{P}\phi_+) - b(\mathcal{P}\phi_-).$$

This definition is consistent because the ambiguity in the fields $\mathcal{P}\phi_\pm$ which occurs because of the group $H^1(\mathbf{P}^I, \mathcal{O}(-n-2))$ is the addition of a holomorphic field on complex affine Minkowski space to both $\mathcal{P}\phi_+$ and $\mathcal{P}\phi_-$ which has no effect on the difference in boundary values.

The proof that the map is an isomorphism uses hyperfunction theory to show that a hyperfunction massless field can always be expressed as the difference in boundary values of massless fields defined on certain subsets of the future and past tubes. The Penrose transform represents these fields as elements of $H^1(U^\pm, \mathcal{O}(-n-2))$ where $U^\pm \subset \mathbf{P}^\pm$ with $U = U^+ \cup U^-$ being a neighbourhood of P^I in $\mathbf{P}^I$. The connecting homomorphism of the relative cohomology exact sequence is

$$H^1(U^+, \mathcal{O}(-n-2)) \oplus H^1(U^-, \mathcal{O}(-n-2)) \longrightarrow H^2_{P^I}(U, \mathcal{O}(-n-2)),$$

and by the excision theorem (see §3.1), the right hand group above is isomorphic to $H^2_{P^I}(\mathbf{P}^I, \mathcal{O}(-n-2))$. Stringing these together, we obtain an inverse for $\mathcal{P}_\mathcal{B}$. In the process a removable singularity theorem for hyperfunction massless fields on M^I is proved since we have shown that every such field is actually the difference in boundary values of massless fields on the future and past tubes, and the boundary value map can be taken at infinity by choosing another affine co-ordinate patch. This defines an extension of the hyperfunction massless field to the compactification.

Given an open set $V \subset P$, the assignment

$$V \longmapsto H^2_V(\mathbf{P}, \mathcal{O}(-n-2))$$

actually defines a sheaf over P and in contrast to the case of 'non-relative H^1s', there is information locally on twistor space—these groups are nontrivial even for 'small' and 'well-behaved' V. In [37], Penrose speculates as to the connection between the stalk of this sheaf and singularities of massless fields in some sense associated with the corresponding null geodesic. Henkin and Polyakov [29] made this precise by showing that there is an isomorphism between this stalk and the space of hyperfunction massless fields in a neighbourhood of the null geodesic modulo those whose singular spectrum (a hyperfunction generalisation of the distributional notion of wave-front set [28]) does not contain the null geodesic.

In conclusion, we should remark that everything we have said applies equally to the case of left-handed fields also. In most cases this follows immediately by symmetry.

3.4   *Positive and Negative Frequency*

It is well known that fields on affine real Minkowski space which are *positive frequency* in the sense of having their Fourier transform supported on the future light-cone in momentum space, extend holomorphically to fields on

$\mathbf{M}^+$, and similarly for negative frequency and $\mathbf{M}^-$. These classes of fields are important, particularly in quantum field theory.

A hyperfunction massless field on affine or compactified real Minkowski space is said to be *future (resp. past) analytic* [5] if it is the boundary value of a holomorphic field from $\mathbf{M}^+$ (resp. $\mathbf{M}^-$). We have seen that hyperfunction fields on compactified Minkowski space can be written uniquely as a difference (one usually thinks of a sum of course) of future and past analytic solutions, and that hyperfunction solutions on affine Minkowski space can be so written modulo the ambiguity of the addition of holomorphic solutions on complex affine Minkowski space to both components. A positive frequency plane wave solution of the wave equation (for example) on affine Minkowski space is both future and past analytic.

The splitting of hyperfunction solutions on affine Minkowski space into a sum of future and past analytic solutions thus depends on deciding on a definition of the field at infinity—there is a natural choice if the field has a Fourier transform which is sufficiently well behaved at the origin, but this is not the case even for the constant field.

3.5 Ingredients for Field Theory

One of the areas of physics in which the twistor description of fields in terms of free holomorphic data might be expected to play a decisive role, is quantum field theory. There are many unresolved difficulties in the translation of standard (massless) QFT into the twistor picture, let alone the development of an intrinsically twistorial procedure for introducing interacting quantum fields. Nonetheless, the basic structures (symplectic forms, polarizations, etc.) of the 'classical phase space' (the space of L^2 massless fields on M) have a very satisfactory twistor description, a feature of which is the manifest conformal invariance of the constructions involved. We wish to give a brief account of such matters in this section; closely related material is to be found in the article by Dunne & Eastwood in this volume.

We saw in §3.3 that the Penrose transform gives an identification

$$H^2_P(\mathbf{P}, \mathcal{O}(-2s-2)) = \{\text{hyperfunction fields on } M \text{ of helicity } s\}.$$

Although of considerable importance, this space is too large for the purpose of field theory since it contains states of infinite norm. A more appropriate starting-point is the space

$$H^1(P, \mathcal{O}(-2s-2)) = \{\text{real-analytic fields on } M \text{ of helicity } s\}. \tag{13}$$

Since P is not a complex manifold, the definition of the left-hand side requires some care. We adopt the sheaf-theoretic approach to its definition,

$$H^1(P, \mathcal{O}(-2s-2)) = \varinjlim H^1(U'', \mathcal{O}(-2s-2))$$

where the limit is taken over the directed system of all open neighbourhoods U'' of P. It should be noted that the symbol on the LHS of (13) is given a different meaning by other authors, particularly those with a background in analysis.

In §3.3, it was mentioned that standard polarization (positive/negative-frequency decomposition) of massless fields on M corresponds to writing the relative cohomology class as a sum of boundary values of classes defined on $\mathbf{P}^+$ and $\mathbf{P}^-$. Restricting to real-analytic fields, the standard polarization comes from the Mayer-Vietoris splitting

$$H^1(P, \mathcal{O}(-2s-2)) = H^1(\bar{\mathbf{P}}^+, \mathcal{O}(-2s-2)) \oplus H^1(\bar{\mathbf{P}}^-, \mathcal{O}(-2s-2)).$$

This result allows one to identify many examples of massless fields via the inclusion

$$H^1(\mathbf{P} \setminus L, \mathcal{O}(-2s-2)) \longrightarrow H^1(\bar{\mathbf{P}}^+, \mathcal{O}(-2s-2)), \tag{14}$$

which is valid for any (complex projective) line $L \subset \mathbf{P}^-$. There are classes in the LHS of (14) with poles of some definite order on L. These are the *elementary states* based on L and can be represented in Dolbeault terms in the following way. We choose coordinates (o^A, ι^A) on $\mathbf{T}$, in such a way that the real structure is given by the form

$$\Phi(Z) = T_{AA'} \iota^A \bar{\iota}^{A'} - T_{AA'} o^A \bar{o}^{A'};$$

here T is a positive-definite matrix with $T^{AA'} T_{AA'} = 2$, and we shall find it convenient to write

$$\hat{o}_A = T_{AA'} \bar{o}^{A'} \text{ and } \hat{\iota}_A = T_{AA'} \bar{\iota}^{A'}.$$

Then $o^A = 0$ defines a line in $\mathbf{P}^+$ and $\iota^A = 0$ defines a line in $\mathbf{P}^-$. (This is the reason for our non-standard conventions; the usual twistor coordinates $(\omega^A, \pi_{A'})$ are chosen so that both $\omega^A = 0$ and $\pi_{A'} = 0$ are *real* lines.)

For $s \geq 0$, set

$$E_+(s, n; A) = A^{A...EF...S}_{F_0...S_0} \frac{o^{F_0} \ldots o^{S_0} \hat{\iota}_A \ldots \hat{\iota}_S \hat{\iota}_T \, d\hat{\iota}^T}{(\iota^U \hat{\iota}_U)^{n+2s+2}} \tag{15}$$

and

$$E_+(-s, n; B) = B^{F_0...S_0}_{A...EF...S} \frac{o_{F_0} \ldots o_{S_0} \hat{\iota}^A \ldots \hat{\iota}^S \hat{\iota}_T \, d\hat{\iota}^T}{(\iota^U \hat{\iota}_U)^{n+2}} \tag{16}$$

where $A...E$ is a collection of $2s$ indices and $F...S$ is a collection of n indices. Application of the standard contour integral formulae shows that the fields represented by (15) and (16) are respectively

$$\frac{1}{n+2s+1} A^{A...EF...S}_{F_0...S_0} T_{AA'} \ldots T_{EE'} x_F^{F_0} \ldots x_S^{S_0}$$

and

$$\frac{(n+2s)!}{(n+1)!} B^{F_0\ldots S_0}_{A\ldots E F\ldots S} x^F_{F_0} \ldots x^S_{S_0}.$$

The cohomology classes defined by (15) and (16) are important because, relative to certain standard topologies on the cohomology groups involved, the system of elementary states based on L spans a dense subspace of the cohomology groups $H^1(\bar{\mathbf{P}}^+, \mathcal{O}(-2s-2))$ and $H^1(\mathbf{P}^+, \mathcal{O}(-2s-2))$ (cf. [19]). Thus the space $H^1(P, \mathcal{O}(-2s-2))$ is a closure of the span of the elementary states E_+ based on $\iota^A = 0$ and the elementary states E_- (defined by interchanging o and ι in (15) and (16)) based on $o^A = 0$.

Let us now focus attention on (13); we wish to give it a natural (i. e. conformally invariant) pre-Hilbert space structure, so that its completion is the space of L^2 massless fields. Notice first that there is a conformally invariant pairing

$$H^1(P, \mathcal{O}(-2s-2)) \otimes H^1(P, \mathcal{O}(2s-2)) \longrightarrow \mathbb{C};$$

in terms of the Dolbeault description, this is given by the formula

$$\omega(\alpha, \beta) = \frac{1}{(2\pi i)^3} \int_P \alpha \wedge \beta \wedge \Omega \tag{17}$$

where

$$\Omega = (Z^0 \mathrm{d}Z^1 - Z^1 \mathrm{d}Z^0) \wedge \mathrm{d}Z^2 \wedge \mathrm{d}Z^3 + (Z^2 \mathrm{d}Z^3 - Z^3 dZ^2) \wedge dZ^0 \wedge dZ^1$$

is determined by the identification $\mathcal{O}(-4) \simeq \Omega^3$. Notice that when $s = 0$, the form ω is skew; it then corresponds to i times the standard symplectic form on the phase space of massless scalar fields:

$$\omega(\phi, \psi) = \int (\phi \nabla_a \psi - \psi \nabla_a \phi) \mathrm{d}\sigma^a, \tag{18}$$

where the integral is over a space-like hypersurface. If we grant the equivalence of (17) and (18), the conformal invariance is much easier to see from the twistorial formula.

Applying Stokes' theorem in (17), it is apparent that ω vanishes identically on the subspaces

$$H^1(\bar{\mathbf{P}}^+, \mathcal{O}(-2s-2)) \otimes H^1(\bar{\mathbf{P}}^+, \mathcal{O}(2s-2))$$

and

$$H^1(\bar{\mathbf{P}}^-, \mathcal{O}(-2s-2)) \otimes H^1(\bar{\mathbf{P}}^-, \mathcal{O}(2s-2)).$$

Let us choose the orientation of P so that we have

$$\frac{1}{(2\pi i)^3} \int_P U_p(o) \wedge U_q(\iota) \wedge \Omega = \frac{1}{(p+1)(q+1)} \mathcal{I}_p \mathcal{I}_q,$$

where

$$U_p(o) = \frac{o^{G_0}\ldots o^{L_0}\hat{o}_G\ldots\hat{o}_L}{(o^N\hat{o}_N)^{p+2}}\hat{o}_M\mathrm{d}\hat{o}^M$$

and

$$\mathcal{I}_p = \delta_G^{(G_0}\ldots\delta_L^{L_0)},$$

the indices $G\ldots L$ being p in number. Then we have

$$\omega(E_-(s,n;A), E_+(-s,m;B)) = \frac{\delta_{nm}}{(n+1)(n+2s+1)}A^{A\ldots EF\ldots S}_{F_0\ldots S_0}B^{F_0\ldots S_0}_{A\ldots EF\ldots S},$$

there being a similar formula for the results of pairing $E_-(-s)$ with $E_+(s)$. The non-degeneracy of ω follows from this formula and the density result of Eastwood and Pilato [19].

Let us continue with the construction of the Hermitian form on (13). For this we need the complex conjugation

$$\mathcal{T} : H^1(P, \mathcal{O}(-2s-2)) \longrightarrow H^1(P, \mathcal{O}(2s-2)) \tag{19}$$

which is given by complex conjugation of spinor fields on M. Since the conjugate of a positive frequency field is negative frequency, we see that $\mathcal{T}$ interchanges $\mathbf{P}^+$ and $\mathbf{P}^-$; we have

$$\mathcal{T} : H^1(\mathbf{P}^+, \mathcal{O}(-2s-2)) \longrightarrow H^1(\mathbf{P}^-, \mathcal{O}(2s-2)).$$

In view of the formulae above,

$$\mathcal{T}(E_+(s,n;A)) = \frac{(n+1)!}{(n+2s+1)!}E_-(-s,n;\hat{A}).$$

Now we can combine (17) and (19), to define a conformally invariant Hermitian form (conjugate-linear in the *first* variable) on $H^1(\bar{\mathbf{P}}^+, \mathcal{O}(-2s-2))$ by means of the formula

$$\langle\alpha|\beta\rangle = \frac{1}{(2\pi i)^3}\int_P \mathcal{T}\alpha\wedge\beta\wedge\Omega$$

Then

$$\langle E_+(s,n;A)|E_+(s,m;B)\rangle = \delta_{nm}\frac{n!}{(n+2s+1)^2(n+2s)!}\hat{A}\cdot B,$$

so that $\langle\ |\ \rangle$ is positive-definite on the span of the elementary states based on $\iota = 0$. This completes the construction of the conformally invariant pre-Hilbert space structure on the positive-frequency part of (13).

By passing to the Hilbert-space completion, we construct the space H_s^+ of L^2 positive-frequency massless fields of helicity s on M: we have inclusions

$$H^1(\bar{\mathbf{P}}^+, \mathcal{O}(-2s-2)) \subset H_s^+ \subset H^1(\mathbf{P}^+, \mathcal{O}(-2s-2)),$$

which exhibit H_s^+ as a 'rigged' Hilbert space, in the sense that (in appropriate topologies) each inclusion is dense and the outer two spaces are conjugate-dual.

In conclusion, we have seen that the basic ingredients of four-dimensional quantum field theory have a very satisfactory twistor description. The only object that is hard to define intrinsically is the conjugation $\mathcal{T}$, which is a version of the twistor transform. As we indicated at the outset, the introduction of interactions remains somewhat problematic. Most work in this area so far has been on twistor diagram theory (cf. Hodges and Huggett in this volume). More recently a geometric approach along the lines of two-dimensional conformal field theory has been considered [30, 43]: but the links between such a theory and physics as it is presently understood seem remote. A third, somewhat prosaic possibility is to try to extend the Penrose transform so as to encompass all the standard bits and pieces of standard four-dimensional field theory. A basic problem here is the description of the Feynman propagator where some recent progress has been made. These matters remain very much open to further investigation.

4   Sourced Fields and Global Descriptions

In this section we consider the twistor description of massless fields with source on a world-line and other situations where topology complicates the application of the Penrose transform.

4.1 *Sourced Fields*

If γ is a time-like curve in M^I (parametrized as $y^a(s)$, say, where s denotes proper time) and $\rho(s)$ is a source-density, then the equation

$$\Box\phi = 4\pi\rho_\gamma, \tag{20}$$

where the right-hand side denotes the current

$$\rho_\gamma[u] = \int \rho(s)u(y(s))\,ds \tag{21}$$

is of considerable physical interest. The study of (20) from the point of view of twistor theory was pioneered by Bailey [2].

Each point x of M^I sufficiently close to γ is null separated from two points on γ, $x_{\mathrm{adv}} = y(s_{\mathrm{adv}}(x))$, which is to the future of x, and $x_{\mathrm{ret}} = y(s_{\mathrm{ret}}(x))$, which is to its past. The quantity $r_{\mathrm{ret}}(x)$ (resp. $r_{\mathrm{adv}}(x)$) is defined as the component of $x - x_{\mathrm{ret}}$ (resp. $x_{\mathrm{adv}} - x$) in the direction tangent to γ at x_{ret} (resp. x_{adv}). In terms of r_{ret} and r_{adv}, the basic solution of (20) is

$$\phi_\lambda(x) = \lambda\frac{\rho(s_{\mathrm{ret}}(x))}{r_{\mathrm{ret}}(x)} + (1 - \lambda)\frac{\rho(s_{\mathrm{adv}}(x))}{r_{\mathrm{adv}}(x)} \tag{22}$$

for any value of λ. (Any ϕ_λ can, of course, be obtained as a sum of multiples of the *advanced* field ϕ_0 and the the *retarded* field ϕ_1.)

As far as twistor description is concerned, the first thing to notice about ϕ_λ is that when it is analytically continued into complex Minkowski space (assuming that γ and ρ are real-analytic), it becomes double-valued. The branching locus B consists of those (necessarily complex) points x for which x_{ret} and x_{adv} coincide. If ϕ_λ is analytically continued along a path C which represents the generator of the fundamental group of the complement of B, then the value of ϕ_λ changes to

$$\hat{\phi}_\lambda(x) = -\lambda\frac{\rho(s_{\mathrm{adv}}(x))}{r_{\mathrm{adv}}(x)} - (1-\lambda)\frac{\rho(s_{\mathrm{ret}}(x))}{r_{\mathrm{ret}}(x)}.$$

We note that this is $\phi_{1-\lambda}$ for the source density $-\rho(s)$.

The observation that typical sourced fields are two-valued shows that to extend the Penrose transform to give a description of solutions of (20), one must learn how to describe massless fields defined on the double cover $\tilde{U}$ of $U \setminus B$ (where U is some complex neighbourhood of M^I). We shall proceed to discuss this and related problems in a systematic way §4.2 below; first we want to show how Conway's integral and its generalizations can be used to generate explicit 'twistor functions' for massless fields with given multipole moments on γ. The cohomological interpretation of these functions will lead us to the main conjecture of [2]; the tools needed for the proof will be described in §4.2.

We recall that Conway's integral [15]

$$\phi(x) = \frac{1}{\pi i} \oint \frac{\rho(s)ds}{(x - y(s))^2} \tag{23}$$

yields (22) if the 'contour' used surrounds the pole at $s_{\mathrm{ret}}(x)$ with multiplicity $-\lambda$ and the pole at $s_{\mathrm{adv}}(x)$ with multiplicity $1 - \lambda$. There is a generalization of this formula which defines a massless field of arbitrary helicity having a prescribed multipole structure: this is described on p.152 of [2]. From (23) one obtains a twistor function which represents the field (22) in the following way. Consider the function ξ^A of the parameter s and the twistor $(\omega^A, \pi_{A'})$ defined by

$$\xi^A(s) = \omega^A - iy^{AA'}(s)\pi_{A'}.$$

We can form a twistor function $f(\omega^A, \pi_{A'})$ by performing the contour integral

$$f(\omega^A, \pi_{A'}) = \frac{1}{\pi i} \oint \frac{\alpha_A \beta^A \rho(s)ds}{\alpha_B \xi^B(s)\, \beta_C \xi^C(s)}, \tag{24}$$

where α_A, β_A are arbitrary fixed spinors. Then f is homogeneous of degree -2; and one recovers the Conway integral from it by use of the standard contour integral formula.

In order to arrive at a cohomological interpretation of this we must consider the geometry more carefully. The (complexified) world-line $y^a(s)$ corresponds in twistor space to a piece $\mathcal{L}$ of *ruled surface*, the ruling lines corresponding to the points on the world-line. By using different contours in (24), one can produce twistor functions defined on different subsets of a neighbourhood U'' of $\mathcal{L}$ in **P**. In fact, they precisely give a relative Čech cocycle (see §3.1) for an element of the relative cohomology group $H^1_{\mathcal{L}}(U'', \mathcal{O}(-2))$. The details can be found in [2].

These arguments are highly suggestive. One expects any massless field on $M^I \setminus \gamma$ to be the sum of a source-free field and a field which can be expanded in terms of the multipole fields just described. On the other hand, each multipole field defines explicitly a cohomology class in $H^1_{\mathcal{L}}(U'', \mathcal{O}(-2))$ and one expects to be able to expand the general class in this group in terms of classes which represent multipole fields, i. e. in terms of classes with poles of a definite order on $\mathcal{L}$. Thus it is plausible that there is an extension of the Penrose transform

$$H^1(U'') \oplus H^1_{\mathcal{L}}(U'') \longrightarrow \{\text{fields on } \tilde{U}\} \qquad (25)$$

and that this should be an isomorphism if the standard Penrose transform from U'' to U is an isomorphism. Results of this type do, in fact, hold but the available proof [6] is completely different from this plausibility argument.

4.2 Aspects of Global Twistor Descriptions

It was originally shown by Eastwood, Penrose and Wells [20] that the space of massless fields on an open subset U of **M** is naturally isomorphic to $H^1(U'')$ if U satisfies some mild topological conditions. If U fails to satisfy these conditions, the Penrose transform is affected in two related ways. The primary effect is to upset the isomorphism ('pull-back mechanism')

$$H^1(U'', \mathcal{O}(V)) = H^1(U', \mu^{-1}\mathcal{O}(V)). \qquad (26)$$

The secondary effect appears, for example, in the description of left-handed massless fields: if U has non-trivial topology, then the isomorphism

$$\{\text{potentials}\}/\{\text{gauge}\} = \{\text{fields}\} \qquad (27)$$

as described, for example, in [20], may fail. We shall now discuss these matters in turn, mainly in the context of the standard correspondence between space-time and twistor space. It should be clear how to generalize the discussion to an arbitrary correspondence.

4.2.1 The Pull-back Mechanism

The isomorphism (26) has been investigated by comparing the Dolbeault complex on U'' with its (topological) inverse image by μ. The spectral sequence associated to $\mu^{-1}\mathcal{E}^{0,p}(V)$ starts with

$$E_1^{p,q} = H^q(U', \mu^{-1}\mathcal{E}^{0,p}(V)) \implies H^*(U', \mu^{-1}\mathcal{O}(V)) \tag{28}$$

(where d_1 is induced by $\mu^{-1}\bar{\partial}_V$). We shall refer to this spectral sequence as the *generalized pull-back mechanism*.

To understand the $E_1^{p,q}$ it is helpful to adopt the following point of view. Since μ is a C^∞ map of maximal rank, each fibre $\mu^{-1}(z)$ $(z \in U'')$ is a C^∞ manifold. Thus μ defines, in a natural sense, a C^∞ family U'/U'' of (generally non-compact) manifolds. Then $E_1^{p,q}$ is some measure of the q-th de Rham cohomology of this family (with coefficients on $\mathcal{E}^{0,p}(V)$)—in particular, when U'' is reduced to a point z, $E_1^{p,q}$ becomes the q-th de Rham cohomology of $\mu^{-1}(z)$ with coefficients in the vector space $\mathcal{E}_z^{0,p}(V)$. There is a de Rham theorem, which states that one can calculate $E_1^{p,q}$ by means of the complex of C^∞ μ-relative differential forms on U'. (Recall that a μ-relative q-form is a smooth alternating q-linear map which assigns a number to any collection of q vertical tangent vectors at a point.) From now on, we shall write

$$E_1^{p,q} = H^q(U'/U'', \mathcal{E}^{0,p}(V)) \tag{29}$$

and call the RHS the q-th de Rham cohomology module of the family of manifolds U'/U'' with coefficients in $\mathcal{E}^{0,p}(V)$.

The first general result concerning these modules is due to Buchdahl [14]: if for some $q \geq 1$, the reduced homology in dimensions q and $q-1$ of each fibre of μ is zero,

$$H_q^\sharp(\mu^{-1}(z), \mathbb{C}) = 0, \quad H_{q-1}^\sharp(\mu^{-1}(z), \mathbb{C}) = 0, \tag{30}$$

then we have $E_1^{p,q} = 0$ for all p. And if (29) holds with $q = 0$, then

$$E_1^{p,0} = \Gamma(U'', \mathcal{E}^{0,p}(V)).$$

This result thus gives sufficient conditions for the generalized pull-back mechanism (28) to degenerate to the isomorphism (26) induced by μ^*.

The identification of (29) when μ has topologically non-trivial fibres was studied by Singer in [42] and falls into two parts. First one defines the de Rham homology $H_*(U'/U'')$ by using μ-relative differential forms with compact support in the vertical direction; then the wedge product and integration over the fibres of μ define a $C^\infty(U'')$-linear pairing

$$E_1^{p,q} \otimes H_q(U'/U'') \longrightarrow \Gamma(U'', \mathcal{E}^{0,p}(V)).$$

Then we have the following generalization of the Poincaré duality theorem: for each q, there exists a natural exact sequence

$$0 \to \mathrm{Ext}^1(H_{q-1}(U'/U''), \mathcal{E}^{0,p}(V)) \to E_1^{p,q} \to \mathrm{Hom}(H_q(U'/U''), \mathcal{E}^{0,p}(V)) \to 0 \tag{31}$$

(where Hom and Ext are calculated in the category of $C^\infty(U'')$-modules). The proof of the result involves the reduction of the problem to a combinatorial one by use of a Čech description of the groups involved. The Čech description can be useful in explicit calculation since it shows how to construct (in principle) a two-step projective resolution of $H_q(U'/U'')$, from which everything in (31) can (again, in principle) be calculated.

The identification of $H_*(U'/U'')$ is aided by the other main result of [42]. To describe this recall that one can localize any $C^\infty(U'')$-module M at any point z in U'' to get a kind of 'fibre' $M_z = M/M \otimes \mathcal{I}_z$ where $\mathcal{I}_z$ is the ideal of functions which vanish at z. Then we have

$$(H_q(U'/U''))_z = H_q(\mu^{-1}(z)).$$

In contrast, the cohomology groups of a general family are not at all well behaved under localization (cf. the example in [14]). We note that Buchdahl's theorem is contained in these two results.

4.2.2 The Isomorphism of Fields and 'Potentials Modulo Gauge'

Let us now consider the other way in which non-trivial topology of U enters, in the possible failure of (27). The failure of this isomorphism is relevant because the Penrose transform naturally identifies the cohomology of $\mathcal{O}(n-2), n \geq 1$ (for example) with 'potentials modulo gauge' (see [20], but it is an easy exercise in the use of the methods of §2.3). In his thesis [13], Buchdahl used what is now recognized as the BGG resolution on $\mathbf{M}$ to analyze the relation between potentials and fields. He showed that if U is Stein there is an exact sequence

$$0 \longrightarrow H^1(U) \longrightarrow \{\text{potentials}\}/\{\text{gauge}\} \longrightarrow \{\text{fields}\} \longrightarrow H^2(U) \longrightarrow \cdots,$$
$$(32)$$

where the coefficients in the cohomology groups are $\odot^{n-2}\mathbf{T}^*$ (for fields with n un-primed indices) and $\odot^{n-2}\mathbf{T}$ (for fields with n primed indices). If U is not Stein then the BGG resolution is not acyclic and a more complicated result, involving the analytic cohomology of U, holds. The map from fields to $H^2(U)$ which is an obstruction to the existence of potentials can be interpreted physically as giving the *charge* of the field.

4.2.3 Examples

Now that we have described the machinery of the pull-back mechanism and have indicated how topology can also affect the analytic part of the Penrose transform let us indicate how to set up the Penrose transform for any complex conformally flat four-manifold R with complex spin structure. The correspondence space R' is defined as the total space of the dual projective primed spin bundle and the twistor space R'' is the space of leaves of the foliation generated by $\pi^{A'}\nabla_{AA'}$: this will be a complex manifold but it may very well be non-Hausdorff. We illustrate the use of this construction and the generalized pull-back mechanism with some examples.

Example 1

If R is a subset of $\mathbf{M}$ then R'' will be the corresponding subset of $\mathbf{P}$.

Example 2

We have seen that to develop a twistor description of sourced fields, we want to take $R = \tilde{U}$, the double cover of $U \setminus B$, where it is assumed that U is Stein and topologically trivial. One finds that each α-plane in R maps in two-to-one fashion to the corresponding α-plane in $U \setminus B$. When the α-plane corresponds to a point z not on $\mathcal{L}$, this is an honest double-cover, but when z moves on to $\mathcal{L}$, the α-plane in R breaks into two connected pieces. Thus the space of leaves R'' is the non-Hausdorff complex manifold obtained from U'' by adding an extra copy of $\mathcal{L}$. (This is not the way in which the non-Hausdorff twistor space R'' was first arrived at: cf. [2].)

Once analytic cohomology of non-Hausdorff twistor spaces has been appropriately defined, it is not particularly difficult to construct an extension of the Penrose transform

$$H^1(R'', \mathcal{O}(-n-2)) \longrightarrow \{\text{fields on } \tilde{U}\} \tag{33}$$

in terms of the standard contour integral formulae. Bailey showed, moreover, that the LHS of (33) is canonically isomorphic to the LHS of (25). In [6], it was proved that (33) is in fact an isomorphism for all $n \geq 0$ and is nearly so for $n < 0$. The problem for the positive homogeneities comes from the 'secondary difficulty', the failure of (27) (cf. (32)). It turns out that (33) becomes an isomorphism when $\mathcal{O}(n-2)$ is replaced by the quotient sheaf $\mathcal{C}(n-2)$ defined by the short exact sequence

$$0 \to \odot^{n-2}\mathbf{T}^* \to \mathcal{O}(n-2) \to \mathcal{C}(n-2) \to 0 \tag{34}$$

where $\odot^{n-2}\mathbf{T}^*$ is identified with the polynomial functions in $\mathcal{O}(n-2)$. We shall see later that the passage from $\mathcal{O}$ to $\mathcal{C}$ can be regarded as a twistorial version of the passage from potentials to fields. One should think of $\mathcal{C}$ as a minor modification whose cohomology differs from that of $\mathcal{O}$ by a finite-dimensional piece which consists of the simplest sourced fields having non-zero conserved quantities (generalized charges).

Example 3

In [41], the twistor transform from cohomology of U'' to cohomology of $''U$ was investigated by use of the correspondence between twistor space and dual twistor space. If U'' and $''U$ are the open sets in $\mathbf{P}$ and $\mathbf{P}^*$ corresponding to the same convex open set U of $\mathbf{M}$, one obtains a correspondence

$$\mathbf{P} \xleftarrow{\ \alpha\ } \hat{U} \xrightarrow{\ \beta\ } {}''U$$

in which the fibres of β are compact (copies of $\mathbf{CP}^2$, in fact) and the topological type of the fibre $\beta^{-1}(z)$ is that of a point if $z \notin U''$ and is that of S^2 if $z \in U''$. This results in an identification

$$H^1(U'', \mathcal{O}(-n-2)) = H^3(\hat{U}, \alpha^{-1}\mathcal{O}(-n-2))$$

and it turns out that the RHS can be identified with $H^1(''U, \mathcal{O}(n-2))$. The latter identification involves significant use of the BGG resolution on twistor space, a topic to be dealt with in §4.3.

We conclude this section with the observation that the generalized pullback mechanism (28) should be of considerable use in other contexts where a global twistor description is needed. Such contexts include the global description of the 'complex propagator' $(x-y)^{-2}$ and of other n-point functions which arise in the standard approach to field theory.

4.3 *Miscellaneous Applications of the BGG Resolution on Twistor Space*

We want to gather here some observations concerning various homogeneous vector bundles on twistor space: they will serve to illustrate the machinery described in §1 and will also illuminate the introduction of the sheaves $\mathcal{C}(n-2)$ needed for the twistor description of charged fields. We feel, however, that full use has not yet been made of them, particularly with regard to the twistorial interpretation of the charge integrals.

The prototypical example is the holomorphic de Rham sequence on $\mathbf{P}$

$$0 \longrightarrow \mathbb{C} \longrightarrow \Omega^\bullet. \tag{35}$$

This is evidently closely linked to the twistor description of Maxwell theory; the first and last sheaves are $\mathcal{O}$ and $\mathcal{O}(-4)$ and H^1 of these gives respectively potential modulo gauge for left-handed Maxwell fields and right-handed Maxwell fields. Moreover, the constant sheaf $\mathbb{C}$ which is resolved by (35), has significance in that electromagnetic charges live in $\mathbb{C}$ (as opposed to some more elaborate representation space of the conformal group). It is of interest to consider the Penrose transform of (H^1 of) the other bundles which appear in (35). In §2.3.1, it was shown that the Penrose transform of $H^1(U'', \Omega^1)$ consists of two pieces,

$$\{\text{gauge-restricted gauge for right-handed potentials on } U\}$$

and

$$\{\text{left-handed fields on } U\}.$$

The Penrose transform of $H^1(U'', \Omega^2)$ is, as we also remarked in §2.3.1,

$$\{\text{gauge-restricted right-handed potentials on } U\}.$$

In these identifications, 'gauge-restricted' refers to the imposition of the (conformally invariant) conditions

$$\Box^2 f = 0 \text{ and } \Box \nabla^a \Phi_a = 0,$$

as discussed above. Thus the Penrose transform of the complex (35) contains all the spaces of fields that one is interested in for Maxwell theory: potential modulo gauge *and* fields for both the left-handed and the right-handed parts of the field. Moreover, the Penrose transform of the differential d of the complex effects the various maps $\{\text{gauge}\} \rightarrow \{\text{potentials}\} \rightarrow \{\text{fields}\}$. In particular, the two pieces in the Penrose transform of Ω^1 are identified as

$$\{\text{gauge for right-handed potentials}\} = \ker d$$

and

$$\{\text{left-handed fields}\} = \operatorname{coker} d.$$

Thus the part of Ω^1 that corresponds to left-handed fields can be identified with $\mathcal{C}(0) = \mathcal{O}/\mathbb{C}$. From this point of view, the introduction of $\mathcal{C}(0)$ to describe left-handed fields on our double-cover $\tilde{U}$ is very natural. In that connection, note that we have the connecting homomorphism from (34)

$$H^1_{\mathcal{L}}(U'', \mathcal{O}/\mathbb{C}) \longrightarrow H^2_{\mathcal{L}}(U'', \mathbb{C}) \simeq \mathbb{C} \tag{36}$$

and the 'iterated connecting homomorphism' from the de Rham sequence (cf. [2])

$$H^1_{\mathcal{L}}(U'', \mathcal{O}(-4)) \longrightarrow H^4_{\mathcal{L}}(U'', \mathbb{C}) \simeq \mathbb{C} \tag{37}$$

which evaluate the charge of a left-handed or a right-handed Maxwell field. It is easy to see from (36) that the charge of ϕ is precisely the obstruction to the representation of ϕ by a class in $H^1_{\mathcal{L}}(U'', \mathcal{O})$ (as opposed to a class in $H^1_{\mathcal{L}}(U'', \mathcal{C}(0))$).

For other helicities we need to consider the BGG resolutions on twistor space. These exact resolutions are given by

$$0 \to (0, p, p+q, p+q+r)(\mathsf{T}) \to \mathcal{O}(p, q, r) \to \mathcal{O}(-p-2, p+q+1, r)$$
$$\to \mathcal{O}(-p-q-3, p, q+r+1) \to \mathcal{O}(-p-q-r-4, p, q) \to 0$$

where the notation $(0, p, p+q, p+q+r)(\mathsf{T})$ denotes the constant sheaf with values in a certain irreducible part of a tensor power of the non-projective twistor space T as described in the appendix. It is instructive to write out these sequences with twistor indices (obeying certain restrictions) as described just before (8). Some of the maps are given easily in terms of powers of $\partial/\partial Z^{\alpha}$, whereas some can only be written down implicitly.

We are particularly concerned with the resolutions of $\odot^{n-2}\mathsf{T}^{*}$ and $\odot^{n-2}\mathsf{T}$ which take the form

$$0 \to \odot^{n-2}\mathsf{T}^{*} \to \mathcal{O}(n-2) \to \mathcal{O}(-n, n-1, 0)$$
$$\to \mathcal{O}(-n-1, n-2, 1) \to \mathcal{O}(-n-2, n-2, 0) \to 0 \tag{38}$$

and

$$0 \to \odot^{n-2}\mathsf{T} \to \mathcal{O}(0, 0, n-2) \to \mathcal{O}(-2, 1, n-2)$$
$$\to \mathcal{O}(-3, 0, n-1) \to \mathcal{O}(-n-2) \to 0; \tag{39}$$

they play an analogous role in the twistor description of massless fields of spin $n/2$. Indeed, both (38) and (39) reduce to the holomorphic de Rham sequence (35) when $n = 2$. Thus for example $\mathcal{C}(n-2)$ can be identified with a subsheaf of $\mathcal{O}(-n, n-1, 0)$ and the Penrose transform of its cohomology is the space of left-handed fields of spin $n/2$. Also, $\mathcal{O}(-2, 1, n-2)$ and $\mathcal{O}(-3, 0, n-1)$ give gauge-restricted gauge and gauge-restricted potentials for right-handed fields of spin $n/2$. The twistorial charge integrals (36, 37) have obvious modifications

$$H^{1}_{\mathcal{L}}(U'', \mathcal{C}(n-2)) \longrightarrow H^{2}_{\mathcal{L}}(U'', \odot^{n-2}\mathsf{T}^{*}) \simeq \odot^{n-2}\mathsf{T}^{*}$$

and

$$H^{1}_{\mathcal{L}}(U'', \mathcal{O}(-n-2)) \longrightarrow H^{4}_{\mathcal{L}}(U'', \odot^{n-2}\mathsf{T}) \simeq \odot^{n-2}\mathsf{T}. \tag{40}$$

Note that when $n = 4$, (40) gives a cohomological interpretation to the twistor integral formula

$$\oint Z^{\alpha} Z^{\beta} f_{-6}(Z) \Omega$$

for the 'gravitational charges' due to a right-handed spin-2 field.

It is of interest to note that each of the vector spaces which appear in the 'moment sequences' of [39] can be resolved using a BGG resolution and then the two $n = 2$ sequences are separated by the BGG resolution for $\mathsf{T} \otimes \mathsf{T}^{*}$. The implications of these observations remain obscure.

A   Notations for Homogeneous Bundles

The purpose of this notation is to provide a dictionary relating the notation for homogeneous bundles and their curved analogues which we have used to that used in other parts of the literature.

Table 3: Dynkin notation for homogeneous bundles

Space	Dynkin notation	Our notation	restrictions
$\mathcal{M}$	$\overset{p}{\bullet}\!-\!\overset{q}{\times}\!-\!\overset{r}{\bullet}$	$\mathcal{O}^{\overbrace{A'\cdots L'}^{p}\overbrace{A\cdots L}^{r}}[q]$	$p, r \geq 0$
$\mathcal{P}$	$\overset{p}{\times}\!-\!\overset{q}{\bullet}\!-\!\overset{r}{\bullet}$	$\mathcal{O}(p,q,r)$	$q, r \geq 0$
$\mathcal{A}$	$\overset{p}{\times}\!-\!\overset{q}{\bullet}\!-\!\overset{r}{\times}$	$\mathcal{O}(p,q,r)$	$q \geq 0$
$\mathcal{G}$	$\overset{p}{\times}\!-\!\overset{q}{\times}\!-\!\overset{r}{\times}$	$\mathcal{O}(p,r)[q]$	None
$\mathcal{F}$	$\overset{p}{\times}\!-\!\overset{q}{\times}\!-\!\overset{r}{\bullet}$	$\mathcal{O}^{\overbrace{(A\cdots L)}^{r}}(p)[q]$	$r \geq 0$

A.1   Dynkin Diagram Notation

The dynkin diagram based notation is used in [18, 4, 9] (the last of these contains a full explanation of the general case) and there is also a brief note by Dunne explaining the general case in this volume. It is related to our notation according to table 3.

A.2   Flag Bundle Notation

There is another notation that has been used to describe the homogeneous bundles on flag varieties. It has the advantage that it can be used when the primed and unprimed conformal weights are not identified.

To begin, let T be an n-dimensional vector space. The irreducible parts of the tensor powers of T are then classified by *Young tableaux* as described in [46] or from a different viewpoint in [38]. We write $(a_1, a_2, \cdots, a_n)$ for the Young tableaux with rows of length $a_n, \cdots, a_1$ Thus, for example

$$(2, 2, 5, 6) =$$

and we write $(2, 2, 5, 6)(\mathsf{T})$ for the corresponding space of irreducible tensors. The numbers a_i are of course necessarily non-decreasing and non-negative.

To continue, let us for definiteness set $n = 4$. We note that $(1, 1, 1, 1) = \Lambda^4(\mathsf{T})$ and thus we can include all the irreducibles constructed from T and T^* if we use the symbols (a, b, c, d) for *any* non-decreasing sequence of integers a, b, c, d together with the rule

$$(a, b, c, d) \otimes (\Lambda^4(\mathsf{T}))^n = (a + n, b + n, c + n, d + n).$$

The dual of (a, b, c, d) is then $(-d, -c, -b, -a)$.

Minkowski space M is the Grassmannian of 2-dimensional subspaces of T. As such, it has a naturally defined *tautological bundle* L whose fibre over any point is just the 2-plane defining that point (this is actually $\mathcal{O}_{A'}$ of course). There is also the quotient bundle $L' = \mathsf{T}/L$ where T is the trivial product bundle over M (which is $\mathcal{O}^A$).

All the homogeneous bundles on M are given as

$$(a, b | c, d) \stackrel{\text{def}}{=} (a, b)L \otimes (c, d)L'$$

for integers a, b, c, d which clearly must obey $b \geq a, d \geq c$. Note that we have extended our notation so that (a, b) etc. act on vector bundles fibre by fibre as well as on vector spaces. Similarly, the homogeneous bundles on P are specified by symbols of the form $(a | b, c, d)$ with $d \geq c \geq b$ and those on A by symbols of the form $(a | b, c | d)$ with $c \geq b$. It is easy to see that duality in this notation is given by, for example,

$$\begin{aligned} (a, b | c, d)^* &= (-b, -a | -d, -c) \\ (a | b, c, d)^* &= (-a | -d, -c, -b) \end{aligned}$$

on M and P respectively.

Translating to our notation, we have

$$(a, b | c, d) = \mathcal{O}^{\overbrace{A \cdots L}^{d-c} \, \overbrace{A' \cdots L'}^{b-a}}[c - b]$$

but if this notation is used in full generality, keeping track of primed and un-primed conformal weights separately, the conformal weight above is actually $[c][-b]'$. We will not make this distinction here, and this also makes itself felt in the notation by the equivalences

$$\begin{aligned} (a, b | c, d) &= (a + n, b + n | c + n, d + n) \\ (a | b, c, d) &= (a + n | b + n, c + n, d + n). \end{aligned}$$

It is then easy to convert these symbols to the equivalent Dynkin form and back again (with the above ambiguity of course) using the formulae

$$\text{On } \mathsf{M}: \quad (a, b | c, d) = \overset{\text{b-a} \quad \text{c-b} \quad \text{d-c}}{\bullet\!\!-\!\!\times\!\!-\!\!\bullet}$$
$$\text{On } \mathsf{P}: \quad (a | b, c, d) = \overset{\text{b-a} \quad \text{c-b} \quad \text{d-c}}{\times\!\!-\!\!\bullet\!\!-\!\!\bullet}.$$

As an application, we will check that $\left(\overset{1 \quad 0 \quad 1}{\times\!\!-\!\!\bullet\!\!-\!\!\bullet}\right)^* = \overset{-2 \quad 1 \quad 0}{\times\!\!-\!\!\bullet\!\!-\!\!\bullet}$ as is implicit in the results of §2.3.2. Thus we have (making a choice for simplicity) $\overset{1 \quad 0 \quad 1}{\times\!\!-\!\!\bullet\!\!-\!\!\bullet} = (-1 | 0, 0, 1)$. Taking duals we get $(-1 | 0, 0, 1)^* = (1 | -1, 0, 0)$ and converting back we get $(1 | -1, 0, 0) = \overset{-2 \quad 1 \quad 0}{\times\!\!-\!\!\bullet\!\!-\!\!\bullet}$ as required.

References

[1] Arnold, V.I., *Mathematical methods of classical mechanics*, Springer, New York (1978).

[2] Bailey, T.N. *Twistors and fields with sources on world-lines*, Proc. R. Soc. Lond. **A397** (1985), 143–155.

[3] Bailey, T.N. *Relative cohomology and multipole expansions*, Preprint (1988).

[4] Bailey, T.N. & Eastwood, M.G. *Complex paraconformal manifolds, their differential geometry and twistor theory*, Forum Math. (1990), to appear.

[5] Bailey, T.N., Ehrenpreis, L. & Wells, R.O., Jr. *Weak solutions of the massless field equations*, Proc. R. Soc. Lond. **A384** (1982), 403–425.

[6] Bailey, T.N. & Singer, M.A. *On the twistor description of sourced fields*, Proc. R. Soc. Lond. **A422** (1989), 367–385.

[7] Bailey, T.N., *Some geometry associated with the Einstein bundle*, Class. Quantum Grav. (1989), in press.

[8] Baston, R.J., *The algebraic construction of invariant differential operators*, D. Phil thesis, Oxford University 1985.

[9] Baston, R.J. & Eastwood, M.G., *The Penrose transform: its interaction with representation theory*, O.U.P., Oxford (1989).

[10] Baston, R.J. & Mason, L.J., *Conformal gravity, the Einstein equations and spaces of complex null geodesics*, Class. Quantum Grav. **4** (1987), 815–826.

[11] Berstein, I.N., Gelfand, I.N. & Gelfand, S.I., *Differential operators on principal affine spaces and investigation of g-modules*, in *Lie groups and their representations* (Ed. I.M. Gelfand), Hilgar, London, 1975, pp. 21–64.

[12] Bredon, G., *Sheaf theory*, McGraw-Hill (1967).

[13] Buchdahl, N.P. 1982 *Applications of several complex variables to twistor theory* D.Phil Thesis, Oxford University.

[14] Buchdahl, N.P. *On the relative de Rham sequence*, Proc. A.M.S. **87** (1983) 363–366.

[15] Conway, A.W., *The field of force due to a moving electron*, Proc. Lond. math. Soc. series 2, **1** (1903), 154–165.

[16] Dighton, K., *An introduction to the theory of local twistors*, Int. J. Theor. Phys. **11** (1974), 31–43.

[17] Eastwood, M.G., *A duality for homogeneous bundles on twistor space*, J. Lond. Math. Soc. **31** (1985), 349–356.

[18] Eastwood, M.G., *The Penrose transform for curved ambitwistor space*, Quart. J. Math. **39** (1988), 427–441.

[19] Eastwood, M.G. & Pilato, A.M., *On the density of twistor elementary states*, Pac. Math. J., to appear.

[20] Eastwood, M.G., Penrose, R. & Wells, R.O., Jr. , *Cohomology and massless fields*, Commun. Math. Phys. **78** (1981) 305–351.

[21] Eastwood, M.G. & Rice, J.W. *Conformally invariant differential operators on Minkowski space and their curved analogues*, Commun. Math. Phys. **109** (1987) 207–228.

[22] Eastwood, M.G. & Singer, M.A. *A conformally invariant Maxwell gauge*, Phys. Lett. **107A** (1985), 73–74.

[23] Eastwood, M.G. & Tod, K.P. *Edth—a differential operator on the sphere*, Math. Proc. Camb. Phil. Soc. **92** (1982), 317–330.

[24] Godement, R., *Topologie algébrique et théorie des faisceaux*, Hermann (1964).

[25] Gover, R. *Conformally invariant operators of standard type*, Quart. J. Math. **41** (1989).

[26] Gover, R., *A geometrical construction of conformally invariant differential operators*, D.Phil. thesis, Oxford University (1989).

[27] Griffiths, P. & Harris, J. *Principles of algebraic geometry*. New York: Wiley (1978).

[28] Guillemin, V.W., Kashiwara, M. & Kawai, T., *An introduction to microlocal analysis*, Ann. Math. Stud. 93, Princeton, (1979).

[29] Henkin, G.M. & Polyakov, P.L., *Homotopy formulas for the $\bar{\partial}$ operator on $\mathbb{CP}^n$ and the Radon-Penrose transform*, Isvestja Akad. Nauk. SSSR **50** (1986), 566-597 (in Russian). Math. USSR Izvestiya **28** (1987), 555–587 (in English).

[30] Hodges, A.P., Penrose, R. & Singer, M.A., *A twistor conformal field theory for four space-time dimensions*, Phys. Lett. **216B** (1989), 48–52.

[31] Huggett, S.A. & Tod, K.P. *An introduction to twistor theory*, London Math. Soc. Student Texts 4, Cambridge University Press (1985).

[32] Komatsu, H., Two articles in *Hyperfunctions and pseudo-differential equations*, Lecture notes in mathematics, vol. 287. Berlin: Springer (1973).

[33] LeBrun, C.R., *Ambi-twistors and Einstein's equations*, Class. Quantum Grav. **2** (1985) 555–563.

[34] LeBrun, C.R., *Thickenings and conformal gravity*, preprint (1989).

[35] Lepowsky, J., *A generalization of the Bernstein-Gelfand-Gelfand resolution*, J. Algebra **49** (1977), 496–511.

[36] Mason, L.J., *Local twistors and the Penrose transform for homogeneous bundles*, Twistor Newsletter **23** (1987), 36–44.

[37] Penrose, R., *Local H^1s and propogation*, in *Advances in twistor theory* (eds. Hughston, L.P. & Ward, R.S.), Pitman (1979).

[38] Penrose R. & Rindler, W. *Spinors and space-time, Vol. 1*, C.U.P., Cambridge, 1984.

[39] Penrose, R. & Rindler, W. *Spinors and space-time, Vol. 2*, C.U.P., Cambridge, 1986.

[40] Polking, J.C. & Wells, R.O., Jr., *Hyperfunction boundary values and a generalized Bochner-Hartogs theorem*, in 'Several complex variables' (Proc. Sympos. Pure Math., Vol. XXX, Part 1, Williams Coll. Williamstown, Mass., 1975), AMS (1977).

[41] Singer, M.A., *Duality in twistor theory without Minkowski space*, Math. Proc. Cam. Phil. Soc. **98** (1985), 591–600.

[42] Singer, M.A., *A duality theorem for the de Rham theory of a family of manifolds*, Preprint (1988).

[43] Singer, M.A., *Flat twistor spaces, conformally flat manifolds and four-dimensional conformal field theory*, Preprint (1989).

[44] Wells, R.O., Jr., *Hyperfunction solutions of the zero-rest-mass field equations*, Commun. Math. Phys. **78** (1981) 367–600.

[45] Woodhouse, N.M.J. *Real methods in twistor theory*, Class. Quantum Grav. **2** (1985), 257–291.

[46] Wybourne, B.G., *Symmetry principles and atomic spectroscopy*, Wiley-Interscience (1970).

Twistor Diagrams and Feynman Diagrams

A.P. Hodges

1 Introduction

The purpose of twistor diagram theory remains as first envisaged by Penrose [18]. It is to enable us to identify a new fundamental structure in twistor geometry for the description of quantum field theory (QFT) in flat space-time. This new structure should eliminate the unsatisfactory features of standard QFT, particularly the divergences and renormalisation schemes. The further hope is that a twistor-geometric formalism could explain the spectrum of observed physical particles and provide the starting point for the unification of QFT with gravity.

The essential character of twistor diagram theory also remains Penrose's. The idea is that twistor diagrams themselves are to play the rôle in a twistorial QFT of Feynman diagrams in conventional QFT. That is, they are to supply a perturbation expansion calculus for physical scattering amplitudes. The guiding principle is that in analogy to Feynman diagrams they should be defined by the systematic combination of very simple elements. These elements and the rules for combining them should be expressible in a form which is essentially combinatorial (hence appropriately represented by a diagrammatic formalism). The elements and the rules should also be manifestly finite. In the original 'classical' theory they were also to be manifestly conformally invariant. As such the theory was at first necessarily restricted to the description of massless fields, and indeed was envisaged originally as to be applied to massless quantum electrodynamics (QED). However, manifest finiteness has always been the more fundamental demand. The task, from the inception of the theory to the present moment, has been to locate such finite elements and combinatorial rules.

2 Strategy of the Theory

It was originally hoped to develop twistor diagrams from an autonomous twistor dynamical principle [18], and indeed the very first twistor diagrams written down by Penrose were based on some ideas drawn from twistor Hamiltonians. But establishing the correctness or otherwise of these diagrams meant in effect showing their equivalence to standard Feynman diagram calculations, and during the 1970s the emphasis of the theory became focussed upon the work of finding a direct correspondence between twistor diagrams and Feynman diagrams (for massless QED). According to this approach, our immediate object is basically to translate term by term the Feynman perturbative expansion, which in conventional theory is determined uniquely by the Lagrangian for the interaction, and gives finite predictions provided a renormalisation scheme is incorporated. In effect we recognise this expansion and its finite predictions as embodying observed truth, and test candidate twistor calculations for agreement with that truth.

Although such a term-by-term 'translation' may appear somewhat unambitious as a strategy, it does in fact demand some subtle innovative element. For the twistor diagram calculus must not remain so close to the standard theory that it merely reproduces every feature of the Feynman integrals, with their well-known problems. We are actually looking for a calculus that keeps close enough to space-time calculations to ensure agreement with their finite conclusions, yet which possesses new twistor-geometric features differing sufficiently from space-time constructions for there to be an actual change and improvement where the Feynman integral scheme is unsatisfactory, particularly in its divergences, but also in its treatment of gauge fields.

By emphasising the agreement with finite Feynman calculations we safeguard twistor diagram theory from the wilder speculations to which it was prone in early days. However the 'translation' programme is clearly inadequate as the basis for an actual physical theory. Even if all Feynman diagram calculations could be made manifestly finite in a twistor diagram translation, we should still be borrowing the predictive content of the theory from a space-time Lagrangian. It would not be a fully twistor-geometric theory. At some point we must identify a twistor-theoretic equivalent to a Lagrangian (such as Penrose tried to use in very early days) from which the perturbative expansion of twistor diagrams can be deduced. The general hope has been that such a fully twistorial principle will emerge when the shape of higher-order twistor diagrams becomes more clear. In my view it is still very reasonable proposal that by finding a consistent pattern for twistor diagrams translating Feynman diagrams, we shall discover a fully twistorial generating principle from which those twistor diagrams (and hence a complete predictive calculus for interactions) could be deduced without reference to space-time concepts. The 'skeleton' pattern which will emerge at the conclusion of this review encourages me in this belief. But other recent work has breathed new life into the original, more radical, proposal of formulating a generating principle directly: this is the four-dimensional conformal field theory (CFT4) hypothesis due to Singer [14, 20, 22, 21]. So far it has not been possible to suggest a connection between this principle and the diagram calculus, except in the most general terms, but it is hoped that in future work some version of the CFT4 hypothesis will be the guide in making sense of the diagram calculus as a predictive physical theory.

3 Classical Twistor Diagram Theory

I first offer a very brief review of the core ideas in 'classical' twistor diagram theory as developed in [18, 19, 23, 24, 1, 25, 13, 4, 5, 6]. The results and the problems of this first phase of the theory are still of great importance and motivate the later developments to be described below.

The elements originally proposed by Penrose were essentially:

- External fields $f(Z^\alpha), g(W_\alpha)$ etc. in the projective twistor (dual twistor) representation, these being understood at this stage to be representative singular functions in projective twistor (dual twistor) space. They are indicated diagrammatically by

$$\underset{f}{Z} \qquad , \qquad \overset{g}{W}$$

  These may be either ingoing (positive frequency) or outgoing (negative frequency). For computations we shall always use elementary states, i.e. those for which the representative twistor functions are simple rational functions.

- The natural projective forms

$$DZ = \frac{1}{6}\epsilon_{\alpha\beta\gamma\delta}Z^\alpha dZ^\beta \wedge dZ^\gamma \wedge dZ^\delta, DW = \frac{1}{6}\epsilon^{\alpha\beta\gamma\delta}W_\alpha dW_\beta \wedge dW_\gamma \wedge dW_\delta$$

  indicated by vertices

$$\bullet \, Z \qquad , \qquad \circ \, W \quad \text{respectively}$$

- Singular factors of form $n! \, (-W_\alpha Z^\alpha)^{-n-1}$ indicated by lines

$$W \overset{n}{\rule{2cm}{0.4pt}} Z$$

- The multiplication of functions, forms and singular factors in exterior algebra to yield a form in (a subspace of) a product of projective twistor and dual twistor spaces.

- The extraction of a complex number from this form. This is effected by compact contour integration; more precisely, by the choice of a homology class in the space on which the form is defined. The drawing of a twistor diagram with vertices and lines connected in the obvious way is taken to indicate both the construction of the form and its integration over a suitable contour. The result of the integration is to be interpreted as a functional of the fields. In this respect the rules are severely incomplete: the formalism does not define how such contours are to be chosen.

The object of diagram construction is to build diagrams defining functionals of fields which are interpretable as scattering amplitudes. But Penrose showed that to obtain scattering amplitudes of interest the elements as just described had to be augmented by lines indicated by

$$W \xrightarrow{\quad -1 \quad} Z$$

formally satisfying

$$\frac{\partial}{\partial W_\alpha}\left(W \xrightarrow{\quad -1 \quad} Z\right) = Z^\alpha \left(W \xrightarrow{\quad 0 \quad} Z\right)$$

Roughly speaking, such elements should behave like $\log(W_\alpha Z^\alpha)$, but their proper definition in projective space was (and has continued to be) a source of difficulty. Sparling's work suggested these elements could be consistently defined in the scheme by the rule that a (-1)-line indicates no contribution to the differential form, but prescribes a contour with *boundary* on the relevant subspace of form $\{W_\alpha Z^\alpha = 0\}$.

It is easy to extend this definition to lines labelled by any negative integer but this will not be required here. Sparling also extended the notation to

$$W \xrightarrow{\quad \lambda \quad} Z \;=\; \Gamma(\lambda + 1)(-W_\alpha Z^\alpha)^{-\lambda-1}$$

for complex λ. The limit $\lambda \to -1$ is then consistent when it is defined (a subtlety relevant in discussing Møller scattering below). This extension is to be understood as technically useful and not part of the fundamental theory. The classical rules also specified various factors of $2\pi i$, and some conventions regarding 'double lines' which are irrelevant to this review and will be ignored along with all numerical factors and physical constants.

For present purposes the essential point to note is that a 0-line is a simple pole generally playing the 'evaluating' role that it does in Cauchy's theorem, so that a 1-line (a double pole) plays the role of evaluating a derivative. The vital significance of a (-1)-line is that it defines and evaluates an inverse derivative.

We can now exhibit simple examples of twistor diagrams. Note that we shall always draw both Feynman and twistor diagrams with in-states (positive frequency fields) at the bottom and out-states (negative frequency fields) at the top.

Inner product diagram

Suppose $f(Z^\alpha), g(W_\alpha)$ are both of homogeneity degree $n - 2$, representing positive, negative frequency fields respectively. Then there is a contour for the twistor diagram

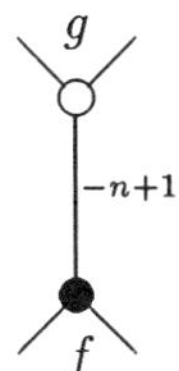

yielding an amplitude which agrees with the inner product of the corresponding fields. This is the simplest possible (i.e. zeroth order) Feynman diagram.

To obtain some intuitive sense for the twistor diagram, note that when the external fields are of spin 0, the internal line is a double pole, corresponding to the derivative in the space-time definition. An inverse derivative (i.e. a (-1)-line) appears when the homogeneity of the external twistor functions is (-4). This corresponds to the fact that a gauge field is involved: in the space-time formula for the spin-1 inner product, an inverse derivative of the field has to be found.

The ϕ^4 integral

Let $f(Z^\alpha), g(W_\alpha), h(X^\alpha), j(Y_\alpha)$ each of homogeneity (-2) correspond to ϕ_1, ϕ_2, ϕ_3, ϕ_4 massless scalar fields; ϕ_1, ϕ_3 positive frequency, ϕ_2, ϕ_4 negative frequency. Then there is a contour for

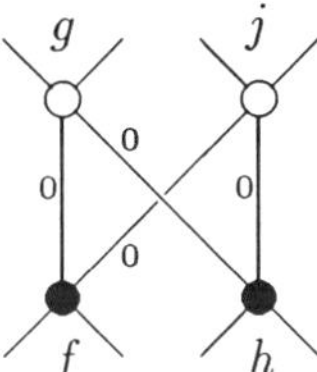

such that the result equals

$$\int d^4x \; \phi_1(x) \; \phi_2(x) \; \phi_3(x) \; \phi_4(x)$$

A twistor diagram of this form is commonly called a 'box' diagram; and I shall refer to this as the 0000 box diagram. The space-time functional of fields can be regarded as the first-order scattering amplitude in massless ϕ^4 theory, with Feynman diagram

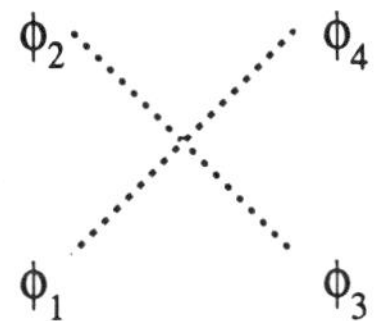

but it has a more general use as the starting-point for translating other first-order scattering amplitudes, as we shall shortly demonstrate.

Note that neither the inner product diagram, nor the 0000 box diagram, is the *unique* twistor representation of the corresponding functionals of fields. It is easy to see that there are infinitely many possible twistor diagram representations of these functionals, since *twistor transform* lines may be freely added. For the spin-0 inner product, for instance, a twistor diagram formed as a chain of any number of 1-lines will also serve. For spin-1/2 a diagram formed as a chain of alternating 0-lines and 2-lines will serve; for spin-1 a diagram of alternating (−1)-lines and 3-lines. Such chains can also be added to link the vertices of the 0000 box to the external fields; moreover there are alternatives to the 0000 box structure as representations of the the ϕ^4 integral, some of which will be detailed below. Thus at this stage the idea of 'translating' Feynman diagrams to twistor diagrams is only very loosely defined.

We may now progress to the study of first-order scattering amplitudes which involve Feynman propagators. The essential feature is the translation of *inverse derivatives* into the twistor picture, and not surprisingly this has the effect of introducing (−1)-lines into the twistor diagram. The simplest possible example of this is given by:

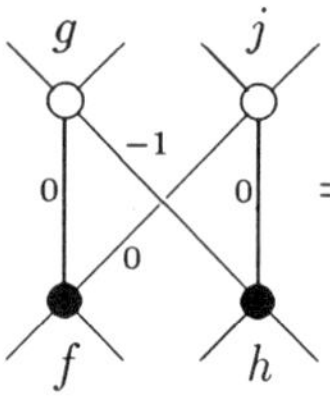

$$= \iint d^4x\, d^4y\, \phi(x)\, \psi_A(x)\, \Delta_F^{AA'}(x-y)\, \psi_{A'}(y)\, \theta(y) \qquad (1)$$

where $f(Z^\alpha)$ of degree −2 represents ϕ a positive frequency massless field; $g(W_\alpha)$ of degree −3 represents ψ^A a negative frequency massless field; similarly for $h(X^\alpha) \leftrightarrow \psi^{A'}$ and $j(W_\alpha) \leftrightarrow \theta$, and $\Delta_F^{AA'}(x-y)$ is the spin-1/2 Feynman propagator satisfying

$$\Box_x \Delta_F^{AA'}(x-y) = \nabla^{AA'}\delta(x-y)$$

The essential idea used here is that the 0000 box is a representation of the ϕ^4 integral which allows the solution of the differential equation for the massless spin 1/2 Feynman propagator. We can think of this as the application of an inverse derivative operator:

$$(I^{\alpha\beta}Y_\alpha W_\beta)(I^{\alpha\beta}Y_\alpha \frac{\partial}{\partial X^\beta})^{-1}$$

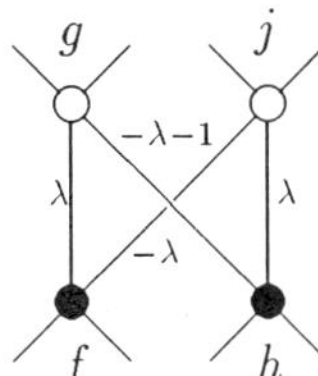

It needs only straightforward spinor algebra to verify that the interior of
the twistor diagram is a formal solution of the essential differential equation.
Much more work is needed to show that a contour with boundary actually
exists, and that the result of the contour integral then agrees exactly with
the space-time calculation. The necessary arguments are given in [5]. As a
technique, it is very useful to study

and the limit $\lambda \to 0$.

Similar inverse derivative operations relevant to other helicities may readily be found; in particular we shall shortly apply the same techniques to
obtain diagrams for massless QED. However, before going further we must
address a question that underlies all these claims.

4   Diagrams and Cohomology

The classical twistor diagram rules specify compact contour integration of
representative functions. But how do we know that the result will be independent of the representative? For—as has been known since the earliest
days—there are certainly contours which yield representative-dependent results. Some extra rules are therefore required. Originally the non-uniqueness
of the twistor representation of fields, and the lack of linear structure on the
space of fields, was very puzzling. It only became clear in 1976, when several
years of work on twistor diagrams had been done, that massless free fields
were to be identified with sheaf cohomology group elements; in particular
positive (negative) frequency fields with elements of

$$H^1(\mathbf{PT}^\pm, \mathcal{O}(-n-2)), \quad H^1(\mathbf{PT}^{*\mp}, \mathcal{O}(n-2))$$

(which we will refer to as '1-functions'). At last some sense could be made of
the ill-defined singularity structures in the diagram contour integrals.

Amplitudes are multilinear maps from the spaces of positive (negative)
frequency fields into $\mathbf{C}$; so in the light of this identification, the diagrams

could be considered as defining functionals on 1-functions. For instance, the integration of the 0000 box diagram should correspond to a map

$$H^1(\mathbf{PT}^+, \mathcal{O}(-2) \otimes H^1(\mathbf{PT}^+, \mathcal{O}(-2))$$
$$\otimes H^1(\mathbf{PT}^{*+}, \mathcal{O}(-2)) \otimes H^1(\mathbf{PT}^{*+}, \mathcal{O}(-2)) \to \mathbf{C}$$

In principle it should be possible to reformulate the twistor diagram rules so that the concept of representative function is never introduced. The whole diagram formalism should be defined to yield convolutions of 1-functions directly. Unfortunately this complete reformulation has not been achieved. More modestly (and rather less satisfactorily) one might retain the concept of contour integration with representative functions, but add some rule which will guarantee that the contour used has the effect of seeing the 1-functions and not their representatives. Even this has proved no easy task, but progress has been made. These problems have been addressed principally by Huggett, Ginsberg, Eastwood, and Singer. At this point the reader should refer to Huggett's article in this volume which explains the results that have been achieved, and in particular to §5 which gives the criterion found by Singer and Huggett for a contour to be cohomological.

From the point of view of the translation programme, one drawback in this new work is that it does not actually characterise the contours which yield amplitudes, i.e. functionals of positive and negative frequency states. Huggett and Singer's formalism treats the external fields as 1-functions relative to $\mathbf{CP}^1$'s in $\mathbf{PT}$; but there will in general be many contours which are cohomological in this sense, and yet which fail to be amplitudes. (For instance there is a 'period' contour of the 0000 box which has this character.)

However, perhaps the most striking difficulty is that no mathematically satisfactory analysis has yet incorporated the (-1)-lines which are essential to the Feynman diagram correspondence. Thus even for the inner product, Singer and Huggett's theory does not yet extend to the case of spin 1. (Ginsberg's work also failed to achieve this satisfactorily.) Nor does it apply to the box diagrams with (-1)-lines, of which many examples will appear below. For this reason the only actual scattering diagram to which this analysis has been applied is the 0000 box. And although it is certainly important to have a satisfactory treatment of this ϕ^4 integral, it cannot be said to have any real QFT content. In momentum-space language, it is just the delta-function. It is in introducing (-1)-lines to the box diagram through inverse derivative operators that the content of QFT begins to emerge.

This means that there are still many constructions in the 'classical' diagram theory which are not yet properly interpreted as functionals of 1-functions. The situation can be summarised as follows: there are contour constructions in which each of the two separated singularities representing the external elementary states is surrounded by an S^1; in these cases the new work provides a proof that the result of the integral is cohomological. There are other constructions in which only *one* of these singularities is surrounded by an S^1; and for which the new work provides the proof that although the explicit result of the integration may look as though it is a functional of the fields, in fact the contour is not cohomological. (This is the situation with the 'hard' contours discussed below.) However, when (-1)-lines are present the phenomenon arises of contours which certainly do yield functionals of the fields but which cannot be fibred in such a way that an S^1 surrounds each of the external singularities. The new work is silent about these cases.

Although this aspect of twistor diagram theory is still unsatisfactory, there is a good prospect that the Huggett and Singer's approach will in fact make complete sense of the classical diagram formalism, and indeed it has already played a very important rôle in clarifying the situation. The rest of this review continues on the assumption that the naïve contour integral approach will suffice for the purpose of finding out which twistor diagram structures are of physical interest. Another reason for adopting this attitude is that the contour-integral approach will lead, as we shall see, to a modification of the theory in which the interpretation of twistor functions and the connection with cohomology is changed. One should not hold up work until the classical theory is made properly cohomological, if the classical theory must in any case be superseded.

5   Problems Found in the Classical Theory

Leaving aside the problems of mathematical definition, we now turn to the application of the classical diagram theory to massless QED. We shall discover two major problems in identifying a correspondence with physical theory: (1) the appearance of an infra-red divergence and (2) the apparent absence of relations between 'crossed' processes.

First we consider Møller scattering for massless electrons, with Feynman diagram

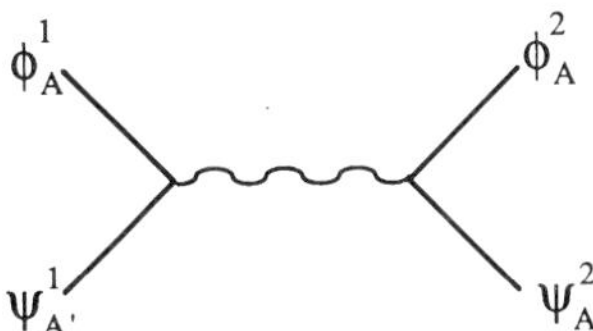

By techniques equivalent to the translation of Feynman propagators described above, Penrose and Sparling arrived at a corresponding twistor diagram

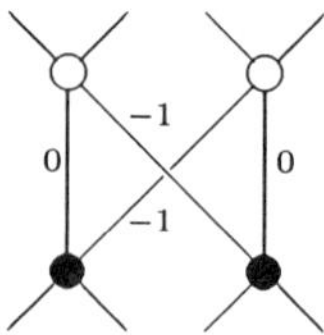

It was however noticed that the limit as $\lambda \to 1$ of

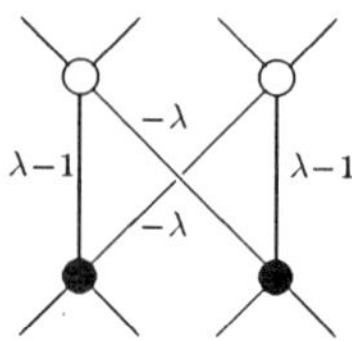

is infinite, thus presenting the theory with a serious puzzle.

Sparling hoped to resolve the problem by finding suitable contours with boundary from first principles, avoiding the use of limiting techniques. For this purpose he pioneered the use of singular homology theory to investigate the classical diagrams [23, 24]. Unfortunately the conclusion drawn from this analysis was erroneous: the contour said to yield the amplitude for Møller scattering did not in fact yield an amplitude at all, nor a solution of the essential differential equation.

It soon transpired [1, 5] that the Feynman integral, as defined by the standard rules, is in fact infinite. There is an infra-red divergence in the forward direction which is not usually taken seriously because evaluation of amplitudes is conventionally left in momentum-space rather than carried out in the Hilbert space of finite-normed states. The puzzle we faced was not therefore a problem in twistor theory but a problem in QED: how can we make sense of its prediction in this case? We cannot have a manifestly finite formalism for infinite answers, so unless finiteness is abandoned, some modification must to be introduced.

Leaving this question unsolved, we now address the problem of crossing relations. Conventional quantum field theory has the powerful consequence that the interior of any Feynman diagram defines a single kernel which can be analytically continued to different regions of momentum space and thus evaluated for different allocations of positive and negative-frequency external states. One effectively describes all the related processes at once. What is the twistor analogue of this statement?

Rather than pursue this question in relation to Møller scattering, where we are faced by the problem of divergences, it is useful to examine it in the context of the Compton scattering amplitude. Here everything is finite. Thus consider first the process:

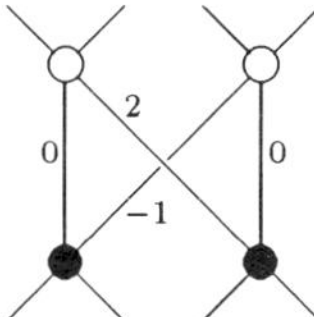

where only the sum has gauge-invariant meaning. Explicit calculation [6] shows that

gives the correct amplitude. An impressive point is that the twistor integral is not only manifestly conformally invariant but also manifestly gauge invariant; this hints at the possibility of a twistor integral theory which improves upon the Feynman diagram expansion. But what is distinctly *unimpressive* is that the twistor integral is restricted to this allocation of incoming and outgoing fields, and tells us nothing about the other related processes.

Now certainly the amplitude for one other 'channel' can be represented by a 'box' twistor diagram, namely

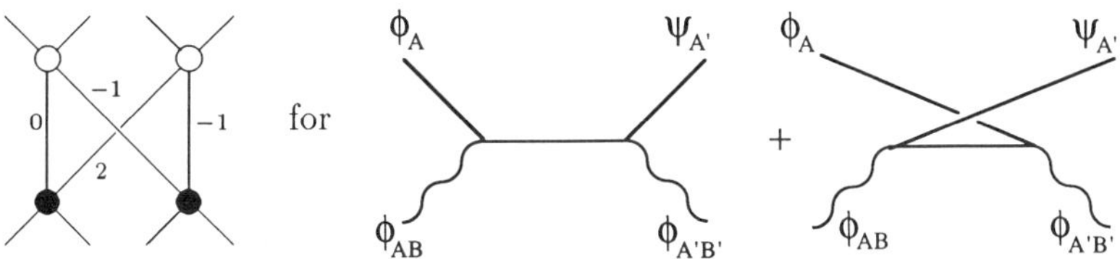

but this does not solve our problem. It only presents it more clearly. For what is the connection between these two diagrams? Must we always translate each channel independently? In general, when there are n external fields, we have up to 2^n related amplitudes. We do not want to have 2^n independent twistor diagrams corresponding to these; the analogy with Feynman diagrams would suggest that we should have just one, interpretable in 2^n ways. Without some such feature we have lost a key element of quantum field theory. It would be a particularly strange part to lose, since analyticity plays so central a role in twistor geometry. We cannot possibly abandon so vital an aspect of Feynman diagrams in our translation programme. We also have, even for this simple first-order process, a third channel for which the amplitude cannot be represented by a 'box' diagram; indeed the same question arises for the crossed 'annihilation-creation' channel of Møller scattering. How are these to be represented?

At one time it appeared that the resolution of these puzzles lay within the structure of the box diagram itself. The claim was that box diagrams such as those described above actually possessed more contours than had hitherto been identified, in fact one for *every* allocation of positive and negative frequency fields to the four vertices. This claim implied in particular that the 0000 box, with positive frequency states attached at two adjacent vertices, and negative frequency states at the other two, had a contour yielding the ϕ^4 integral. So investigation focussed on this simplest version of the claim.

In 1980 this claim seemed valid. There was good reason for this belief. Penrose [19] had sketched a proof that the 0000 box diagram represented the ϕ^4 integral, by following an explicit translation from space-time to twistor space. Huggett [15] completed this work, studying in detail not only the form but the contour induced by the translation. This indicated the existence of contours for every allocation of positive and negative frequency states, at least for elementary states. These could also be constructed directly in the twistor diagram [4]. These contours became known in folklore and *Twistor Newsletter* articles as 'hard contours'. With more work, they could indeed be generalized to Compton scattering [6].

At the time this construction seemed quite a triumph and it certainly stimulated further work, but in retrospect it was a blind alley, because the new contours constructed were *not cohomological.* In a sense this was known from the start because when applied to the 0001 box diagram, the 'hard contour' explicitly gave a representative-dependent answer. However at that time we lacked a clear conception (such as Huggett and Singer have now supplied) of what it means for a contour to be cohomological. Later papers [7, 8] clung to idea that this 'hard contour' could be the basis of a satisfactory formalism, even though one could not see how to generalise away from elementary states. It now seems clear that this view was mistaken: this construction *cannot* be identified with a functional of 1-functions. A box diagram has a cohomological contour for only one channel, and therefore cannot translate the full content of one of these first-order Feynman diagrams.

We thus have two major areas of difficulty with formulating a twistor diagram theory even for first order massless QED. Progress in the 1980s came through seeing how to extend and modify the theory to deal with them.

6   Inhomogeneous Twistor Diagrams

A mechanism for resolving the first of these problems, that of divergent amplitudes, was seen in 1983. It involves shifting the diagram structure into *non-projective* twistor space. So far we have always assumed the framework of projective twistor space. However there were always reasons to think that non-projective twistor space might prove to be the more appropriate framework, and the mechanism to be described exploits this idea.

The development actually arose somewhat indirectly. It originated in

Sparling's homology calculation for the Møller scattering diagram. This analysis restricted attention to contours which surrounded both singularities of every external function with an S^1, but Huggett's work [15] removed this restriction. The effect was to show the existence of a new contour for the Møller scattering diagram, with the property of surrounding only *one* of the external singularities of each external function with an S^1. In 1983 it was seen how to evaluate the Møller scattering diagram on this contour [7], thus exhibiting the fact that the answers are representative-dependent (i.e. the contours are non-cohomological).

Nevertheless it was striking fact that this contour did formally yield a finite solution to the essential differential equation for the Feynman propagator, and this suggested a new idea. This was that the same Huggett contour could be applied to a related integral in *non-projective* twistor space. This could be done by re-defining the integral ito make each internal line *inhomogeneous*. Thus the 0-lines are re-defined to take the form

$$(W_\alpha Z^\alpha - k_1)^{-1}, \ (Y_\alpha X^\alpha - k_3)^{-1}$$

and the boundaries re-defined to lie on

$$\{W_\alpha X^\alpha - k_2 = 0\}, \ \{Y_\alpha Z^\alpha - k_4 = 0\}$$

The external fields must be represented by functions in non-projective twistor (dual twistor) space and the forms associated with vertices are

$$D^4 Z = \frac{1}{24}\epsilon_{\alpha\beta\gamma\delta}dZ^\alpha \wedge dZ^\beta \wedge dZ^\gamma \wedge dZ^\delta, \ D^4 W = \frac{1}{24}\epsilon^{\alpha\beta\gamma\delta}dW_\alpha \wedge dW_\beta \wedge dW_\gamma \wedge dW_\delta$$

The extra dimension in non-projective twistor space makes room for an S^1 round each of the external function singularities and the Huggett contour becomes cohomological. The result of the integration then takes the form of a sum of two terms:

$$\left\{ \begin{array}{l} \textit{finite conformally invariant functional of the fields, satisfying the} \\ \textit{essential differential equation} \end{array} \right\}$$
$$+ \ \log(k_1 k_3 / k_2 k_4) \ \times \ \{\textit{zeroth order amplitude}\}.$$

Full details are in [7]. The result has been stated for the general case where k_1, k_2, k_3, k_4 are distinct in order to exhibit the functional dependence of the result on these parameters; but we might expect them all to take a universal constant value k. The finite functional of the fields agrees with the standard Feynman calculation for every scattering which has no component in the forward direction; i.e. it contains exactly the same information about differential cross-section. The k-dependent term is of course entirely in the forward direction, so makes no contribution to differential cross-section.

Although this modification was first noted as a way of eliminating the divergence in this specific diagram, it is in fact a change that can be made to the formalism throughout. The essential point we have used is that the 'inhomogeneous' 0000 box diagram

$$\oint \frac{D^4 Z \wedge D^4 X \wedge D^4 W \wedge D^4 Y \; f(Z^\alpha) g(W_\alpha) h(X^\alpha) j(Y_\alpha)}{(W_\alpha Z^\alpha - k)(W_\alpha X^\alpha - k)(Y_\alpha Z^\alpha - k)(Y_\alpha X^\alpha - k)}$$

is a perfectly good representation of the ϕ^4 integral. But it is one susceptible to strictly more inverse derivative operations than is the original projective box diagram; in particular the inverse derivative operator corresponding to the spin-1 Feynman propagator can be applied. This argument is not specific to the box diagram; any projective diagram could be made inhomogeneous in this way. If a projective diagram has a genuine contour then it will have a corresponding contour when made inhomogeneous, and the result of the integration will be unchanged; but the modification will in general allow more inverse derivative operations to become well-defined.

The shift to non-projective twistor space also has the general effect of simplifying the treatment of the representative functions for the external fields: *all* contours known can now be fibred with an S^1 surrounding each singularity of each external function. It is for this reason that it was indicated earlier that supplying a completely cohomological account of classical diagrams might not be essential for the development of the theory.

A further aspect of the inhomogeneous framework is that there is now a straightforward solution of the equation

$$\frac{\partial}{\partial W_\alpha} W \overset{-1}{\underset{}{\longrightarrow}} Z = Z^\alpha \, W \overset{0}{\underset{}{\longrightarrow}} Z$$

namely

$$c + \log(W_\alpha Z^\alpha - k),$$

where c is any constant. Thus in the Møller scattering diagram we could use such logarithmic terms instead of the boundary prescriptions. The use of a logarithm rather than a boundary means however a strict increase in the scope of the diagram formalism, since there will generally be more contours available if the logarithm is used. The additional contours have the feature that they do not contract when the logarithm is replaced by 1. Contours of this nature will 'see' the constant c, so we must specify a choice; there is only one possible consistent choice, namely Euler's constant. The definition can be extended consistently to $(-n)$-lines.

This further extension does not have any clear application in the theory of massless fields as known at present, but it is an important possibility for future development of the theory. For instance, it means that we may now meaningfully multiply (-1)-line factors together. This hints at how twistor diagrams might supply a mechanism for the regularisation of ultraviolet divergences, since these arise when inverse derivatives are multiplied.

Furthermore it is known that the logarithmic form of the (-1)-line, complete with Euler's constant, is needed in the formula for generating massive fields, as will be discussed below.

It has to be stressed that the introduction of k is not equivalent to introducing a mass; indeed it does not break the manifest conformal invariance of the formalism. Nor is it to be understood that a limit $k \to 0$ has to be taken; the integrals are well defined for finite k.

The introduction of inhomogeneity gives an example of how our strategy for the twistor diagram programme need not amount to mere copying of space-time Feynman diagram prescriptions, even though the answers it yields agree with finite QFT statements. Flat space-time structure corresponds to projective twistor space, so to utilise the extra dimension offered by twistor scale is to add something new. In the classical projective 0000 box diagram, the internal lines may be interpreted as Cauchy poles restricting twistors to a space-time point. It is not surprising therefore that this formalism reproduces the divergence of the space-time theory whern we try to extend it to Møller scattering. But in the inhomogeneous 0000 box, we 'enforce' $W_\alpha Z^\alpha = k$, which has no interpretation in space-time. It is however interestingly reminiscent of Penrose's original motivating idea of 'fuzzing out' points while retaining causal structure [18].

The physical dimension of $W_\alpha Z^\alpha$, and hence of k, is that of angular momentum. As to the value of k, we have no sensible interpretation at present. (One tantalising thought, however, is that a real value for k would subtly introduce a time-reversal asymmetry into the formalism.) Another difficulty is that we still do not have a clear picture of how the infra-red divergence is to be understood in conventional quantum field theory, where it points up a difficulty in defining the Fock space when long-range interaction is present. Because the interpretation is so obscure, the inhomogeneous diagrams cannot be as yielding a definitive theory at this stage, but as showing a powerful property of the twistor integral calculus with the potential to regularise divergences. Only in a more complete theory of Feynman integrals, in which mass appears and other divergences are treated coherently, could we expect to interpret it.

7 Skeleton Diagrams

We now take up the problem of crossing-related amplitudes. First note that the preceding work has all been based on the 0000 box diagram as a representation of the ϕ^4 integral, a representation allowing certain inverse derivative operations to be performed. However, there do exist many *other* representations of the ϕ^4 integral. In fact if we were *solely* interested in the first-order ϕ^4 scattering integral we should have no trouble at all with the 'crossing' property: it has been known since the earliest days that it *can* be expressed in a form allowing all allocations, namely by

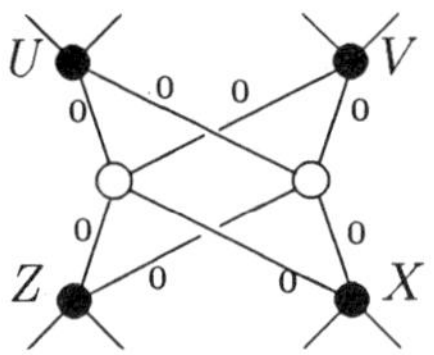

(as noted by Penrose [19]). By integrating out the dual twistors we can reduce the interior of the diagram to

$$(\epsilon_{\alpha\beta\gamma\delta}Z^\alpha X^\beta U^\gamma V^\delta)^{-2}$$

which, again, allows all allocations. The kernel

$$(W_{[\alpha}Y_{\beta]}Z^\alpha X^\beta)^{-2}$$

studied by Huggett and Singer [16] is very similar. Thus we might say that the problem of crossing relations for the ϕ^4 integral is 'solved' by any of these.

However, what we require is not just an isolated statement about one particular integral but a structure that can be generalised to all the other processes we are interested in (Møller and Compton scattering, for instance), and which has the capacity for generalisation beyond first order amplitudes.

It turns out that the key to locating such structure is given by the *double box*, first written down long ago by Penrose [18]. Pioneering work by Sparling [23] has been corrected and extended recently [10] and its properties are now more clear. It turns out that this diagram does no 'better' than the box diagram, in that it only allows one allocation of positive and negative frequencies. But it has the useful property of being susceptible to a *different* set of inverse derivative operators.

In particular the double box satisfies:

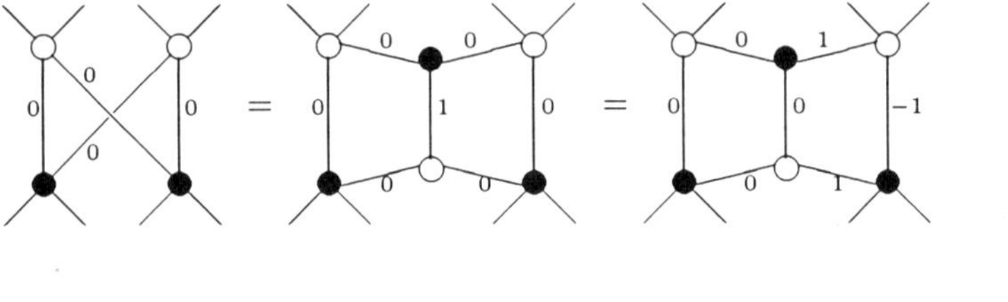

$$(2)$$

thus presenting two alternative representations of the ϕ^4 integral, and

$$(I^{\alpha\beta}Y_\alpha W_\beta)(I^{\alpha\beta}\frac{\partial}{\partial X^\alpha}\frac{\partial}{\partial Z^\beta})^{-1}$$

These facts enable us to give twistor diagram representation of all the first-order QED processes in all channels: in particular for the 'crossed' version of Möller scattering, which cannot be represented by a single box:

and the 'missing' channel for Coulomb scattering:

This result also vindicates Penrose's original guessed diagram for Coulomb scattering [18].

As pointed out above, we cannot be satisfied with having independently specified diagrams for crossing-related processes. The double-box diagrams do not in themselves solve the crossing-relation problem. But this expansion of our repertoire does allow us to observe and verify a remarkable connection between the crossing-related processes. The idea goes back to one of the earliest guesses in twistor diagram theory. Penrose was led to propose his double-box representation of Coulomb scattering partly by drawing a sequence of lines to carry the spin-$1/2$ field (the lines labelled 0 or 2 in our notation) and then attaching (-1)-lines and spin-1 fields in an appropriate manner. But inspection reveals that *all* the first-order QED diagrams can be described in just this way: the original guess seems to embody a general principle. We may express this in terms of 'skeleton diagrams'.

Suppose we define the *skeleton diagram* of a twistor diagram by omitting its (-1)-lines. (That is, the skeleton represents only the form to be integrated and not any boundary-defining data.) Then skeleton diagrams for the crossing-related processes are either already identical or can be made identical by adding twistor transform lines.

In a sense this is only a small modification to our original expectation of how crossing relations should appear: the idea of locating different contours for a single diagram. We retain the principle of the same differential form being integrated in each case, and the only change is that the choosing of different contours for different channels now includes the obligation to choose different boundaries as well. However, in another sense the change is considerable because it means that the (-1)-lines have a different status from that of the other lines. This has the important implication that gauge fields have a different status in the calculus from that of the other fields.

Lastly, the 0000 box itself can be fitted into this new 'skeleton diagram' scheme, by extending it to the skeleton

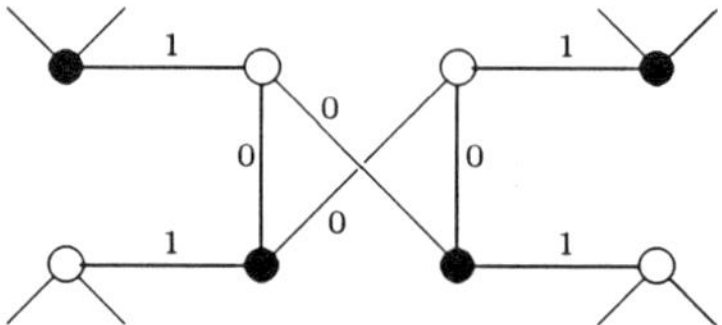

This possesses contours for all channels by virtue of (2) above.

8 Diagrams and the Standard Model

So far we have looked at amplitudes for first-order massless QED and ϕ^4 interactions, finding corresponding twistor diagrams which can be fitted into a 'skeleton' pattern. But it is easy to write down many other twistor 'box' diagrams which certainly correspond to various conformally invariant functionals of positive and negative frequency massless fields. Do these have any connection with physical theory? In particular, consider the twistor diagram (1) written down in §3 as the simplest example of the application of an inverse derivative to the 0000 box diagram. There it was identified simply with a mathematical expression involving fields and a Feynman propagator; but in fact this expression can be given a physical interpretation as the amplitude for

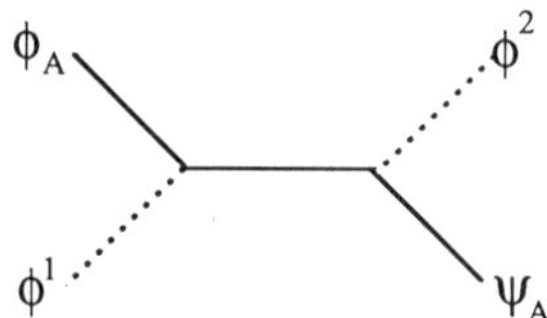

where the vertices arise from the 'Yukawa' interaction Lagrangian $\phi\psi^{A'}\bar{\psi}_{A'}$ In the other channels we may establish the correspondences

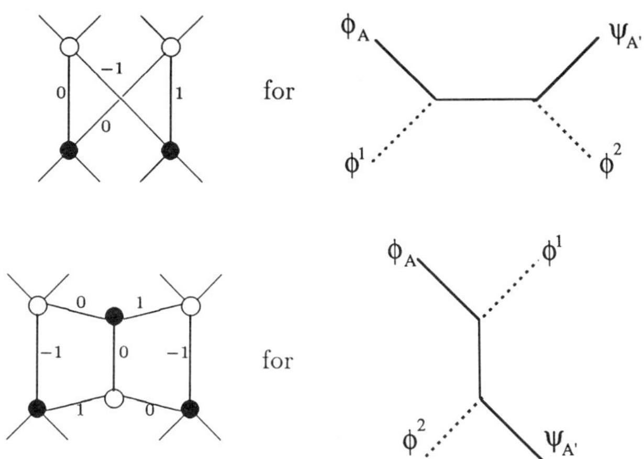

so that the twistor diagrams for all three channels are consistent with the 'skeleton diagram'

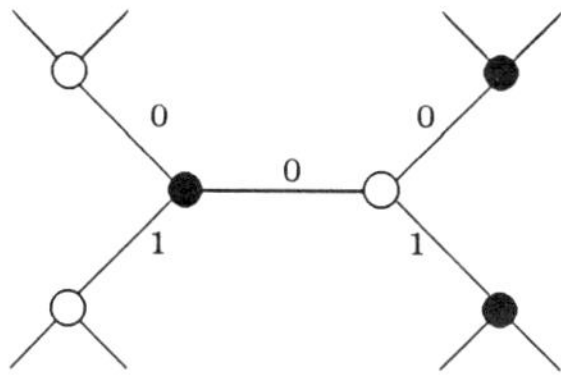

The Yukawa vertex also gives rise to the Feynman diagrams and corresponding twistor diagrams

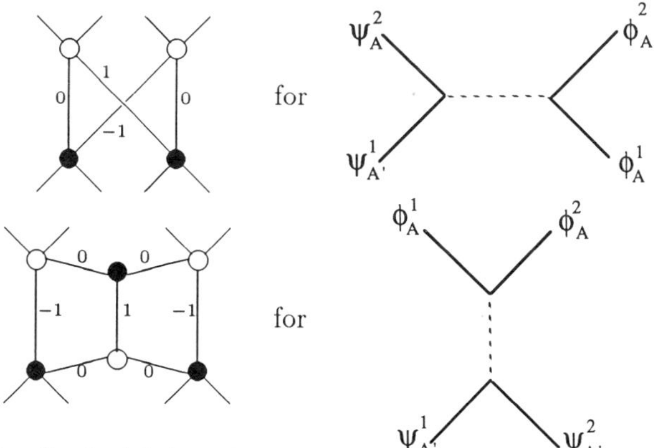

Note these also fit the 'skeleton' pattern:

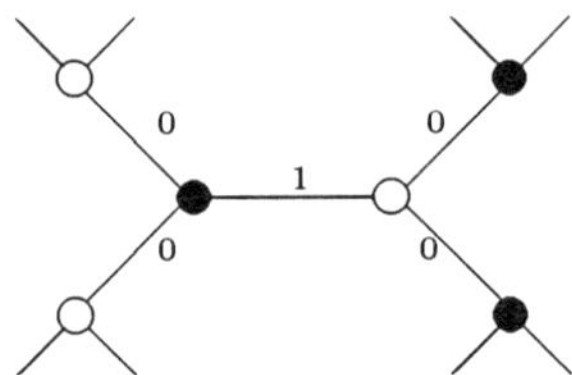

As another striking example of how twistor diagrams naturally express the content of these massless field theories, we can study the gauge-invariant sum of three diagrams which involve both Yukawa and massless QED vertices:

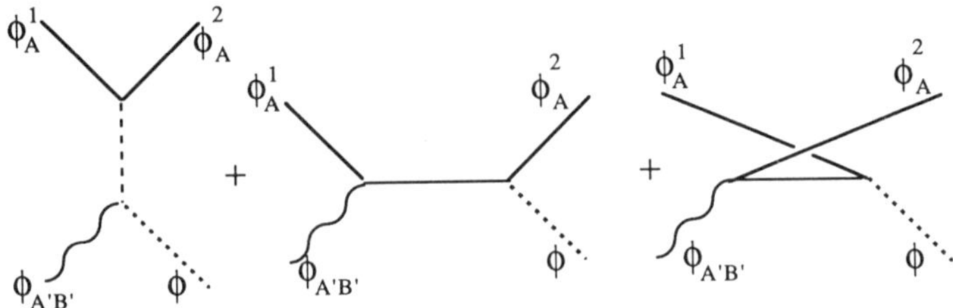

This sum translates into twistor diagrams such as

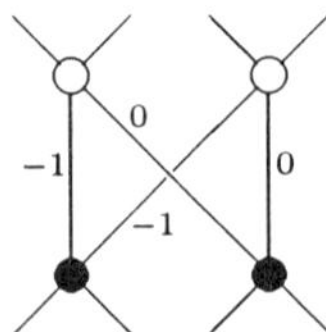

which again fit the 'skeleton' pattern.

These observations indicate that the twistor diagram formalism does not relate specifically to massless QED: it is applicable to all the interactions of fundamental massless fields that appear in the 'standard model' of leptons and the electroweak forces.

This observation marks something of a departure from the original spirit of the twistor programme. In the 1970s much work was done (e.g. [17]) on investigating the internal symmetry groups implied by the twistor representation, in the hope that this would lead to a alternative to the mainstream account of the particle spectrum. However, it seems at the present time a useful idea to borrow what insight we can from the 'standard model' picture— in much the same spirit as we borrow the Feynman diagram expansion—as a reasonably consistent working scheme which certainly embodies a great deal of experimental information. An encouraging feature is that it describes the electron (for instance) as arising from the more fundamental massless fields (as we shall discuss below), a viewpoint close to that of the twistor programme. In adopting this strategy, one does not have to go as far as believing in the mechanism of 'dynamical symmetry-breaking'; we shall merely use the general description of fields offered by the standard model as a guideline, expecting new features in the twistor-geometric description to shed new light on how it can be interpreted.

In particular the earlier work done on internal symmetries in the twistor representation might re-emerge in this new context. At this stage the connection between the gauge symmetries of the standard model and twistor diagram structure is unclear. Tantalising connections however do exist. For instance, in studying the first-order pure scattering of SU(2) gauge fields we find a correspondence between the gauge-invariant sum of Feynman diagrams

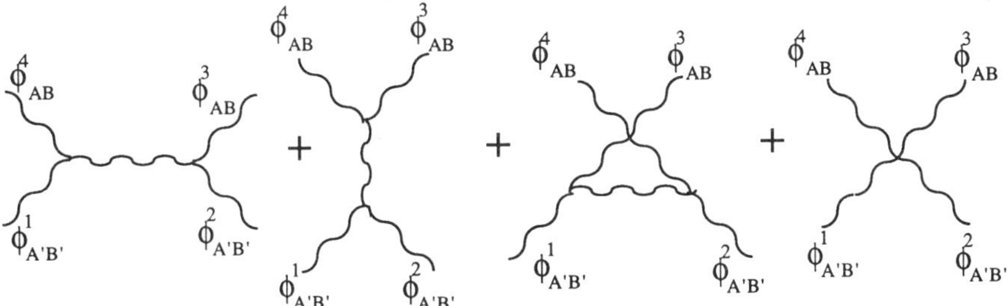

and the sum of twistor diagrams

where the Λ_i are the SU(2) matrices associated with each external field. Note the connection between the order in which the SU(2) traces are taken, and the order in which the (-1)-lines connect the vertices.

9 Higher Order Diagrams: Breaking Conformal Symmetry

Studying first-order processes is not enough. We wish to study general Feynman diagrams. From 1976 onwards I studied the possibility of attacking

Feynman diagram structure directly—seeking a twistor integral formula for the convolution of three general fields (i.e. fields not satisfying any equation). This could be done by studying the convolution of six massless scalar fields. The idea was that given such a 'ϕ^6 integral' formula, a general Feynman vertex (i.e. the convolution of three Feynman propagators) would be related to it by a differential equation, just as the various first order amplitudes are related by differential equations to ϕ^4 integral formulae. One such ϕ^6 integral formula was indeed found [2] and others followed [3]: but of course there are many representations and the difficulty one faces (just as in studying the ϕ^4 integral, but with an even greater variety of possibilities to consider) is that of finding a representation relevant to the differential equations satisfied by the Feynman propagators. These ideas did give some useful results on higher order ϕ^4 theory amplitudes and their twistor diagram representations [9], but not a systematic translation principle.

In studying these difficult multiple integrals, with many possible twistor representations, one is easily lost without some guiding light, and the earlier work lacked any such illumination. Recent work has benefited greatly from two ideas. The first of these is the observation that study of higher-order diagrams is much simpler within the ϕ^4 and the Yukawa interaction theories than it is in massless QED; we do not have the requirement of a summation over Feynman diagrams to obtain a gauge-invariant amplitude. The second is the recognition of the 'skeleton diagram' pattern in the first-order amplitudes, which suggests a picture of how the higher order diagrams might appear in a twistor diagram translation.

Guided by these thoughts we can look first at the simplest (non-zero) amplitude arising from the interactions in the standard model which involves more than four external fields: this arises from composing a ϕ^4 vertex with a Yukawa vertex and has the Feynman diagram

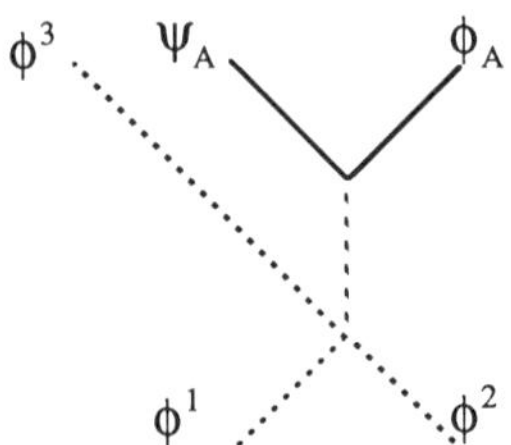

where $\phi_1 = \phi_2 \leftrightarrow (Z^\alpha A_\alpha)^{-1}(Z^\alpha B_\alpha)^{-1}, \quad \phi_3 \leftrightarrow (Z^\alpha E_\alpha)^{-1}(Z^\alpha F_\alpha)^{-1}$

$\psi_A \leftrightarrow (W_\alpha C^\alpha)^{-1}(W_\alpha D^\alpha)^{-2}, \quad \phi_A \leftrightarrow (W_\alpha C^\alpha)^{-2}(W_\alpha D^\alpha)^{-1}$

A simple calculation yields the amplitude:

$$\frac{I^{\alpha\beta} A_\alpha B_\beta}{(A_{[\alpha} B_{\beta]} C^\alpha D^\beta)^2 (\epsilon^{\alpha\beta\gamma\delta} A_\alpha B_\beta E_\gamma F_\delta)(I_{\gamma\delta} C^\gamma D^\delta)} \tag{3}$$

and mere observation of this result is enough to reveal the vital fact that although scale-invariant it is not conformally invariant.

This tells us that in any twistor diagram for this amplitude there must be some mechanism for breaking conformal symmetry. Indeed it is not hard to see that conformal symmetry-breaking will in general occur in any amplitude involving more than four external fields. Such a breaking of conformal symmetry is in any case to be expected on general QFT grounds. It was only while our attention was restricted to first-order amplitudes that this feature remained unobserved.

Inevitably, therefore, new elements must be introduced into the diagram formalism, elements which break conformal symmetry. It means that instead of demanding a formalism which is conformally invariant, we can only demand one in which the breaking of conformal invariance is always explicit. But this can be achieved without sacrificing manifest finiteness, and without sacrificing the 'skeleton' diagram feature: the breaking of conformal symmetry can go entirely into contour *boundaries*. We simply augment the rules to allow the contours to have boundaries on subspaces defined by $I_{\alpha\beta}X^\alpha Z^\beta$, $I^{\alpha\beta}W_\alpha Y_\beta$; in the projective diagram picture these will be $\{I_{\alpha\beta}X^\alpha Z^\beta = 0\}, \{I^{\alpha\beta}W_\alpha Y_\beta = 0\}$ but the same arguments as above tell us that these could be generalised to inhomogeneous boundaries of form $\{I_{\alpha\beta}X^\alpha Z^\beta = m\}, \{I^{\alpha\beta}W_\alpha Y_\beta = m\}$ in non-projective diagrams.

We can now try to find a twistor diagram to represent the particular Feynman diagram amplitude thus calculated. In fact the skeleton diagram thesis tells us what to try: by composing the first-order skeletons we are led to consider

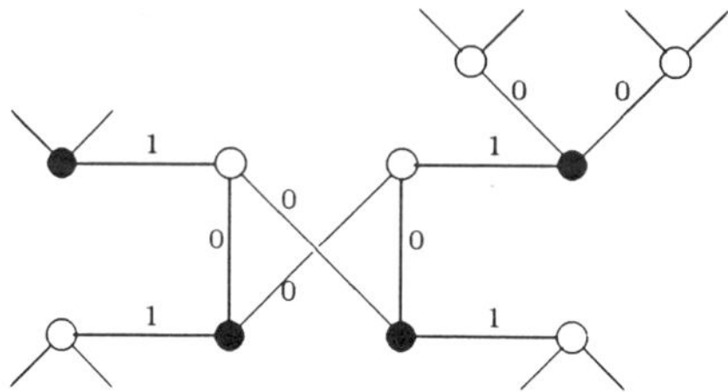

This skeleton can indeed be furnished with boundaries and integrated to yield the amplitude (3). Indeed it turns out to have contours for *all* allocations of positive and negative frequency fields, provided we allow contours to have boundaries of the new type.

These recent calculations mark the frontier of current research. The principal point is that all the results known now fit a quite simple pattern of skeleton diagrams: spin-1/2 propagators translate to a chain of alternating 0-lines and 2-lines; spin-0 propagators to a chain of 1-lines; a Yukawa vertex corresponds to the meeting of two 0-lines and a 1-line; a ϕ^4 vertex to a 0000 box attached to 1-lines. Gauge fields make *no* contribution to the skeleton except in isolated 3-lines; their properties are translated into the prescription of contour boundaries. There is much to be done both to confirm this picture, and to establish rules for how the skeleton should be 'fleshed out' with (-1)-lines and contour prescriptions to yield the amplitudes in its various channels. At present this is only understood on an *ad hoc* basis, and the boundary contours used are certainly not uniquely specified.

10   New Directions

These observations suggest three particular directions for further research:

(1) We can study the Feynman diagram

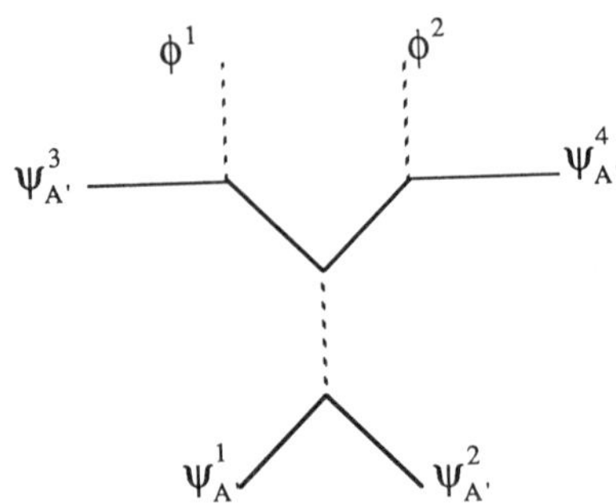

for which at least in some channels the skeleton diagram

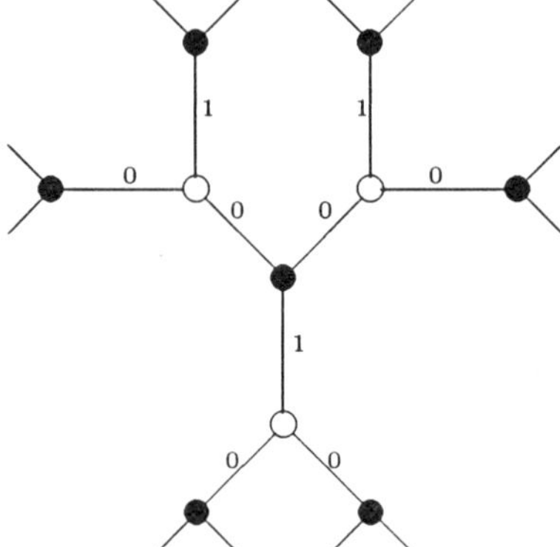

is certainly valid, although the analysis is not yet complete. The point of looking at this diagram is that the six external fields, taken in all possible allocations of positive and negative frequencies, are equivalent to test-functions for the full content of the internal Yukawa vertex. Thus if this diagram is fully understood we shall have translated a Feynman vertex in its most general position. It should then be possible to compose such vertices and thus derive a theory of general Feynman diagram translation. In particular one might study the Feynman diagrams which give rise to ultra-violet divergences. Note that in doing this we can use boundaries on subspaces of form $\{I_{\alpha\beta}X^\alpha Z^\beta = m\}$, which have no direct interpretation in space-time. This means the twistor diagram representation has the potential to regularise divergences in a quite new way.

(2) Progress with the programme outlined above would naturally lead to study of the corresponding diagrams for higher order massless electroweak interactions. These are computationally much more difficult, because of the gauge-field aspects. However, they would be worth studying if only from the point of view of finding a link with the CFT4 hypothesis. In particular, if we consider pure SU(2) gauge field scattering, it appears that the 'skeleton' diagram must disappear completely, or rather be reduced to isolated 3-lines. In effect the differential form evaporates into pure volume, and all that is left is a prescription of a region of integration. In the first-order case the resulting structure seems remarkably close to the formulas appearing in bosonic string theory [12], suggesting that it is with pure gauge field scattering that the connection of twistor diagrams with conformal field theories may be closest. It would be worth knowing whether this pattern extends to a higher order. Apart from this CFT4 connnection, it would in any case be particularly valuable to know if the manifestly gauge-invariant formulation of first-order interactions also extends to higher order gauge-field interactions. The Feynman description, although a complete predictive calculus, becomes extremely complicated for SU(2) gauge-field theory, and any reformulation would be of great interest.

(3) So far we have considered only *massless* fields, thought of as appearing in the standard model with unbroken symmetry. However, an important aspect of the standard model is that it relates massless fields to massive fields in a way that is both systematic and motivated by experimental findings. Its principle is that the scalar field happens to be in a ground state $\phi = \text{constant}$; and that when we do perturbation theory we must redefine the zeroth order to include all interactions with the ground state. The Yukawa interaction then implies that the massless spin-1/2 field propagator must be supplemented by an infinite series of 'zigzag' diagrams

in which the scalar field is always: ϕ = constant. The (massive) electron propagator can then be identified with such a summation. It is certainly worth studying the twistor-diagrammatic equivalent to this analysis of massive fields as a guide to constructing a consistent twistor description. Note that the constant scalar field appears in the twistor description as the elementary state based at $\mathcal{I}$. A first look at the twistor representation of these zig-zag Feynman diagrams shows immediately a feature of contour 'pinching' due to the singularities at $\mathcal{I}$ thus introduced. Correspondingly the space-time integrals are divergent. However, we know the properties of the massive fields and propagators to which these divergent integrals formally sum, and it should be possible to regularise the divergent integrals and make this summation manifestly finite within the twistor representation. In attempting this we have a guide: there is a twistorial formula for massive fields, developed as one of the applications of the inhomogeneous twistor diagram formalism. This will be found described in [8]. This formula has the remarkable feature that it requires the use of the logarithmic definition of the inhomogeneous (-1)-line complete with Euler constant. This is employed in combination with inhomogeneous poles of form $(I_{\alpha\beta}X^\alpha Z^\beta - m)^{-1}$. The resulting integral formula automatically generates finite-normed massive states, i.e. those which converge exponentially at infinity. At present the connection between formula and the standard model 'zigzags' is not quite clear, but it does suggest that the geometry of non-projective twistor space, with special reference to the structure at $\mathcal{I}$, may be a crucial aspect of the twistor reformulation of field theory.

References

[1] A. P. Hodges, *The description of mass within the theory of twistors,* Ph. D. thesis, University of London (1975)

[2] A. P. Hodges, The integrated product of six massless fields, *Twistor Newsletter* **4**, Mathematical Institute, Oxford University (1977);

reprinted in *Advances in Twistor Theory,* eds. L. P. Hughston and R. S. Ward, Pitman (1979)

[3] A. P. Hodges, Recent progress in twistor diagrams, *Twistor Newsletter* **12**, Mathematical Institute, Oxford University (1981)

[4] A. P. Hodges, Twistor diagrams, *Physica* **114A** 157–175 (1982)

[5] A. P. Hodges, Twistor diagrams and massless Møller scattering, *Proc. R. Soc. Lond.* **A 385** 207–228 (1983)

[6] A. P. Hodges, Twistor diagrams and massless Compton scattering, *Proc. R. Soc. Lond.* **A386** 185–210 (1983)

[7] A. P. Hodges, A twistor approach to the regularization of divergences, *Proc. R. Soc. Lond.* **A 397** 341–374 (1985)

[8] A. P. Hodges, Mass eigenstates in twistor theory, *Proc. R. Soc. Lond.* **A 397** 375–396 (1985)

[9] A. P. Hodges, A twistor diagram for second order ϕ^4 scattering, *Twistor Newsletter* **25**, Mathematical Institute, Oxford University (1987)

[10] A. P. Hodges, Double box diagrams, *Twistor Newsletter* **28**, Mathematical Institute, Oxford University (1989)

[11] A. P. Hodges, Pochhammer contours in twistor diagrams, *Twistor Newsletter* **28**, Mathematical Institute, Oxford University (1989)

[12] A. P. Hodges, String amplitudes and twistor diagrams: an analogy, in *The Interface of Mathematics and Particle Physics*, eds. D. G. Quillen, G. B. Segal and Tsou S. T., Oxford University Press (1990)

[13] A. P. Hodges and S. A. Huggett, Twistor diagrams, *Surv. High Energy Phys.* **1**, 333 (1980)

[14] A. P. Hodges, R. Penrose, M. A. Singer, A twistor conformal field theory in four dimensions, *Phys. Letters* B **216** 48–52 (1989)

[15] S. A. Huggett, Sheaf cohomology in twistor diagrams, D. Phil. thesis, University of Oxford (1980)

[16] S. A. Huggett, M. A. Singer, Relative cohomology and projective twistor diagrams, to appear in *Trans. Amer. Math. Soc.* (1990)

[17] L. P. Hughston, Twistors and particles, *Springer Lect. Note. Phys.* **97** (1979)

[18] R. Penrose and M. A. H. MacCallum, Twistor theory: an approach to the quantisation of fields and space-time, *Phys. Reports* **6** 241-315 (1972)

[19] R. Penrose, Twistor theory: its aims and achievements, in *Quantum Gravity, an Oxford Symposium,* eds. C. J. Isham, R. Penrose and D. W. Sciama, Oxford University Press (1975)

[20] R. Penrose, Twistors, Particles, Strings and Links, in *The Interface of Mathematics and Particle Physics*, eds. D. G. Quillen, G. B. Segal and Tsou S. T., Oxford University Press (1990)

[21] R. Penrose, Article in this volume.

[22] M. A. Singer, Twistors and Four-dimensional Conformal Field Theory, in *The Interface of Mathematics and Particle Physics*, eds. D. G. Quillen, G. B. Segal and Tsou S. T., Oxford University Press (1990)

[23] G. A. J. Sparling, Ph. D. thesis, University of London (1974)

[24] G. A. J. Sparling, Homology and twistor theory, in *Quantum Gravity, an Oxford Symposium,* eds. C. J. Isham, R. Penrose and D. W. Sciama, Oxford University Press (1975)

[25] A. Qadir, Penrose graphs, *Phys. Reports* **39**, 131–167 (1978)

Cohomology and Twistor Diagrams

S.A. Huggett

1   Introduction

The description of massless fields in terms of the analytic cohomology of twistor space opened up a new line of enquiry in the study of twistor diagrams. Traditionally, these diagrams represent holomorphic differential forms on products of twistor spaces, in which representative cocycles are written down for the fields (instead of the actual cohomology classes), and the diagrams are then evaluated using a (compact) contour integral. We review the development of a reinterpretation of twistor diagrams, in which the interior of a diagram determines a finite-dimensional family of functionals on the tensor product of the twistor cohomology classes representing the interacting fields. Our principal example is the twistor description of the first order massless scalar ϕ^4 vertex. We conclude by considering the question of which of the traditional contour integrals are in fact cohomological functionals.

We take as our starting point the theory of twistor diagrams as described in [17, 8, 9]. In particular, we do not explore their relationship with the Feynman diagrams they are intended to mimic. No detailed knowledge of twistor diagrams is assumed; indeed we now summarise their most basic features.

Each vertex in a diagram represents projective twistor space P (or its dual P^*) and contributes the differential form DZ (or DW), where

$$DZ = \epsilon_{\alpha\beta\gamma\delta} Z^\alpha dZ^\beta \wedge dZ^\gamma \wedge dZ^\delta$$

(and $\epsilon_{\alpha\beta\gamma\delta}$ is a given element of $\Lambda^4 T$). An internal edge joining vertices labelled Z and W contributes $(Z^\alpha W_\alpha)^{-1}$, unless the edge has a non zero integer on it, in which case $Z^\alpha W_\alpha$ is raised to a different power. External edges come in pairs, each pair being a representative cocycle such as

$$\frac{1}{Z^\alpha A_\alpha Z^\beta B_\beta}$$

for an 'elementary state'. (We will see how to generalise this to, for example, positive frequency fields in due course). Given a diagram (with v vertices) one takes the cup product of its various components and obtains a differential form

$$\omega \in H^0(\Pi - \Upsilon; \Omega^{3v})$$

where Π is the product of all the twistor (and dual twistor) spaces and Υ is the union of all the subspaces on which the $(Z^\alpha W_\alpha)^{-1}$ and $(Z^\alpha A_\alpha)^{-1}$ factors are singular. This differential form is integrated over a contour

$$\kappa \in H_{3v}(\Pi - \Upsilon; \mathbb{C})$$

(which we refer to as a traditional contour).

The question is, given *cohomology classes* instead of representative cocycles, how do we replace

$$\oint_\kappa \omega.$$

We more or less follow the historical development in answering this question, so it is only in §4 that we arrive at a general procedure for describing all the cohomological functionals on a given collection of fields. §5 then describes an approach to the more modest, but perhaps as useful, question of how to tell if a traditional contour corresponds to a cohomological functional.

The extent to which we lag behind the *theory* of twistor diagrams is evident in that throughout we concentrate on only two examples: the scalar product and the first order massless scalar ϕ^4 vertex. Prospects for extending the work are discussed in §6.

2   The Dot Product

Early ideas [7] on the cohomological interpretation of twistor diagrams emphasised the importance of being able to construct arbitrarily large diagrams out of elementary pieces according to certain rules. (These rules—whatever they turn out to be—are intended to be a twistor version of the Feynman rules for constructing Feynman diagrams. Early versions can be seen in [17, 8]).

The simplest thing to do is to take each internal edge in a diagram to be an element of an appropriate H^0 group. (If the edge has the non-negative integer r on it the coefficient sheaf is $\mathcal{O}(r)$). Pairs of external edges, of course, represent the interacting fields and so are taken as elements of appropriate H^1 groups. As one puts the various edges together to construct a diagram, so one multiplies the various H^0 and H^1 elements in some way.

It would have been possible (as later developments showed) to use the *cup product* on cohomology to multiply the elements together. Instead though, an even more 'atomic' approach was taken. We can regard the elementary state

$$\left[\frac{1}{Z^\alpha A_\alpha Z^\beta B_\beta}\right] \in H^1(P - L; \mathcal{O}(-2))$$

(where A_α and B_β are planes through the line L) as having been made out of the two pieces

$$\frac{1}{Z^\alpha A_\alpha} \in H^0(P - A; \mathcal{O}(-1)) \quad \text{and} \quad \frac{1}{Z^\beta B_\beta} \in H^0(P - B; \mathcal{O}(-1)).$$

In general, we can define a product

$$H^p(U;S) \times H^q(V;T) \longrightarrow H^{p+q+1}(U \cup V; S \otimes T) \qquad (1)$$

as follows. We use the Čech description of cohomology classes. Suppose that $f \in H^p(U;S)$ is represented by $f_{i_0\ldots i_p}$ on $U_{i_0\ldots i_p}$ (with respect to the covering $\{U_i\}$ of U), and that $g \in H^q(V;T)$ is represented similarly with respect to the covering $\{V_j\}$ of V. Let $\{W_\alpha\}$ be the disjoint union of the covers $\{U_i\}$ and $\{V_j\}$, so that $\{W_\alpha\}$ is an open covering of $U \cup V$. Then on those $(p+q+2)$-fold intersections from $\{W_\alpha\}$ in which $p + 1$ of the sets are from $\{U_i\}$ and $q + 1$ from $\{V_j\}$ put the corresponding function ($f_{i_0\ldots i_p} g_{j_{p+1}\ldots j_{p+q+1}}$ say). It is easy to check that this defines a map in cohomology, called the *dot product*.

We usually need the dot product in the more general setting in which $U = X - F$ and $V = Y - G$, where X and Y are disjoint complex manifolds containing the closed subsets F and G. We use the projections π_X and π_Y from $X \times Y$ to X and Y to pull back the elements in $H^p(X - F; S)$ and $H^q(Y - G; T)$ to the product, obtaining elements in $H^p((X - F) \times Y; \pi_X^* S)$ and $H^q(X \times (Y - G); \pi_Y^* T)$. Then (1) yields an element of $H^{p+q+1}(X \times Y - F \times G; \pi_X^* S \otimes \pi_Y^* T)$ (but we often omit the π_X^* and π_Y^*).

This enables us to think of *all* the (single) edges in a twistor diagram (whether internal or external) as elements of H^0 groups. The multiple edges have to be 'lumped together' and counted as single edges of a different homogeneity (because dotting something with itself yields zero). Of course, the multiple edge notation was originally [17] intended to indicate a higher sphere (usually S^2) contribution to the contour, although in practice this rule was often broken. We will see that it has an echo in cohomology, where a multiple edge indicates the necessity for an extra element (see (2) below).

Now consider a twistor diagram with e edges and v vertices, and let d ($= 3v$) be the complex dimension of the underlying space Π. Then if all the edges have zeroes on them and all the parameters are in general position the dot product yields an element of

$$H^{e-1}(\Pi - \Upsilon''; \Omega^d)$$

(where Υ'' is the codimension e intersection of the subspaces in Υ). The rule that four edges meet each vertex implies that in the case of a *tree* diagram (in which there are no multiple edges or closed circuits)

$$e = d + 1$$

so that we have an element of

$$H^d(\Pi; \Omega^d)$$

which by Serre duality is isomorphic to $\mathbb{C}$.

The fact that we have a unique evaluation procedure here reflects the fact that in making use of the dot product we *exclude* contours which allow any of the edges in the diagram to coincide. In the case of tree diagrams that leaves exactly one contour, as can be seen from a purely topological study of the diagram.

Interesting though this is, it doesn't help us much with physically relevant twistor diagrams, none of which is a tree. The scalar product diagram (see [9]) is our first example. Here $e = 5$ and $v = 2$, so our naive method yields an element of

$$H^4(\Pi - \Upsilon''; \Omega^6)$$

which we have to find some way of 'evaluating'. The first approach to this [5] maintained the 'atomic' viewpoint and reinterpreted the double edge in the middle of the diagram.

For simplicity, suppose the two H^1 elements are elementary states representing *scalar* fields. They are then elements of $H^1(P - L; \mathcal{O}(-2))$ and $H^1(P^* - L'; \mathcal{O}(-2))$ for some lines L and L' in P and P^*, and their dot product is an element of $H^3(P \times P^* - L \times L'; \mathcal{O}(-2, -2))$. We need to interpret the double edge, then, as an element of $H^2(N; 0(-2, -2))$ (where N is a neighbourhood of $L \times L'$ in $P \times P^*$) so that when everything is dotted together we obtain an element of

$$H^6(P \times P^*, \Omega^6) \cong \mathbb{C}$$

as before.

Ginsberg [5] first noted that an appropriate element of $H^2(N; \mathcal{O}(-2, -2))$ *had* to exist, because the (two-variable) Penrose transform of this group is the space of two-point massless scalar fields, one of which is

$$\frac{1}{(x - y)^2}.$$

The significance of this field lies in its role as a positive frequency reproducing kernel for massless scalar fields:

$$\int \int \theta(x) \frac{\partial}{\partial x^{AA'}} \frac{1}{(x - y)^2} \frac{\partial}{\partial y^{BB'}} \phi(y) d^3 x^{AA'} d^3 y^{BB'} = k \int \theta(x) \nabla_{AA'} \phi(x) d^3 x^{AA'}.$$

This last integral, when skew-symmetrised on θ and ϕ and taken over some spacelike hypersurface, is the scalar product of the two massless scalar fields θ and ϕ (and k is some uninteresting constant). The notion of a reproducing kernel is considerably developed by Eastwood and Ginsberg in [4]. As well as using the two-variable Penrose transform

$$H^2(P^- \times P^{*-}; \mathcal{O}(-2, -2))$$
$$\cong \{\phi \in \Gamma(M^- \times M^+; O[-1][-1]') : \Box_x \phi = 0 = \Box_y \phi\}$$

in the (twistor, dual twistor) case they consider the corresponding isomorphism in the (twistor, twistor) case. Here the left hand side is

$$H^2(P^- \times P^+; \mathcal{O}(-2,-2)).$$

We return to this, and to Ginsberg's construction of a twistor kernel for the first order massless scalar ϕ^4 vertex, a little later.

There is another method of getting into $H^2(N, \mathcal{O}(-2,-2))$. Take our naive interpretation of the double edge:

$$\frac{1}{(Z^\alpha W_\alpha)^2} \in H^0(P \times P^* - A; \mathcal{O}(-2,-2))$$

(where $A = \{(Z^\alpha, W_\alpha) \in P \times P^* : Z^\alpha W_\alpha = 0\}$) and dot it with an (as yet unidentified) element

$$\mu \in H^1(U; \mathcal{O}) \tag{2}$$

where U is a neighbourhood in N of A. It would obviously be preferable if this element μ could be identified with some *topological* object. This was achieved by Ginsberg in [6] who used the following commutative diagram of exact sequences

$$
\begin{array}{ccccccc}
 & & & & & \downarrow & \\
 & & & & \rightarrow & H^1(U; \mathcal{O}) & \rightarrow \\
 & & & \downarrow & & \downarrow r & \\
 & & & \rightarrow & H^2(N,U;\mathbb{Z}) \xrightarrow{j} & H^2(N,U;\mathcal{O}) & \rightarrow \\
 & & \downarrow & & \downarrow i & \downarrow & \\
 & & \rightarrow H^1(N;\mathcal{O}^*) \xrightarrow{\delta} & H^2(N;\mathbb{Z}) & \rightarrow & \\
 & \downarrow & \downarrow \rho & \downarrow & & \\
\rightarrow & H^1(U;\mathcal{O}) \xrightarrow{\exp} & H^1(U;\mathcal{O}^*) & \rightarrow & & \\
 & \downarrow r & & & & \\
\rightarrow & H^2(N,U;\mathcal{O}) & \rightarrow & & & \\
 & \downarrow & & & & \\
\end{array}
$$

in which the rows are induced from the exponential sheaf sequence while the columns are relative exact sequences. Ginsberg started with an element (the generator, in fact) $\tau \in H^2(N,U;\mathbb{Z})$ and defined

$$\mu = \log(\rho\delta^{-1}i(\tau)) \in H^1(U; \mathcal{O}).$$

He then went on to show that the element

$$\frac{1}{(Z^\alpha W_\alpha)^2} \cdot \mu \tag{3}$$

corresponds to the kernel referred to earlier. In fact he managed to achieve all this not just for the case of elementary states (which we have used in our description above) but for general massless fields in $H^1(P^+; \mathcal{O}(-2))$ and $H^1(P^{*+}; \mathcal{O}(-2))$. The commutative diagram of exact sequences is the same except that the space N is replaced by $P^- \times P^{*-}$.

Furthermore, Ginsberg used the same commutative diagram to construct a twistor kernel for the first order massless scalar ϕ^4 vertex. Here the underlying space Π of the diagram is

$$P_W^* \times P_X \times P_Y^* \times P_Z,$$

where the subscripts refer to the variables used below. The interior edges form the well-known 'box' (see [9]):

$$\frac{1}{W_\alpha X^\alpha \, Y_\beta X^\beta \, W_\gamma Z^\gamma \, Y_\delta Z^\delta},$$

and the exterior edges comprise the four interacting fields, which we take to be elements of

$$H^1(P_W^{*+}; \mathcal{O}(-2)), \ H^1(P_X^+; \mathcal{O}(-2)), \ H^1(P_Y^{*+}; \mathcal{O}(-2)), \ H^1(P_Z^+; \mathcal{O}(-2)).$$

Ginsberg's kernel is of the form

$$\frac{1}{W_\alpha X^\alpha} \cdot \frac{1}{Y_\delta Z^\delta} \cdot \frac{1}{Y_\beta X^\beta W_\gamma Z^\gamma} \cdot \tau$$

(where τ, which is an element of an H^2 group, is obtained by chasing through the commutative diagram). This kernel, being an element of

$$H^4(P_W^{*-} \times P_X^- \times P_Y^{*-} \times P_Z^-; \mathcal{O}(-2, -2, -2, -2))$$

enables us to assemble the complete diagram using the dot product and evaluate it via Serre duality, as planned. Moreover, the structure of the kernel (in which two of the opposite sides of the box are individually dotted into the extra element τ) is similar to the structure of the Sparling contour [20] normally used in the integration of the box diagram. (The phrase 'is similar to' can be made precise. There is a relationship between taking a residue on a codimension one submanifold defined, say, by $s = 0$, and dotting with $1/s$. See [6] or [13] for details).

However, a very long-standing problem with the box diagram has been that of crossing symmetry. The original hope had been that *one* twistor diagram (the box diagram) would represent the Feynman vertex for first order massless scalar ϕ^4 scattering, but that there would be *three* contours for the corresponding twistor integral, one for each of the three channels in the Feynman diagram. One would expect, then, there to be *three* kernels, whereas Ginsberg's procedure only led to one.

There is a clue in Ginsberg's kernel, though, as to how one might proceed. Note that two of the edges of the box ($1/Y_\beta X^\beta$ and $1/W_\gamma Z^\gamma$) are merely *cupped* together, not dotted. So the 'rule' that the cohomological version of a twistor diagram is assembled using the dot product has already been broken, which suggests we re-examine it.

3   Cohomological Contours

By way of introduction to the next phase of development of the subject, we give an alternative description of the dot product. In fact right at the start [7] Eastwood pointed out that the dot product could be thought of as the cup product followed by the Mayer-Vietoris map: given $\alpha \in H^p(X; S)$ and $\beta \in H^q(Y; T)$ their cup product is $\alpha \cup \beta \in H^{p+q}(X \cap Y; S \otimes T)$ and the connecting map in the Mayer-Vietoris sequence:

$$\partial^* : H^{p+q}(X \cap Y; S \otimes T) \to H^{p+q+1}(X \cup Y; S \otimes T)$$

takes us to the dot product of α and β:

$$\alpha \cdot \beta = \partial^*(\alpha \cup \beta).$$

This description of the dot product was very useful in a new approach to the cohomology of twistor diagrams initiated by Singer [11]. The idea (whose roots lie in Penrose [18]) is almost the opposite of the procedure we have been discussing so far, in that we assemble a given twistor diagram by *cupping* everything together. The interior edges are again taken as elements of H^0 groups (the extra elements such as μ or τ having been abandoned) and the exterior edges form the fields (assumed now to be elementary states) in various H^1 groups. If there are f of these H^1 elements this new procedure gives an element of

$$H^f(\Pi - \Upsilon'; \Omega^d)$$

(where Υ' is the union of all the subspaces defined by the internal edges, together with the $\mathbb{C}P^1$ subspaces on which the f elementary states are singular). There is a mapping from

$$H^f(\Pi - \Upsilon'; \Omega^d) \to H^{f+d}(\Pi - \Upsilon'; \mathbb{C})$$

obtained by using the Dolbeault description of the first group, forgetting the bidegree (d, f) and remembering only the total degree $d + f$. (A Čech description of this map is given in Penrose [18]). Now we simply use Poincaré duality to tell us that in order to evaluate an element of $H^{f+d}(\Pi - \Upsilon'; \mathbb{C})$ we will need a *contour* in

$$H_{f+d}(\Pi - \Upsilon'; \mathbb{C}).$$

This 'cohomological' contour is easy to relate to the traditional ones in

$$H_d(\Pi - \Upsilon; \mathbb{C}),$$

for there is a map

$$H_{f+d}(\Pi - \Upsilon'; \mathbb{C}) \to H_d(\Pi - \Upsilon; \mathbb{C})$$

given by iterating the Mayer-Vietoris connecting map (in homology) f times, once for each field.

This all works beautifully for the scalar product diagram, and one can show that

$$H_8(\Pi - \Upsilon'; \mathbb{C}) = \mathbb{C}$$

and that the image of the generator of this group under two Mayer-Vietoris maps is the usual physical contour for the scalar product. This tells us that there is only one cohomological contour for the scalar product (as expected) and suggests a method for testing contours to see whether they are cohomological (a point to which we return later).

More interesting, though, is the application of this method to the box diagram. Here $f = 4$ and $d = 12$, so the cohomological contours are in

$$H_{16}(\Pi - \Upsilon'; \mathbb{C}). \tag{4}$$

A calculation by Singer [19] demonstrated that this space has two generators, one of which does not correspond to any of the three physical channels for ϕ^4 (because it does not generalise to positive or negative frequency fields). The other generator is the Sparling contour in its cohomological form. (In other words, four Mayer-Vietoris maps applied to this generator would yield the Sparling contour).

This seems to confirm Ginsberg's work, that is, that there is only one cohomological channel for the box diagram. Indeed, these two approaches (dot product and extra elements on the one hand, and cup product and cohomological contour on the other) were explicitly shown to be equivalent in the scalar product case by Baston using the techniques of cohomology algebra in [3].

4   A General Procedure

There remained, however, the possibility that there might be other cohomological ways of evaluating the box diagram, corresponding to the other two channels. After all, Hodges had succeeded [8] in constructing contours for all three channels, following a suggestion in [10]. Clearly a method was needed which would exhibit *all* the cohomological functionals on a given collection of fields. Some earlier work by Baston provided the starting point.

In [2] another interpretation of the dot product is described. Given a complex manifold $X \cup Y$, closed subsets $F \subset X$ and $G \subset Y$, and elements

$$\alpha \in H^p(X - F; S)$$

and

$$\beta \in H^q(Y - G; T)$$

we can use the connecting maps in the relative cohomology exact sequences

$$H^p(X;S) \;\rightarrow\; H^p(X-F;S) \;\overset{r}{\rightarrow}\; H^{p+1}_F(X;S) \;\rightarrow\; H^{p+1}(X;S)$$

$$H^q(Y;T) \;\rightarrow\; H^q(Y-G;T) \;\overset{r}{\rightarrow}\; H^{q+1}_G(Y;T) \;\rightarrow\; H^{q+1}(Y;T)$$

to obtain elements $r\alpha$ and $r\beta$. Then the cup product on relative cohomology

$$\cup : H^{p+1}_F(X;S) \times H^{q+1}_G(Y;T) \rightarrow H^{p+q+2}_{F\cap G}(X\cup Y; S\otimes T)$$

gives us an element in the image of the connecting map in the following relative cohomology exact sequence:

$$H^{p+q+1}(X\cup Y; S\otimes T) \rightarrow H^{p+q+1}(X\cup Y - F\cap G; S\otimes T) \overset{r}{\rightarrow}$$

$$H^{p+q+2}_{F\cap G}(X\cup Y; S\otimes T) \rightarrow H^{p+q+2}(X\cup Y; S\otimes T)$$

and it can be shown that

$$\alpha \cdot \beta = r^{-1}(r\alpha \cup r\beta).$$

Actually, in order to make use of this new viewpoint we have to be a little more precise (as indeed [2] was). After all, the f interacting fields are given as elements of H^1 groups defined on f *different* spaces. So we need the *cross* product in relative cohomology:

$$\times : H^{p+1}_F(X;S) \otimes H^{q+1}_G(Y;T) \rightarrow H^{p+q+2}_{F\times G}(X\times Y; S\otimes T).$$

Here, strictly speaking, $S\otimes T$ should be $\pi_X^* S \otimes \pi_Y^* T$.

As before, $r\alpha \times r\beta$ is in the image of the connecting map r in

$$H^{p+q+1}(X\times Y - F\times G; S\otimes T) \overset{r}{\rightarrow} H^{p+q+2}_{F\times G}(X\times Y; S\otimes T) \rightarrow H^{p+q+2}(X\times Y; S\otimes T),$$

with

$$\alpha \cdot \beta = r^{-1}(r\alpha \times r\beta).$$

We remark that this explains why Ginsberg was able to stop at μ in his diagram-chasing. Strictly speaking the diagram chase leads to $r(\mu)$, but we then use μ in a dot product, and for any ν

$$\mu \cdot \nu = r^{-1}(r\mu \times r\nu).$$

Before we make use of this description of the dot product we introduce a little more notation. For each $i = 1,\ldots,f$ we let P_i stand for P or P^*, and we let U_i be an open subset of P_i of the correct topology for the Penrose transform on $H^1(U_i; \mathcal{O}(r_i))$ to be an isomorphism. We will also need

$$F_i = P_i - U_i$$

and

$$F = F_1 \times \ldots \times F_f.$$

Finally, we denote by L_i a projective line contained in F_i, and let

$$\Lambda = L_1 \times \ldots \times L_f.$$

So, given our f fields we have an element in

$$H^1(U_1; \mathcal{O}(r_1)) \otimes \ldots \otimes H^1(U_f; \mathcal{O}(r_f))$$

and our first result [14] is that we lose nothing by choosing to *dot* all these fields together. In fact the Künneth formula for relative cohomology implies that

$$H^1(U_1; \mathcal{O}(r_1)) \otimes \ldots \otimes H^1(U_f; \mathcal{O}(r_f)) \cong H^{2f}_F(\Pi; \mathcal{O}(\mathbf{r})), \tag{5}$$

where $\mathbf{r} = (r_1, \ldots r_f)$. Every continuous linear functional on these fields is therefore an element of the compact relative cohomology group

$$H^{3f,f}_c(\Pi, \Pi - F; \mathcal{O}(-\mathbf{r})) \tag{6}$$

(but (5) and (6) are not in general dual to each other [16]).

We now have to decide how the interior of the diagram picks out some of these functionals. We regard the interior as a holomorphic kernel

$$h \in H^{3f,q}(\Pi - \Sigma; \mathcal{O}(-\mathbf{r})).$$

For example, in the scalar product (spin zero)

$$h = \frac{DWZ}{(W_\alpha Z^\alpha)^2} \in H^{6,0}(\Pi - A; \mathcal{O}(-2,-2))$$

(here and hereafter, we use DWZ for $DW \wedge DZ$, etc.) while in the box

$$h_0 = \frac{DWXYZ}{W_\alpha X^\alpha \, W_\beta Z^\beta \, Y_\gamma X^\gamma \, Y_\delta Z^\delta} \in H^{12,0}(\Pi - \Sigma_0; \mathcal{O}(-2,-2,-2,-2)). \tag{7}$$

Usually, q is equal to zero. In these cases h can in principle be calculated by integrating out the interior vertices of the twistor diagram, although it is not always easy to see how to do this in practice. If q is non-zero the determination of h from the diagram is even less clear.

Suppose we had an element

$$\alpha \in H^{o,f-q}_c(\Pi - \Sigma, \Pi - \Sigma \cup F).$$

Then

$$\alpha \cup h \in H^{3f,f}_c(\Pi - \Sigma, \Pi - \Sigma \cup F; \mathcal{O}(-\mathbf{r})).$$

There is a map induced by inclusion

$$i : H_c^{3f,f}(\Pi - \Sigma, \Pi - \Sigma \cup F; \mathcal{O}(-\mathbf{r})) \to H_c^{3f,f}(\Pi, \Pi - F; \mathcal{O}(-\mathbf{r}))$$

so $i(\alpha \cup h)$ is a functional picked out by the interior of the diagram (i.e. h) as required.

But as it stands α does not look very much like a contour. To overcome this we note first that the embedding of the constant sheaf $\mathbb{C}$ into $\mathcal{O}$ induces a map

$$H_c^{f-q}(\Pi - \Sigma, \Pi - \Sigma \cup F; \mathbb{C}) \to H_c^{f-q}(\Pi - \Sigma, \Pi - \Sigma \cup F; \mathcal{O}) \qquad (8)$$

and second that the groups

$$H_{5f+q}(\Pi - \Sigma, \Pi - \Sigma \cup F) \quad \text{and} \quad H_c^{f-q}(\Pi - \Sigma, \Pi - \Sigma \cup F; \mathbb{C})$$

are isomorphic. Now all we have to do is insist that α is in the image of the map (8), and then it can be regarded as a contour. When α *is* a contour we call $i(\alpha \cup h)$ a functional 'associated to' the kernel h, and remark straight away that there are *none* if $F \subset \Sigma$, because then

$$H_{5f+q}(\Pi - \Sigma, \Pi - \Sigma \cup F) = 0$$

In this case we refer to the problem as 'ill-posed'.

It should be noted that so far in this section our fields are perfectly general. If in fact the fields are elementary states then $F_i = L_i$ and F is equal to the closed *submanifold* Λ (of real codimension $4f$ with orientable normal bundle). Now we can use the Thom isomorphism

$$H_{f+q}(\Lambda - \Sigma) \xrightarrow{\cong} H_{5f+q}(\Pi - \Sigma, \Pi - \Sigma \cup \Lambda)$$

to deduce that the contours we seek are in $H_{f+q}(\Lambda - \Sigma)$. Indeed, this result still holds even when the fields are *not* elementary states, as long as

$$(\Pi - \Sigma, \Pi - \Sigma \cup F) \quad \text{is homotopic to} \quad (\Pi - \Sigma, \Pi - \Sigma \cup \Lambda). \qquad (9)$$

To summarise, then:

Theorem 4.1 *If* (9) *is satisfied then the functionals on*

$$H^1(U_1, \mathcal{O}(r_1)) \otimes \ldots \otimes H^1(U_f, \mathcal{O}(r_f))$$

associated to the kernel

$$h \in H^{3f,q}(\Pi - \Sigma; \mathcal{O}(-\mathbf{r}))$$

are given by elements of the homology group

$$H_{f+q}(\Lambda - \Sigma).$$

Now we consider some applications. For the usual scalar product we note only that all works well, as usual! Instead let us look at the scalar product on $P \times P$ (instead of $P \times P^*$). In this case we can choose Σ to be the diagonal Δ in $P \times P$, which defines for each r what Atiyah [1] calls the 'Serre class'

$$h \in H^{6,2}(\Pi - \Delta; \mathcal{O}(r-2, r-2)).$$

Our theorem now tells us that functionals associated to this kernel are in

$$H_4(\Lambda - \Delta),$$

which again has exactly one generator if $\Lambda \cap \Delta = \emptyset$, that is, if the problem is well-posed. It is ill-posed if L_1 and L_2 intersect, but then the corresponding points in Minkowski space are null-separated, so no functional is expected. This Serre class h was also identified by Eastwood and Ginsberg [4] as the (twistor, twistor) description of their kernel.

However, the big question is what happens in the box diagram? Here $f = 4$, h_0 is as in (7), and in the case of (spin zero) elementary states we are interested in functionals on

$$H^1(P_1 - L_1; \mathcal{O}(-2)) \otimes \ldots \otimes H^1(P_4 - L_4; \mathcal{O}(-2))$$

associated to h_0. In particular, of course, we want to know whether functionals exist corresponding to the three channels for the first order massless scalar ϕ^4 vertex. These functionals will have the property that under a continuous motion of the lines L_i into one of the following special configurations exactly one survives.

(i) $L_1 = L_2$ and $L_3 = L_4$
(ii) $L_1 = L_4$ and $L_2 = L_3$
(iii) $L_1 = L_3$ and $L_2 = L_4$

In cases (i) and (ii) the problem is ill-posed, so there are *no* associated functionals. In case (iii) and for the L_i in general position, the problem is well-posed and we need to study the group

$$H_4(\Lambda - \Sigma_0).$$

But this is exactly the group studied by Sparling [20]. It has two generators (for the L_i in general position), only one of which can be generalised to positive or negative frequency fields. This functional corresponds to the channel given by the Sparling contour.

So our general procedure shows that the box diagram does *not* have 'crossing symmetry', and explains why the earlier attempts at a cohomological understanding of this diagram only led to one functional. The contours for the other channels constructed by Hodges remain a bit of a mystery. One reasonable guess is that in our scheme they correspond to elements α *not* in the image of the map (8). (These generalised contours were first considered in [15].)

Before we leave our discussion of the applications of this general procedure we should note (omitting the details) that it can be used to show that it *is* possible to describe all three channels for the first order massless scalar ϕ^4 vertex in twistors. We abandon the 'box' part of the box diagram (i.e. the kernel h_0) and choose a new kernel

$$h = \frac{DW\,XY\,Z}{(W_\alpha X^\alpha Y_\beta Z^\beta - W_\alpha Z^\alpha Y_\beta X^\beta)^2}.$$

This choice of kernel is justified from the space-time point of view in [14], where it is also shown that there are three associated functionals, one for each channel.

5 Traditional Contours Regarded as Cohomological Functionals

Finally, we consider the question of deciding which contours are cohomological. We take our fields to be elementary states, and choose representative cocycles for them. That is, for each field we choose functions of the form

$$\frac{1}{Z^\alpha A_\alpha Z^\beta B_\beta} \tag{10}$$

and omit to use the Mayer-Vietoris map which would make this into an element of H^1. Constructing a diagram in this way yields a differential form

$$\omega \in H^0(X - A - B; \Omega^d). \tag{11}$$

Here X is the rest of the diagram, so that

$$X = \Pi - \Sigma \cup S$$

where

$$S = \bigcup_{i=1}^{2f-2} S_i,$$

the S_i being (in pairs) the planes defining representative cocycles for the other $f - 1$ fields. In other words, in writing ω as an element of (11) we are merely focusing attention on the one field (10). The Mayer-Vietoris maps in cohomology and homology fit together in the following way:

$$
\begin{array}{ccc}
(\omega \in)\ \ H^0(X - A - B; \Omega^d) & \xrightarrow{\partial^*} & H^1(X - A \cap B; \Omega^d) \\
\times & & \times \\
(\kappa \in)\ \ H_d(X - A - B; \mathbb{C}) & \xleftarrow{\partial_*} & H_{d+1}(X - A \cap B; \mathbb{C}) \\
\downarrow & & \downarrow \\
\mathbb{C} & & \mathbb{C}
\end{array}
$$

The traditional contour κ evaluates ω, but as far as the field (10) is concerned we should be considering $\partial^*\omega$, which means that there should be a contour λ such that

$$\kappa = \partial_*\lambda$$

If κ passes this test for *each* of the fields in the diagram, then it is a cohomolgical functional.

Recent work [13] analyses this test in more detail. Consider the two Leray sequences

$$
\begin{array}{ccccccc}
\to & H_{p+1}(X - A \cap B) & \overset{\cap_1}{\to} & H_{p-1}(A - B) & \overset{\delta_1}{\to} & H_p(X - A) & \to \\
 & \uparrow & & \| & & \uparrow & \\
\to & H_{p+1}(X - B) & \overset{\cap_2}{\to} & H_{p-1}(A - B) & \overset{\delta_2}{\to} & H_p(X - A - B) & \to
\end{array}
$$

(where $\cap$ is intersection and δ is Leray's cobord map). It is shown in [13] that

$$\delta_2\cap_1 = \partial_*$$

so that if $\kappa = \partial_*\lambda$ then (by Leray's Residue Theorem)

$$\int_\kappa \omega = \int_{\partial_*\lambda} \omega = \int_{\delta_2\cap_1\lambda} \omega = \int_{\cap_1\lambda} res(\omega)$$

which confirms that these contours always involve treating at least one of the two singularities in (10) as a Cauchy pole. In fact we believe that they always treat *both* of these singularities as Cauchy poles (and therefore make an $S^1 \times S^1$ contribution to the overall contour for each field) but so far this remains conjectural.

6   Concluding Remarks

A limitation of the general procedure in §4 (not shared with the work in §5) is that as it stands it only applies to *projective* twistor diagrams. The only result we know of here is that it can be shown from (4) that the non-projective box diagram also lacks three cohomological contours. So the new kernel is still needed.

Also, the reader will be aware that almost without comment we have restricted our attention to *scalar* fields. What can be done for other spins? Obviously the kernel in (3) can be modified to give a (twistor, dual twistor) scalar product

$$H^1(U_1; \mathcal{O}(n - 2)) \otimes H^1(U_2; \mathcal{O}(n - 2)) \to \mathbb{C}$$

for any $n > -2$, but it is not so clear what to do with the other cases. One traditional interpretation [17, 20] of diagram edges of the form

$$\frac{1}{(Z^\alpha W_\alpha)^{2+n}}$$

for $n \leq -2$ has been to insist that the contour have *boundary* in the subspace $Z^\alpha W_\alpha = 0$. It would be of interest to discover how far the methods of §4, 5 generalise to contours with boundary. Indeed it seems likely that these methods will provide a good foundation for future work.

I would like to thank all the members of the twistor group, but especially Roger Penrose, for many discussions and constant encouragement. I would also like to thank Andrew Hodges and Michael Singer for our very enjoyable and fruitful collaboration.

References

[1] Atiyah, M.F. *Green's Functions for self-dual four-manifolds* Adv. Math. Supp Studies **7A**, 129–158 (1981)

[2] Baston, R.J. *Local cohomology, elementary states, and evaluation* Twistor Newsletter (Oxford preprint) **22** 8–13 (1986)

[3] Baston, R.J. *Twistor diagram 'magic' and cohomology algebra* Twistor Newsletter (Oxford Preprint) **23** 31–35 (1987)

[4] Eastwood, M.G. & Ginsberg, M.L. *Duality in twistor theory* Duke Math J. **48** 177–196 (1981)

[5] Ginsberg, M.L. *A cohomological scalar product construction*, in *Advances in twistor theory* (eds L.P. Hughston and R.S. Ward), Pitman, San Francisco (1979)

[6] Ginsberg, M.L. *Scattering theory and the geometry of multi-twistor spaces*, Trans. Amer. Math. Soc **276**, 789–815 (1983)

[7] Ginsberg, M.L. & Huggett, S.A. *Sheaf cohomogy and twistor diagrams* in *Advances in twistor theory* (eds L.P. Hughston and R.S. Ward), Pitman, San Francisco (1979)

[8] Hodges, A.P. *Twistor diagrams* Physica **114A** 157–175 (1982)

[9] Hodges, A.P., Article in this volume.

[10] Huggett, S.A., & Penrose R., *Three channels for the box diagram* Twistor Newsletter (Oxford preprint) **10** 18–22 (1980)

[11] Huggett, S.A., & Singer, M.A., *Two philosophies for twistor diagrams* Twistor Newsletter (Oxford preprint) **23** 20–30 (1987)

[12] Huggett, S.A., & Singer, M.A., *Cohomology and projective twistor diagrams* Twister Newsletter (Oxford preprint) **25** 33–40 (1987)

[13] Huggett, S.A., & Singer, M.A., *Cohomological residues* Twistor Newsletter (Oxford preprint) **28** 1–9 (1989)

[14] Huggett, S.A., & Singer, M.A., *Relative cohomology and projective twistor diagrams* to appear in Trans. Amer. Math. Soc.

[15] Josza, R.O., *Sheaf homology and contour integrals* in *Advances in twistor theory* (eds L.P. Hughston and R.S. Ward), Pitman, San Francisco (1979)

[16] Laufer, H.B., *On Serre duality and envelopes of holomorphy* Trans. Amer. Math. Soc. **128** 414–436 (1967)

[17] Penrose, R., *Twistor Theory: its aims and achievements*, in *Quantum Gravity, an Oxford Symposium* (eds C.J. Isham, R. Penrose & D.W. Sciama), Oxford University Press, Oxford (1975)

[18] Penrose, R., *On the evaluation of twistor cohomology classes* Twistor Newsletter (Oxford preprint) **10** 14–15 (1980)

[19] Singer, M.A., & Huggett, S.A., *On the homology of the box diagram*, to appear in *Further advances in twistor theory* (ed L.J. Mason), Pitman (1990)

[20] Sparling, G.A.J., *Homology and twistor theory* in *Quantum Gravity, an Oxford Symposium* (eds. C.J. Isham, R. Penrose & D.W. Sciama), Oxford University Press, Oxford (1975).

Authors' Addresses

T. N. Bailey, Department of Mathematics, University of Edinburgh, J. C. Maxwell Building, The King's Buildings, Mayfield Road, Edinburgh, EH9 3JZ, U.K.

R. J. Baston, Mathematical Institute, 24–29 St. Giles, Oxford, OX1 3LB, U.K.

F. E. Burstall, School of Mathematics, University of Bath, Claverton Down, Bath, BA2 7AY, U.K.

C. J. Cutler, Department of Physics and Astronomy, University of Pittsburgh, Pittsburgh, PA 15260, U.S.A.

E. G. Dunne, Department of Mathematics, Oklahoma State University, Stillwater, Oklahoma, 74078–0613, U.S.A.

M. G. Eastwood, Department of Pure Mathematics, University of Adelaide, G.P.O. Box 498, Adelaide, 5001, South Australia.

J. Fletcher, Mathematical Institute, 24–29 St. Giles, Oxford, OX1 3LB, U.K.

J. Frauendiener, Department of Physics and Astronomy, University of Pittsburgh, Pittsburgh, PA 15260, U.S.A.

S. G. Gindikin, Molecular Biology Laboratory, Building A, Moscow University, GSP-2BY, Moscow, 119899, U.S.S.R.

A. P. Hodges, Mathematical Institute, 24–29 St. Giles, Oxford, OX1 3LB, U.K.

S. A. Huggett, Department of Mathematics and Statistics, Polytechnic South West, Plymouth, PL4 8AA, U.K.

L. P. Hughston, Robert Fleming Securities Ltd., 25 Copthall Avenue, London, EC2 7DR, U.K.

C. N. Kozameh, Laprida 854, FAMAF, University of Cordobá, Argentina.

C. R. Le Brun, Department of Mathematics, SUNY, Stony Brook, NY 11794-3651, U.S.A.

L. J. Mason, Mathematical Institute, 24–29 St. Giles, Oxford, OX1 3LB, U.K.

E. T. Newman, Department of Physics and Astronomy, University of Pittsburgh, Pittsburgh, PA 15260, U.S.A.

R. Penrose, Mathematical Institute, 25–29 St. Giles, Oxford, OX1 3LB, U.K.

W. T. Shaw, Intera-ECL, Highland's Farm, Grey's Road, Henley on Thames, RG9 4PS, U.K.

M. A. Singer, Lincoln College, Oxford, OX1 3DR, U.K.

K. P. Tod, Mathematical Institute, 24–29 St. Giles, Oxford, OX1 3LB, U.K.

R. S. Ward, Department of Mathematical Sciences, University of Durham, Durham, DH1 3LE, U.K.

N. M. J. Woodhouse, Wadham College, Oxford, OX1 3PN, U.K.